Causality

CAUSALITY

Philosophical Theory Meets Scientific Practice

PHYLLIS ILLARI
University College London

FEDERICA RUSSO
Universiteit van Amsterdam

UNIVERSITY PRESS

Great Clarendon Street, Oxford, OX2 6DP,
United Kingdom

Oxford University Press is a department of the University of Oxford.
It furthers the University's objective of excellence in research, scholarship,
and education by publishing worldwide. Oxford is a registered trade mark of
Oxford University Press in the UK and in certain other countries

© Phyllis Illari and Federica Russo 2014

The moral rights of the authors have been asserted

First Edition published in 2014

Impression: 1

All rights reserved. No part of this publication may be reproduced, stored in
a retrieval system, or transmitted, in any form or by any means, without the
prior permission in writing of Oxford University Press, or as expressly permitted
by law, by licence or under terms agreed with the appropriate reprographics
rights organization. Enquiries concerning reproduction outside the scope of the
above should be sent to the Rights Department, Oxford University Press, at the
address above

You must not circulate this work in any other form
and you must impose this same condition on any acquirer

Published in the United States of America by Oxford University Press
198 Madison Avenue, New York, NY 10016, United States of America

British Library Cataloguing in Publication Data

Data available

Library of Congress Control Number: 2014939866

ISBN 978-0-19-966267-8

Printed in Great Britain by
Clays Ltd, St Ives plc

Links to third party websites are provided by Oxford in good faith and
for information only. Oxford disclaims any responsibility for the materials
contained in any third party website referenced in this work.

To David. A Giuseppe.

FOREWORD

Causality has, of course, been one of the basic topics of philosophy since the time of Aristotle. However during the last two decades or so, interest in causality has become very intense in the philosophy of science community, and a great variety of novel views on the subject have emerged and been developed. This is admirable, but, at the same time, poses a problem for anyone who wants to understand recent developments in the philosophy of causality. How can they get to grips with this complicated and often very technical literature?

Fortunately, Part II of this book provides the solution to this problem. In only 171 pages, the authors provide a survey of recent accounts of causality which is clear, comprehensive and accurate. Considering the breadth and complexity of the material, this is a remarkable achievement. The main emphasis in this part of the book is on the research of others, but the authors do include accounts of some of their own work. In chapter 13 (Production Accounts: Information), there is a discussion of Illari's and Illari and Russo's recent work on an informational account of causality. In chapter 15 (Regularity) and chapter 16 (Variation), there is a good presentation of Russo's criticism of Hume's regularity theory of causation, and an exposition of her view that 'variation' rather than 'regularity' is the underlying rationale for causation.

In the remainder of the book (Parts III and IV), the emphasis shifts to the authors' own research, and we find quite a number of interesting and original developments. The authors begin by outlining their approach, which they call: 'Causality in the Sciences'. This is an approach which emphasizes the need for a continual interaction between philosophy and science. It leads to the discussion in chapter 20 of an important, but rather neglected, question about the methodology of philosophy. Should philosophers consider toy examples or examples from science? The literature of the philosophy of causality is filled with the improbable adventures of Billy and Suzy, and one even finds purely imaginary examples such as Merlin causing a prince to turn into a frog. Are such 'toy' examples really of much use for the analysis of causality? Given the authors' Causality in the Sciences approach, one might expect them to prefer examples from science to toy examples, but, in fact, they provide a nuanced account in which there is room for both sorts of example. What is significant here is the consideration of an important topic, which is rarely discussed.

Part IV of the book draws some general conclusions. The authors favour a pluralist view of causality. There is no universal notion of causality, which applies in all contexts, but each of the various analyses of causality does apply in some particular contexts. In the authors' own words (p. 256):

> All we are saying is that the full array of concepts of causality developed within philosophy could very well be of interest to practising scientists...

So when trying to apply causality to any particular scientific example, there is a need to construct what the authors call, in a striking phrase, a 'causal mosaic'. They then illustrate the construction of a causal mosaic by considering an example from contemporary science. This is the new research area called 'exposomics' which attempts to find out how exposure to objects of various kinds can cause illnesses. Thus the authors' title *Causality* is not a rhetorical flourish, but a programme, which is actually carried out in the course of the book.

Donald Gillies
Emeritus Professor of Philosophy of Science and Mathematics
University College London
March 2014

ACKNOWLEDGEMENTS

We started working on the idea for this book while we were both at the University of Kent, where we shared not only an office but also a flat. This created intensive collaboration between us (and a 'no causality after 8pm!' rule that Lorenzo Casini will also remember). Although we have since moved to different institutions—and countries—we are deeply indebted to the information and communication technologies that allowed our weekly meetings to take place anyway.

We are most grateful to the editors at OUP who have shown enthusiasm about this project from the very beginning: Beth Hannon, Clare Charles and Keith Mansfield.

Several people read the book at several stages, even as an embryonic book proposal. Their comments, suggestions and criticisms have been invaluable to us. We offer profound thanks to: Prasanta Bandyopaday, Christos Bechlivanidis, Neil Bramley, Silvia Crafa, Roland Ettema, Matt Farr, Alex Freitas, Luciano Floridi, Toby Friend, Tobias Gerstenberg, Stuart Glennan, David Lagnado, Bert Leuridan, Michel Mouchart, John Pemberton, Julian Reiss, Attilia Ruzzene, Paolo Vineis, Adam White, Jon Williamson, Guillaume Wunsch, and especially Donald Gillies and Ian McKay who read the first complete draft cover to cover. They often led us to rethink the structure and rhetoric of the book, which are immeasurably better for their dedication. We would also like to thank the 'Causality in the Sciences' network—both steering committee and attendees over the last eight years—for the very valuable discussions that helped form so many ideas here. We apologize if we have missed anyone, and naturally acknowledge that remaining errors are our own.

During the academic year 2012–13, Federica Russo was Pegasus Marie Curie Fellow, funded by the Fonds Wetenschappelijk Onderzoek (Flemish Research Foundation), while Phyllis Illari was supported by the Arts and Humanities Research Council. During the academic year 2013–2014, Federica Russo was Reseacher at the University of Ferrara.

We have worked closely together on the structure of the book and then on the contents; writing and re-writing, commenting, manipulating and polishing each other's work. In the pages you are about to read, it is really hard to identify the work of one or other of us.[1] The result is that writing it has been enormous fun!

Finally, our deepest thanks go to David and Giuseppe. You know why.

---

[1] Naturally, we had to decide who would write preliminary drafts of chapters. Phyllis was in charge of chapters 3, 6, 9, 12, 13, 14, 17, 18, 19, 20, 21. Federica was in charge of chapters 2, 4, 5, 7, 8, 10, 11, 15, 16, 22, 23. Chapters 1, 24, and the tables in the appendix are the result of many revisions and intensive co-writing.

CONTENTS

Part I Prelude to Causality

1 PROBLEMS OF CAUSALITY IN THE SCIENCES — 3
  1.1 Why this book on causality? — 3
  1.2 Five scientific problems — 4
  1.3 The contents of this book — 6

2 A SCIENTIFIC TOOLBOX FOR PHILOSOPHY — 9
  2.1 Methods for finding causes — 9
  2.2 Observational methods — 10
  2.3 Experimental methods — 11
  2.4 Between observation and experiment — 14
  2.5 Beyond observation and experiment — 15
  2.6 How to make a study work — 15

3 A PHILOSOPHICAL TOOLBOX FOR SCIENCE — 19
  3.1 Arguments — 19
  3.2 Methods — 21
  3.3 Levels of abstraction — 22

Part II Causality: Accounts, Concepts and Methods

4 NECESSARY AND SUFFICIENT COMPONENTS — 27
  4.1 Examples: electrical short-circuit and AIDS — 27
  4.2 Component causes — 28
  4.3 INUS causes and related concepts — 30
  4.4 Rothman's pie charts — 32

5 LEVELS OF CAUSATION — 35
  5.1 Examples: personalized medicine and migration behaviours — 35
  5.2 Three parallel literatures — 36
  5.3 Bridging the levels—and the terminology! — 41

6 CAUSALITY AND EVIDENCE — 46
- 6.1 Examples: effects of radiation and smoking causing heart disease — 46
- 6.2 What do we want to know? — 47
- 6.3 Evidence for causal relations — 51
- 6.4 Evidence-based approaches — 56

7 CAUSAL METHODS: PROBING THE DATA — 60
- 7.1 Examples: apoptosis and self-rated health — 60
- 7.2 The need for causal methods — 61
- 7.3 The most widespread causal methods — 64
- 7.4 Key notions in causal methods — 67

8 DIFFERENCE-MAKING: PROBABILISTIC CAUSALITY — 75
- 8.1 Example: smoking and lung cancer — 75
- 8.2 Is causality probability-altering? — 76
- 8.3 Beyond probabilistic causes — 82

9 DIFFERENCE-MAKING: COUNTERFACTUALS — 86
- 9.1 Example: mesothelioma and safety at work — 86
- 9.2 The unbearable imprecision of counterfactual reasoning — 87
- 9.3 Philosophical views of counterfactuals — 88
- 9.4 Counterfactuals in other fields — 93

10 DIFFERENCE-MAKING: MANIPULATION AND INVARIANCE — 99
- 10.1 Example: gene knock-out experiments — 99
- 10.2 The manipulationists: wiggle the cause, and the effect wiggles too — 100
- 10.3 What causes can't we wiggle? — 103

11 PRODUCTION ACCOUNTS: PROCESSES — 111
- 11.1 Examples: billiard balls colliding and aeroplanes crossing — 111
- 11.2 Tracing processes — 112
- 11.3 How widely does the approach apply? — 114

12 PRODUCTION ACCOUNTS: MECHANISMS — 120
- 12.1 Example: how can smoking cause heart disease? — 120
- 12.2 What is a mechanism? The major mechanists — 121
- 12.3 Important features of mechanisms and mechanistic explanation — 127
- 12.4 What is not a mechanism? — 132

13 PRODUCTION ACCOUNTS: INFORMATION — 135
- 13.1 Examples: tracing transmission of waves and of disease — 135
- 13.2 The path to informational accounts — 136
- 13.3 Integrating the informational and mechanistic approaches — 143
- 13.4 Future prospects for an informational account of causality — 146

14	**CAPACITIES, POWERS, DISPOSITIONS**	150
	14.1 Examples: systems in physics and biology	150
	14.2 The core idea of capacities, powers and dispositions	151
	14.3 Capacities in science: explanation and evidence	154

15	**REGULARITY**	161
	15.1 Examples: natural and social regularities	161
	15.2 Causality as regular patterns	162
	15.3 Updating regularity for current science	164

16	**VARIATION**	167
	16.1 Example: mother's education and child survival	167
	16.2 The idea of variation	168
	16.3 Variation in observational and experimental methods	172

17	**CAUSALITY AND ACTION**	178
	17.1 Example: symmetry in physics; asymmetry in agency	178
	17.2 Early agency theorists	179
	17.3 Agency and the symmetry problem	181
	17.4 Agency and action	183
	17.5 Problems for agency theories	184
	17.6 Merits of agency theories	186

18	**CAUSALITY AND INFERENCE**	188
	18.1 Example: combatting the spread of AIDS	188
	18.2 Different sorts of inferences	189
	18.3 Does inferentialism lead to anti-realism?	194
	18.4 The heart of inference	195

Part III Approaches to Examining Causality

19	**HOW WE GOT TO THE CAUSALITY IN THE SCIENCES APPROACH (CITS)**	201
	19.1 A methodological struggle	201
	19.2 Causality and language	202
	19.3 Causality, intuitions and concepts	203
	19.4 Causality in the sciences	206

20	**EXAMPLES AND COUNTEREXAMPLES**	211
	20.1 Examples of examples!	211
	20.2 Toy examples or scientific examples?	214
	20.3 Counterexamples	220

21 TRUTH OR MODELS? — 227
21.1 Two approaches to causal assessment — 227
21.2 Causal assessment using models — 228
21.3 Causal assessment identifying truthmakers — 230
21.4 Truth or models? — 233

22 EPISTEMOLOGY, METAPHYSICS, METHOD, SEMANTICS, USE — 237
22.1 Fragmented theorizing about causality — 237
22.2 Which question to answer when? — 240
22.3 Which question interests me? — 242
22.4 Should we integrate the fragments? — 243

Part IV Conclusion: Towards a Causal Mosaic

23 PLURALISM — 249
23.1 If pluralism is the solution, what is the problem? — 249
23.2 Various types of causing — 250
23.3 Various concepts of causation — 251
23.4 Various types of inferences — 252
23.5 Various sources of evidence for causal relations — 253
23.6 Various methods for causal inference — 253
23.7 The pluralist mosaic — 255

24 THE CAUSAL MOSAIC UNDER CONSTRUCTION: THE EXAMPLE OF EXPOSOMICS — 258
24.1 Making mosaics — 258
24.2 Preparing materials for the exposomics mosaic — 260
24.3 Building the exposomics mosaic — 267

APPENDIX ACCOUNTS, CONCEPTS AND METHODS: SUMMARY TABLES — 273
A.1 The scientific problems of causality — 273
A.2 The philosophical questions about causality — 273
A.3 The accounts: how they fare with scientific problems — 274
A.4 The accounts: how they fare with philosophical questions — 277

References — 281
Index — 303

PART I
Prelude to Causality

In this part of the book we introduce and illustrate five scientific problems of causality that the accounts, concepts, and methods of Part II have to deal with. We also provide a basic scientific and philosophical toolbox to be used throughout the book.

CHAPTER 1

Problems of Causality in the Sciences

1.1 Why this book on causality?

Many of the sciences spend much of their time looking for causes—from causes of disease, to climate change, to earthquakes, to house price bubbles. Finding such causes, and deciding what to do about them, such as creating sustainable sources of energy, is not at all easy. Though these things are sometimes called 'causes', sometimes less loaded names such as 'risk factors', 'determinants', 'associations', and so on are used. Whatever they are called, they are really important, and it is worth everyone thinking about these issues at some point in their lives, academic studies or careers.

It is a good time to think about these issues now, because there has been such intensive progress in thinking about causality in the last decade or so, in both science and philosophy. The capacity of science to collect and process data has expanded enormously in recent years. Powerful computational techniques for searching data sets for correlations have been created, along with methods for identifying causes from correlations. The results of these methods need to be interpreted and explored.

Historically, various philosophers have been sceptical about causality, broadly holding that we can interpret the world without thinking causally. These include Mach (1905), Pearson (1911), and Russell (1913) in the early twentieth century, at least at some point in their work—all prominent figures in philosophy and science. But we see the same view being expressed more recently, such as with Lipton and Ødegaard (2005) in the medical domain, or Norton (2003) in the philosophy of physics. It may turn out that some strategies for eliminating causality are defensible in specific domains, for instance theoretical physics, but that is something that needs to be investigated (see chapters 22 and 23). Most recently, there has been an explosion of literature on causality, developed in connection with the fascinating problems of causality that arise constantly in the practice of the sciences.

If you have avoided thinking or talking about causality because you thought it was tied to determinism, or logical necessity, or some other straightjacket that seemed inappropriate to the fluidity of scientific practice, then hopefully you will find this book useful. Philosophical theorizing about causality has grown, and now allows for great flexibility in reasoning. If you are a student, intrigued by the sophistication of the philosophical notion of causation, you will see how causal notions are also routinely used in scientific reasoning. The philosophical literature now offers an extraordinary range of concepts and accounts of causality. This book will introduce this philosophical literature, and provide many examples of scientific problems of causality. So it is simultaneously an introduction to the philosophical literature and an overview of the problems of finding and using causes in the sciences. While we focus on science, there are of course also humanities disciplines, like history and law, which also try to establish causes. The book might be of use, for example, to law professionals trying to interpret expert scientific testimony. The following provides information on the contents of the book, which can be read in different ways, according to different interests.

1.2 Five scientific problems

In order to link philosophical theory and scientific practice, we first introduce and illustrate five different scientific problems of causality: inference, prediction, explanation, control and reasoning. These five problems will be very important in this book, as we will use them to explain and assess all the views and accounts of causality that we offer in the chapters of Part II.

Inference. In many ways this is the paradigm problem of causality. Many sciences spend a great deal of time trying to find out what causes what. In medicine doctors have to decide what the cause of a patient's illness is. Social scientists study what causes discrimination. In court, lawyers and judges have to decide who is legally responsible for crimes like murder and civil problems like negligence. Scientists study the causes of phenomena as diverse as climate change, cancer, and supernovas.

So the first problem of causality is the problem of inference: Is there a causal relation between X and Y? Does X cause Y? What are the causes of Y? What are the effects of X? How much of X causes how much of Y?

Relevant chapters include: 6, 7, 9, 11, 12.

Prediction. Inference is not the only problem of causality, as we want to know other things too. Prediction is just working out what will happen. We often want to know what will happen, as with climate change, where it is difficult to tell what will happen to global weather; how much it will change, and when. In economics, we have all seen that having much better predictions of market crashes would also be enormously helpful.

So the second problem of causality is prediction: What will happen next, and how do we find out what will happen next? How accurate are our predictions about changes in some population characteristics, such as mortality, morbidity, and so on? What does a physical theory predict will happen in a given experimental setting, such as in the large hadron collider at CERN?

Relevant chapters include: 8, 11, 18.

Explanation. We often want to know not just what happened or will happen, but *how* it happened, and why. How does cancer lead to death? How do carbon emissions lead to climate change? Why did the global market crash of 2008 happen? This is causal explanation.

So the third problem of causality we will use is explanation: How do we (causally) explain phenomena? To what extent is a phenomenon explained by statistical analyses? What level of explanation is appropriate for different audiences, such as fellow specialists, the public, or jurors in a court?

Relevant chapters include: 6, 8, 12, 14, 18.

Control. In many ways, we want to know very much more than what causes what, what will happen, and why. We want to know what to *do*. Causes allow us both to increase our knowledge and understanding of the world and, based on this knowledge, take action, such as developing sustainable energy resources. Policymakers have to interpret messages from science and decide what to do about them. Climate change and cancer are just two of the thousands of concerns clamouring for political attention— and money. Bewildered bystanders innocently listening to the radio or reading the news can struggle to follow these complex debates: shall I avoid milk products to lower cholesterol, thereby avoiding heart disease ... or will that raise the risk of osteoporosis? It can be tempting to give up and ignore all these issues. Control is going beyond prediction, to alter what will happen.

So the fourth problem of causality is control. There are two aspects of control. First, how do we control variables that may confound a relation between two other variables? In what cases is it appropriate to control for confounders (see section 2.3)? Second, how do we control the world or an experimental setting more generally? How do we find out about that? Our interests in control extend all the way from interfering with a system to get a slightly different result, to the construction of entirely new causal systems, such as smartphones, or GPS, or a political system.

Relevant chapters include: 7, 10, 17.

Reasoning. This is the broadest of the scientific problems, lying most clearly at the intersection of science and philosophy. Causal reasoning concerns all the ways we think about causality in science, explicitly and implicitly. We suggest that it is worth pausing at some point to try to make the assumptions underlying practice explicit, for scientists, philosophers, and indeed anyone trying to interpret media messages about newly discovered 'correlations' or 'links'. Note that thinking about causal reasoning more broadly is when we worry about relations between the other four problems. For example, are the assumptions in the models we use to make predictions consistent with the reasoning we use in causal inference?

So the fifth and final problem of causality is reasoning: What reasoning underlies the construction and evaluation of scientific models? What conceptualization of causation underpins causal methods? How do we reason about all aspects of causality? How can we sharpen up that reasoning, making it more precise and so more effective?

Relevant chapters include: 9, 14, 16.

In the chapters of Part II, we will show how different approaches to causality can be used to illuminate the five scientific problems—inference, prediction, explanation, control and reasoning—at least in some fields. Our aim is to introduce the philosophical approaches led by these five problems. Of course, we will also look at how each approach addresses the problems that arise within the philosophical debate. Naturally, philosophical accounts do not make the scientific problems vanish! They can, however, sharpen up reasoning about these problems. Science is also a fascinating resource for philosophy in yielding so many cases where it is worth wrestling with these problems. And helping address these problems in this way demonstrates how the philosophy of causality can be of enormous interest to science.

Ultimately, this book does not argue in favour of one particular approach to or account of causality, although along the way we will of course present the ones we have advocated ourselves. We think that, at least currently, there is no single causal theory that can exhaust the variety and complexity of causal reasoning in the sciences. Instead, we will try to show that, from the perspective of the sciences, the vivid diversity of the philosophical literature is an advantage. The rich mosaic of causal concepts and accounts in philosophy is useful in its entirety.

1.3 The contents of this book

In this book we offer an introduction to philosophy of causality, one that aims to be highly accessible:

- to scientists unacquainted with philosophy,
- to philosophers unacquainted with science, and
- to anyone else lost in the labyrinth of philosophical theories of causality.

We try to make all the debates and views accessible to multiple audiences, but *none* of the topics in this book is easy. Consequently this book requires effort on the part of the reader. We offer the following as a guide to the contents:

- Part I: Prelude to Causality

  We offer an introduction to the alliance of science and philosophy, beginning with an introduction to the five scientific problems of causality we use throughout the book. We then give a scientific toolbox for philosophy, introducing key methods for how the sciences go about finding causes; and then a philosophical toolbox for science, introducing key ideas in philosophy of science.
- Part II: Accounts, Concepts and Methods

  We introduce key philosophical accounts, concepts and methods, giving each major philosophical approach one short chapter. In each chapter we illustrate how to apply the philosophical ideas to the scientific problems presented in section 1.2. This provides a guide to the sprawling philosophical literature. Note that we do not begin with Hume in the traditional way, as we are examining the philosophical theorizing thematically, rather than historically, following threads inspired by the practice of the sciences.

  Chapters in Part II begin with an example section illustrating the problems, give an overview of the views, and then have a discussion section examining the virtues of the views and problems with them. The chapters finish with a summary of 'core ideas', 'distinctions and warnings' and 'further readings'. We also provide extensive references to the more detailed literature we discuss, and cross-references to other relevant chapters. The example sections and summaries of core ideas and distinctions and warnings are a useful way to identify chapters of interest. To the same end, the index offers a comprehensive list of examples used in the book, and specific topics and authors discussed.

- Part III: Approaches to Examining Causality

  This part is more distinctively philosophical, looking at approaches to philosophical method in understanding causality. We explain how we came to form a 'Causality in the Sciences' community and, within this approach, how we came to use examples and counterexamples, to debate the notions of truth and validity, and to specify the scope of causal questions. Our approach in this book is currently very popular, but it is not wholly uncontroversial, and we examine this methodological approach—*philosophical* methodology—throughout Part III.

  If you are approaching philosophy of causality for the first time, Part III can help you get acquainted with the philosophical approaches, in order to pick and choose specific chapters in Part II. This may be useful to scientists dealing with causal questions on an everyday basis and interested in complementing their practice with a philosophical outlook; or to philosophers who wish to find work that falls within a particular approach.

- Part IV: Towards a Causal Mosaic

  We finish by presenting various issues of pluralism in philosophy of causality. We defend the version of pluralism we believe contributes to connecting philosophical theory and scientific practice, and then we apply the material in the rest of the book to an exciting case in current science—exposomics science.

While writing, we had a variety of possible readers in mind, such as undergraduate students taking a philosophy of science course in either a philosophy or science curriculum; a graduate student interested in doing a PhD on causality; a philosopher coming from outside philosophy of causality and in need of a road map to the literature; and any scientists interested in philosophical issues about causality and similarly in need of a road map. So there are many ways in which this book can be read and used. We have included some other features to help you find useful material.

We have written each chapter to be independently intelligible. But so many themes recur throughout the book, and so much can be learned by comparing the philosophical approaches, that we think much can be gained by reading at least a few chapters. To help guide the reader on this, we have included extensive cross-references, so that other places of interest can be easily tracked. We have also set the book up so that it is easy to refer back to it, by using the examples at the beginning of every Part II chapter, also listed in the index, and the 'core ideas' and 'distinctions and warnings' at the end of every Part II chapter. The index includes key abbreviations used in the book and a list of examples used throughout the chapters. Finally, the appendix provides summary tables that give a snapshot view of the contents of the book.

Throughout the book we use examples from natural, social, and biomedical sciences. Sometimes we introduce the same problem or account using examples from very different fields. Methods and specific concerns certainly differ from field to field, but comparison between sciences is often useful. Moreover, similar conceptual problems arise in contexts that are, at first glance, very distant. Our choice of examples also serves to draw attention to areas that have been less discussed in the literature. To help multiple possible audiences, chapters in Parts I and III give an overview of the scientific and philosophical methodology respectively. They can be used in a course in causality or, more generally, philosophy of science. The accounts in Part II can then be selected depending on the focus of the course.

We do our best to guide the reader through vast and complex debates in a way that is as self-contained as possible. We would like to stress that this is an introduction, so the reader will have to engage with the literature we refer to for further details, either philosophical or scientific. We offer a map, not an encyclopaedia. This book systematizes and summarizes the current causality literature, but if you want to take your studies further there is no substitute for the primary sources that we also direct you to.

The variety of perspectives met led us to develop a view of causal theories as a *mosaic*. In chapter 24, we argue that we can't really understand a tile of this mosaic outside its immediate context or the broader context of the mosaic. The accounts, concepts, and methods discussed in Part II are better discussed in relation to the other notions given elsewhere. Therefore we have cross-referenced chapters, providing the 'cement' that fixes the tiles to the base of the mosaic that we gradually lay out throughout the book.

CHAPTER 2

A Scientific Toolbox for Philosophy

2.1 Methods for finding causes

The most obvious scientific problem of causality, the one that scientific practice is riddled with, is how we find out about causes—this is the problem of developing and using suitable and appropriate methods for finding causes; the problem of causal inference. But in order to use such methods to draw inferences, we need to gather data. Drawing inferences may mean either *testing* a causal hypothesis or *formulating* a causal hypothesis, or both. In either case, we need to look at data. So we begin by looking at data-gathering methods in the sciences.

To investigate the natural and social world, we observe what happens around us. We soon realize that some events occur repeatedly, and try to find out what circumstances allow such repetition. We can also isolate events that were unexpected, given what we already knew, and try to find out why they occurred. Whenever we can, we will try to interact with the world, for instance to test a hypothesis coming from theory or to create or recreate a phenomenon in the lab (see e.g. Hacking (1983)).

At least since the scientific revolution, scientists have tried to systematize their methods for observation and for experiment. An important difference between modern science and that of the ancient Greeks, who also worried about the natural world, is that we mess about with the world to understand how it works. Experimental methods at their most basic poke the world to see what happens. What happens when we give people antibiotics? Do they cure disease? Does anything else happen, like increased antibiotic resistance of bacteria? Experimental studies take what the world is doing all by itself, and interfere with it to see what happens next. The decision about where and how and when to intervene is not blind, though. Experimentation is usually carefully planned, although 'serendipitous' discoveries do exist—penicillin is but one famous example in the history of science.

A distinction often made in thinking about the methodology of data-gathering is between experimental and observational studies; namely between studies where data have been 'produced' within the experimental setting by manipulating some factors while holding others fixed, and studies where data have been produced by simply recording observations. For instance, a randomized trial (see below) is an experimental study because data is generated within the specific experimental setting, where scientists decide what data to collect, what treatment or placebo to prescribe to groups, etc. The group's pre-experimental behaviour is thoroughly altered. A study on, say, health perception and wealth is usually considered an observational study because the social scientists running a study will most often analyse data that have been collected, for instance for the national census the previous year, thereby altering the behaviour of the population minimally or not at all.

Needless to say, we cannot interfere with the world just any way we like. There are important restrictions on experiments, very common in social sciences and many areas of medicine. We might be able to show that cigarette smoking causes cancer if we took two groups of people, sorted them according to age, other diseases, and so on, put them in different cages, isolated them from anything else, and forced one cageful to smoke and the other not to smoke for 20 years, to see what happens. Ethical guidelines and ethics boards exist to prevent subtle and unsubtle violations of ethical ways to treat experimental subjects, whether they are humans, animals, or embryos. In this case and others like them, all we can do is observe people who voluntarily choose to smoke, see what happens to them over their lives, and look out for other things they do that may affect whether they get lung cancer.

So there is a need for different methods of causal inference for different situations. In this light, the distinction between 'observational' and 'experimental' is not a distinction between better and worse methods, but a distinction about *how* we get the data. Once we have the data, there are further questions about data analysis, which we also discuss later in this chapter and throughout Part II of the book. In particular, methods for *probing* the data (see chapter 7) have important technical differences. Despite these differences, these methods share the same logic, or rationale, underpinning data analysis, which we also discuss in chapter 16: we learn about causality by comparing different groups, different instances, and different situations.

2.2 Observational methods

Observational studies are widely used in epidemiology, demography, and other social sciences, including, for example, actuarial sciences. What these methods have in common is the *comparison* between instances, groups, categories, circumstances, and so on in order to draw inferences about cause-effect relationships.

Cross-sectional studies. These studies concern a population, or a sample of a population, at a specific point in time. A typical example is a census. Individuals are asked questions, sometimes very specific ones, to get a 'snapshot picture' of the population, including, for instance, information about occupation, the prevalence of a particular disease, or the presence of different ethnic groups in the studied population.

Cohort studies. A cohort is a group of individuals sharing some specific characteristics in a specified time period, such as being women aged 30–40, or having a secondary school education. The purpose is to follow the cohort over a given period of time. Cohort studies, used in social science and medicine, examine the characteristics of the cohort—the group—to establish correlations between variables concerning, for example, environmental exposure, lifestyle habits, social conditions and diseases or other social outcomes such as average income. From these correlations we can possibly draw inferences about causal relations, depending on issues about quality of data, study design, etc.

Case-control studies. These studies are typically used in epidemiology and compare two groups of individuals. For instance: those having a disease (the cases) are compared to those not having a disease (the controls); or those who have been exposed to some factor (the cases) are compared to those who have not been exposed (the controls). In these studies, subjects are not randomized into groups, but are instead observed as part of the groups they already belong to. The *idea* behind these studies—i.e. comparing cases and controls—is not different from randomized controlled trials (RCTs) described below, but the selection of individuals and their allocation to groups is different.

Case series studies. These are qualitative studies mainly used in epidemiology and medicine where a single individual, or a small group of individuals showing similar conditions (for instance exposure to the same substance, situation or treatment for a particular condition) are analyzed through the examination of medical records.

However sophisticated their design is, and however carefully conducted these kinds of study are, they can never 'automatically' establish causal relations. To be sure, there is no method able to do that, and we will discuss reasons why in due course. Most of the time these studies can only establish correlations. More reasoning and more evidence is required to draw causal inferences from statistical analyses of data and to make such conclusions plausible. Chapter 6 discusses evidence for causal relations in detail. This is one reason why we need to understand the different facets of causality, to make appropriate use of the concepts discussed in Part II, so that observational studies can be conducted well, and so that interpretation of the results of such studies can be as nuanced as possible.

2.3 Experimental methods

Scientists like to use experimental methods wherever possible. The literature on experimentation is vast. Weber (2012) and Franklin (2012) offer a useful introduction to general epistemological and methodological issues about experiments, and also to specific issues that arise in biology and physics, respectively. For a historical and philosophical discussion of experiments that marked significant changes in science, see for instance Harré (1981). Experiments vary enormously according to the technical apparatus or the purpose of the experiment. We associate experiments with the idea of working in a lab, but there are other procedures that can count as experiments. One example is randomized trials in biomedical and pharmaceutical contexts. Another is the particle accelerator at CERN.

Lab experiments. Experiments can be very different from each other. Think about the Large Hadron Collider (LHD) at CERN (European Organization for Nuclear Research), which is the largest particle accelerator in the world. Scientists at CERN use this complex and huge experimental apparatus to study particle physics. In a biology lab, we may find instruments for the culture and conservation of cells, or powerful instruments to reveal the structure of DNA. Lab experiments are also done in experimental economics or psychology, where individuals are put in somewhat artificial situations and their behaviour is studied.

Experiments are designed and performed for different purposes. One is to improve measurements, such as the equivalence between inertial mass and gravitational mass, or to control the reliability of an experimental apparatus. Another is to explore the constitution of materials. Yet another is to confirm a theory. Other experiments explore new phenomena. There are also mixed-purpose experiments, or those driven by pure curiosity. Franklin is a classic reference in the philosophy of experiments (see for instance Franklin (1981) and Franklin (1986)). For a general presentation of experiments, and their main features and roles in science, see also Boniolo and Vidali (1999, ch5.3).

A central concept in experiments is *manipulation*. In fact, the goal of a lab experiment is precisely to manipulate the putative cause, i.e. to wiggle some elements, and hold other factors, conditions or features of the experimental apparatus fixed as far as possible, and see what happens to the putative effect. In philosophy of science, Woodward (2003) has spelled out the notion of an intervention for causal assessment and provided stringent conditions to define and identify interventions (for a thorough discussion see chapter 10).

Randomization. Another key concept in experimental methods is randomization. In randomized experiments, subjects need to be allocated to the treatment group and to the control group. But the two groups must be alike in important respects. For example, if the control group is significantly older than the treatment group in a medical trial, the treatment may appear to work, just because young people tend to recover more quickly from illness. We know some important factors like age are potential 'confounders' in such trials. Confounders are variables associated both with the outcome (E) and with the putative cause (C), and for this reason they 'confound' the relation between C and E, i.e. it is unclear how much of E is due to C or to the confounder (see more below). The favourite solution to this problem is randomization, which refers to the random allocation of a treatment (whether a drug, fertilizer, etc.) to the experimental unit in a study or, which amounts to the same thing, the allocation of experimental subjects to treatment or control group. In the original thought of Fisher, randomization is a means for eliminating bias in the results due to uncontrolled differences in experimental conditions (Fisher (1925) and Fisher (1935)). When randomization is used correctly, the treatment group and the control group differ only, to the best of the experimenter's knowledge, in which group gets or does not get the treatment. So the experimenter can conclude that any difference in the outcome between the two groups will be due only to the treatment. If the treatment group improves faster, the treatment works. Randomization is an important tool, but it can be tricky. In laboratory experiments ideal conditions for random allocation are more often met because uncontrolled variations in the environment are much better known, though this is certainly not the case in agricultural studies (where Fisherian randomization originated) nor in social and biomedical contexts where phenomena and environmental conditions are highly complex. Despite these practical difficulties, randomization remains a central concept in many of the causal methods we examine in chapter 7.

Randomization is at the heart of *randomized controlled trials* (RCTs), on which many disciplines, especially those involved in 'evidence-based approaches' (see chapter 6), rely. Today, RCTs are mainly associated with medical trials, for instance to test the efficacy of a drug, but they are increasingly being used in other fields such as in social science to evaluate the effectiveness of various types of policy interventions. Randomized studies started thanks to the work of Fisher and of Neyman, in the field of agriculture, for instance to evaluate the efficacy of a given fertilizer across blocks in a field. RCTs are based on two ideas:

(i) the random allocation of a treatment to experimental units (whether field blocks, mice, patients, or something else) minimizes the problem of bias and confounding, and

(ii) to establish the effectiveness of the treatment, we need to compare the results in a test group (where the treatment has been administered) and in a control group (where the treatment has not been administered, or a placebo or the current alternative best-known treatment has been given instead).

The first historically recognized randomized experiment was run by Peirce and Jastrow (1885) in psychometrics, but the first systematic exposition of randomization is due to Fisher (1925). (For a recent textbook presentation in social science and medicine see, respectively, Solomon et al. (2009) and Jadad and Enkin (2007)).

RCTs can involve several individuals or just one—these are called 'n of 1 trials'. RCTs can be 'blind' or 'double-blind'. This is now standard terminology, although perhaps 'anonymous' would be more appropriate. In blind experiments, individuals don't know whether they have been given the drug under test or the placebo or alternative treatment. This prevents patients from reporting increased improvement due solely to the knowledge that they're getting a new drug, which they may assume is more effective. In double-blind trials, the doctor administering the treatment and assessing the clinical effects also doesn't know which subjects have been given the drug and which the placebo or alternative treatment. Double-blind trials are meant to minimize bias in the design of the trial and in the evaluation of results. For example, it is impossible for the doctor to signal to the patients subliminally whether they are getting the real trial drug, or record patient reports on recovery in a biased way. These measures may sound extreme, but the history of trials has shown that the effects of such biases can be large, and so trial design has improved to avoid such problems.

That the distinction between observational and experimental studies is in a sense artificial becomes clear once we consider other kinds of studies that are not exactly one or the other. In the following sections, we present three: quasi-experiments and natural experiments; qualitative analysis; and simulations.

2.4 Between observation and experiment

The studies described in this section are typically performed in the social sciences or in epidemiology.

Quasi-experiments and natural experiments. Quasi-experiments are empirical studies typically intended to evaluate the impact of an intervention or treatment on the target population. The baseline idea is the same as RCTs, with an important difference: there is no random allocation of experimental subjects. Allocation is decided on other grounds, for instance through techniques called propensity scores as in potential outcome models presented in chapter 9. Sometimes the allocation is done by *nature*, in the sense that subjects just 'naturally' happen to be allocated in way that strongly resembles randomization. If, to the best of our knowledge, the treatment and control groups are the same with respect to all relevant factors, then we treat the study as an experiment, and we take the allocation that nature 'gives' us. The famous 1854 cholera outbreak in London is often discussed as an example of a natural experiment (see for instance Freedman (2005b, ch1) and Di Nardo (2008) on quasi-experiments and natural experiments). So quasi-experiments, or natural experiments, share many features of both experimental and observational studies.

Qualitative analysis. The observational–experimental distinction also masks another type of analysis, typical of the social sciences, which is qualitative analysis. Here the relevant distinction is between quantitative and qualitative methods, rather than observational versus experimental methods. Methods described above are by and large quantitative, i.e. they attempt to gather large amounts of data to analyse using statistical techniques. Qualitative studies, used especially in ethnography or part of sociology, select small groups of individuals and study them in detail, for instance by trying to integrate into their community to observe from within. Don't get sidetracked: this is not just about going to Madagascar to study a small indigenous culture, or learning a different language. These methods are also used to study European or North American cultures, to focus on specific problems. Sociologists of health, for instance, may try to observe the people in an operating theatre to understand the dynamics of the operating group. There are also focus groups to study people's perceptions of new technologies and thus increase public participation in some decision-making processes. Qualitative analysis can be as rigorous as quantitative analysis (see for instance Cardano (2009)), as rigour is not an intrinsic property of studies but depends on how scientists set up and carry out the studies. Clearly, the distinction between observational and experimental methods cuts across other important distinctions, such as the difference between quantitative and qualitative methods.

2.5 Beyond observation and experiment

There are some methods, increasingly important to science, that defy classification as either observational, experimental, or mixed. These are known as simulations.

Simulations in some way mimic the system of interest, often called the 'target system'. An example of a physical simulation is a small model car, or the wing of an aeroplane, which can be placed in a wind tunnel to test its behaviour in circumstances that are difficult to create for the full-sized vehicle (which may not yet have been created).

A computer simulation is basically a computer programme that mimics, or simulates, a target system. The system simulated can be a natural system, whether in biology, physics, or astrophysics, or a social system. In the simulation, the scientist, usually in collaboration with computer scientists, designs a process based on available knowledge of the phenomenon. Simulation models are built on observational data, or data gathered from a few real experiments. The model is designed to mimic, in some way, the behaviour of the target system over time. Simulations are then run on the model to see what would happen for experiments we haven't actually tried!

Such simulations can be very illuminating about various phenomena that are complex and have unexpected effects. Simulations can be run repeatedly to study a system in different situations to those seen for the real system, or when the system itself, like a binary star system, is quite impossible to re-run, or to study in any circumstances other than its actual circumstances.

If the parameters of the simulation are physically realistic representations of the real system, then often the results of the simulation are taken to be genuinely illuminating about the real system. This is particularly so if at least some interesting results obtained from the simulations are observed in the real system, such as when studying distant star systems, or can be replicated in the original system, such as in the biological 'wet lab'. For further discussion of various philosophical issues arising in computer simulation, see, for example, Hartmann (1996), Frigg and Hartmann (2012).

So simulations, whether run on computer or physical models, are neither experimental nor observational, but something much more complex. Scientists create an imitation system that is capable of having something like experiments run on it, and use that imitation system to probe features of the target system; it can then be tried out in some way on the target system—perhaps by suggesting new experiments, or by directing new observations.

There is a longstanding distinction between studies '*in vivo*' and those '*in vitro*', to mark the difference between studying phenomena in the environment of the living organism, versus in an artificial environment such as a test tube. Many scientists have now added the expression 'in virtuo' to mark simulation, a study in a virtual environment.

2.6 How to make a study work

Experimental and observational methods have their own strengths and weaknesses. For each of these methods, the references we have given above contain in-depth discussions, highlighting what kind of data they are best suited for, what inference techniques are available for them, or what software, and what their limitations are. Here, we select two important problems all studies share: the problem of confounding, and the problem of validity.

2.6.1 The problem of confounding

The first problem, the problem of confounding, might also be called the 'third-variable' problem. It is the problem of moving from a correlation or association between two variables, to a causal conclusion. The problem is that the correlation might not be due to a causal connection between the two variables, but instead to the presence of a third variable which is a cause of both.

A stock example of this from the philosophical literature is: coffee drinking and lung cancer are associated, but coffee drinking doesn't cause lung cancer, nor vice versa. They are associated because both are associated with cigarette smoking. Smokers happen to drink more coffee on average than non-smokers. Yet, while there is a direct link between cigarette smoking and lung cancer, cigarette smoking and coffee drinking are linked indirectly, perhaps because other factors interfere such as stress or habitual behaviour. A slightly different issue concerns spurious relations: yellow fingers and lung cancer are also associated, but again lung cancer doesn't cause yellow fingers, nor vice versa. They are associated because both are *effects* of cigarette smoking. Once the factor 'cigarette smoking' is introduced, the correlation between yellow fingers and lung cancer vanishes. The 'third-variable problem' illustrates why *correlation does not imply causation*.

Hans Reichenbach pioneered the study of probabilistic structures like these, in the context of the philosophy of spacetime. Firstly, his work led to the formulation of the 'common cause principle', which says that if two events are correlated, either one causes the other, or there is a third event causing both of them. The epistemological, methodological, and metaphysical implications of the common cause principle have been extensively discussed in the philosophical literature (for a compendium presentation, see for instance Arntzenius (2008)). Secondly, Reichenbach's work led to the formulation of the screening-off relation: in the presence of a common cause S (such as Smoking), of two other variables Y (such as Yellow fingers) and L (such as Lung cancer), S screens off Y from L. That is, if you hold S fixed, the association between Y and L disappears. This would appear correct: for smokers, Y and L are associated, while for *non*-smokers they are not. So once S is held fixed, the association disappears, showing that it is due to S, rather than being a symptom of a genuine causal relation between Y and L, in either direction. The causal structure represented in Figure 2.1 is often called a 'conjunctive fork'.

Fig 2.1 Common cause

Confounding is probably one of the most pervasive methodological problems in science, occurring in experimental and observational studies in the natural, biomedical, and social sciences alike. Understanding the idea of screening off is useful, because it allows us to grasp the point of a great deal of scientific method: to control potential common causes (S in figure 2.1) and so get rid of correlations that do not indicate causes—spurious correlations such as that between Y and L. This can explain the obsession with 'controls', which arose in many of the study designs we have described. These controls are attempts to fix possible common causes, and while broad strategies of the kinds we have described are often used, they are adapted to the specific design of each study. There are methods of control *ex ante*, i.e. variables are controlled *before* data are generated and collected, as in the setting up of an experiment, or in the division of patients taking part in a trial into subpopulations. There are methods of control *ex post*, i.e. variables are controlled *after* data are generated and collected. For instance, the population can be stratified during the stage of data analysis, depending on what data have been collected. It is worth noting that when we do *not* know what the relevant causal factors may be, randomization is a good strategy to avoid confounding. However, if we *do* know what the relevant factors may be, we may try to control *ex ante*, or, if that's not possible, at least control *ex-post*.

It is worth noting that the problem of confounding arises in other ways. In physics, for instance, a famous example is the experiments run by Penzias and Wilson in the 1960s. They were using a massive antenna at Bell Labs, and kept finding 'background noise' they didn't understand. They attempted to understand the source of the noise in order to control or minimize it. Several possibilities were tried experimentally, including the possibility that the noise came from pigeons on the big antenna, or urban interference such as from New York City. Unable to find the source in this way, they then turned to more theoretical possibilities. This led to the study of cosmic background microwave radiation (see Guth (1998)). Here, the question is not about solving the 'third-variable' problem in the sense explained above. Yet this can be considered a genuine case of confounding in science.

2.6.2 The problem of validity

The second problem is to show that a particular study is *valid*—this is about the level of confidence we have in the results of a study. Validity has been especially discussed in the context of quasi-experiments in social science. (The classic sources, which generated a lot of subsequent discussion, are Campbell and Stanley (1963) and Cook and Campbell (1979).) The validity of a study depends on a coherent, plausible, convincing story from data collection to interpretation of results. In general, statistical tests are applied to assess the validity of studies. But statistical tests are not enough. Vitally, the validity of a study depends on what particular result you have in mind. For different purposes, we might need to examine different aspects of the study more or less stringently. One particularly urgent question is: is your purpose to test a causal structure in a population? Or is it to export an explanation or treatment or intervention to *another* population?

Frequently, we are interested in using the results of a particular study to tell us about alternative populations, perhaps in a different country, or a different age group. You have to do a great deal to ensure that your results are valid for your study population, particularly dealing with the problem of confounding, by controlling adequately. This is often called the problem of 'internal validity'. Another big problem is to ensure that your results will tell you something interesting about other populations. This is often called the problem of 'external validity' or the 'extrapolation problem'. You might also see it informally referred to as the question of whether study results are 'exportable' or whether they 'generalize'.

This problem, or problems, has been extensively challenged and discussed, and there is no programmatic solution. Jimenez-Buedo and Miller (2010), for instance, discuss the use and meaning of the concept of 'external validity'. Cartwright (2011), to give another example, talks about the 'long road from "it-works-somewhere" to "it-will-work-for-us" ', to indicate precisely how difficult it is to establish whether, and to what extent, the results of a study will be applicable outside the study itself. The crucial points are, first, that statistical tests of validity can help, but ultimately the whole narrative of a study, from how data have been collected to interpretation of results, is vital in meeting challenges of validity (see Russo (2011a)). Second, validity is relative to the planned use of the study results, and so there are as many types of validity as there are objectives of different studies. Guala (2005, p142, n1) identifies the following:

- Internal versus external validity as we have discussed: the extent to which a relation between two variables is causal *within* the sample (internal) and *outside* the sample (external);
- Population validity: generalizing to different populations of subjects (Christensen, 2001, ch14);
- Ecological validity: generalizing to the behaviour of the same subjects in different circumstances ((Christensen, 2001, ch14), (Brunswik, 1955));
- Temporal validity: generalizing to the same population, in the same circumstances, but at different times (Christensen, 2001, ch14).

For different types of studies, having different scopes and purposes, we could no doubt identify further forms of validity.

The problem of confounding and the challenge of validity are probably the two most invidious problems for science, affecting almost all studies, and driving an enormous amount of methodology.

A final note: causality and time is a hotly disputed question. We will not address the problem of time separately, but we do comment on it when specific accounts or concepts or methods discussed in Part II prompt it. It suffices here to lay down the most general terms of the discussion. On one hand, we, as epistemic agents in everyday or scientific contexts, experience causal relations in time. On the other hand, this strong intuition clashes with the content of some part of physics, where causal relations seem to be symmetric with respect to some time frame (see particularly chapter 17). Yet several scientific domains, most typically the social sciences or epidemiology (broadly construed), assume that causation happens in time, and focus on how a model should grasp, measure, or take into account temporal information.

CHAPTER 3

A Philosophical Toolbox for Science

The causal results of a study, whether it is experimental, or quantitative, or of any other kind are not easy to decide. We still have to interpret the data, in the light of the whole setup of the study. To get to causes from data requires a good deal of causal *reasoning*. We need to be careful about what we infer from the data, and think about whether our evidence supports whatever we want to do with the new knowledge. Indeed, more and more scientific fields require an immense amount of advanced data processing before any results are reached. For a fascinating case to study, the data processing required by various brain-imaging techniques is eye-opening (Uttal, 2001, ch2). In order to think carefully about all this, we need to be equipped with a 'conceptual toolbox'; we present this now.

It is worth distinguishing between arguments (as they are developed in critical thinking or logic) and the abstract names for methods we introduce below (as they are developed in science for data analysis). The take-home message is that a deductive argument is not the same as a deductive method, and an inductive argument is not the same as an inductive method, as we will explain; this is why it is important to draw such distinctions. We finish by looking at levels of abstraction, which relate to the modelling practices we examine further in chapter 7.

3.1 Arguments

An argument is a set of propositions, usually just a set of sentences expressing claims, from which we infer another proposition or set of propositions, which is the conclusion or conclusions. Three kinds of arguments are worth distinguishing when thinking about science.

Deductive arguments. Deductive arguments allow us to infer a conclusion with certainty from a set of premises, given that they employ a valid inference rule. If the premises are true, the conclusion *must* also be true. But a deductive argument does not lead to 'new' knowledge: the content of the conclusion is already contained, albeit implicitly, in the premises. So, in a sense, deductive arguments allow us to make explicit what is already known and expressed in the premises. Deductive arguments can go from the general to the general, from the general to the particular, and from the particular to the particular:

- A syllogism from the general to the general:
  All humans are mortal.
  All philosophers are humans.
  Therefore, all philosophers are mortal.
- A syllogism from the general to the particular:
  All men are mortal;
  Socrates is a man;
  Therefore, Socrates is mortal.
- A modus ponens from the particular to the particular:
  If Socrates is a philosopher, then Socrates reasons about first principles;
  Socrates is a philosopher.
  Therefore Socrates reasons about first principles.

Inductive arguments. Inductive arguments allow us to infer a conclusion from a set of premises, but not with certainty. This means that the conclusion can be false, even though the premises are true. The reason is that, in inductive inferences, we move from content known in the premises to content in the conclusion that is not already known in the premises. So these arguments go beyond what is already contained in the premises. Inductive arguments are therefore called 'ampliative' and they are fallible. Inductive arguments can go from particular observations to generalizations, or to predict another observation:

Alice observes ten black ravens.

- Inductive inference to a generalization: *all* ravens are black.
- Inductive inference to the next case: the *next* raven Alice observes will be black.

Abductive inferences. Abductive inferences share with inductive inferences the fact that the conclusion does not follow from the premises with certainty and that, in a sense, the conclusion expands on what is stated in the premises. However, in abduction, from a set of premises, we 'abduce' the conclusion, which is a proposition that purportedly best explains the premises. The definition of abduction is much less clear-cut than the definition of deduction or induction—for a discussion, see e.g. Douven (2011). Here is an example of an abductive inference:

You happen to know that Tim and Harry have recently had a terrible row that ended their friendship. Now someone tells you that she just saw Tim and Harry jogging together. The best explanation for this that you can think of is that they made up. You conclude that they are friends again. (Douven 2011)

3.2 Methods

The types of inferences presented above have broad counterparts in the methods most commonly used in the sciences, but the argument forms are not the same as the methods. On one hand, an argument is a set of propositions, from which we infer another proposition or set of propositions. The important question about arguments is whether the truth of the conclusions does or does not follow from the truth of the premises. On the other hand, the *methods* derive consequences from axioms, explain a phenomenon using laws of nature, infer empirical generalizations from observations, and so on. The point of the methods described below is to expand our knowledge, whether in mathematics, or sociology, or biology. The methods describe a scheme of reasoning, used in scientific practice.

Deductive method. Generally speaking, a method is deductive when a system, such as a piece of logic or a physical theory, is generated using axioms or postulates and the inferences used in the generation are all deductive. Thus, for instance, Euclidean geometry is considered axiomatic because all of its geometrical propositions are deduced from its axioms. In mathematics, there have been attempts to 'axiomatize' the whole discipline (famously, by Russell and Whitehead). The 'deductive-nomological' model of explanation developed by Hempel is also an example of a deductive method, as particular events that have occurred are explained by deducing a statement of their occurrence from premises stating an appropriate set of laws of nature and initial conditions. The explaining argument shows that the occurrence of the event was to be expected.

Inductive method. Inductive methods are widely used in the empirical sciences in order to make predictions or to generate empirical generalizations that account for our observations. We use some form of inductive method any time we wish to draw conclusions about the world beyond what we have observed. In statistical analyses, we talk about 'model-based induction'. Inferences are inductive because from a sample we wish to draw conclusions about the whole population of reference, or about an altogether different population. For instance, we can randomly sample members of staff from philosophy departments in the UK, find out that, say, 10% are philosophers of science and infer that 10% of the *whole* population of philosophers in the UK consists of philosophers of science. Such inferences are 'model-based' because the modelling of the data (for instance, how we construct a variable), the sampling methods, the choice of measure of dependence, etc. all play a role in the quality and reliability of the inferences drawn.

Hypothetico-deduction (H-D). Hypothetico-deductive methods, despite the name, are more closely related to inductive methods than deductive methods. The goal of using them is to gain new knowledge, based on the available data and on background knowledge. There are three crucial steps in an H-D method. First, study the context (for instance the socio-demographic-political background) and the data (what data have been collected? how? what variables represent the concepts?). Second, hypothesize a causal structure, based on the preliminary analysis of data, the background context, similar studies performed, etc. Third, build and test a model for this causal structure, where consequences of the hypotheses are 'deduced' from the model; here deduction is to be taken in a loose sense. For instance, if the causal structure under test works for one sub-population but not another, we draw conclusions about the non-adequacy of the model for the *whole* population. There may have been problems in measuring the variables, or there may be an important difference in the population that is not captured by the model, and so extra work is needed. H-D methods are of course part of the legacy of Popperian falsificationism, which, crudely, holds that science tests the empirical consequences of theories, and if we find they do not obtain, we have falsified the theory. But there are important differences. One is the less strict meaning of 'deduction', allowing the consequences of hypotheses to be derived more loosely. Another is that, while Popper believed that the *whole* theory or model was to be rejected if its empirical consequences are not found, present-day methodologists advocate a milder view whereby we can update the theory or model instead and so build upon previous studies.

Inference to the best explanation (IBE). A core business of science is to explain phenomena. There are often several competing explanations for the same phenomenon. Which explanation should we choose? The obvious answer is to choose the *best* explanation. But what counts as the best explanation, and how to make such inferences, is a matter for debate. IBE follows abductive reasoning, so much so that 'IBE' and 'abduction' are often—but not always—used interchangeably,[2] because what may count as the best explanation does not follow deductively from the observations and the theory, nor is it merely an inductive step. For example, many authors have argued that features such as simplicity, or beauty, or intelligibility are important to choosing the best explanation. These kinds of features clearly go beyond deductive or inductive inferences.

3.3 Levels of abstraction

The last tool that we will be using in the course of the book is Levels of Abstraction (LoAs), in the form that derives originally from computer science, and is much discussed in the philosophy literature by Floridi (for example, in Floridi (2011*b*)). It allows us to state precisely problems that arise in our modelling practices in science.

Floridi notices that we pay attention to some features of any system we are interacting with, and ignore other features. This is true whether we are trying to model a system formally, or whether we are just looking at it, talking about it, or jumping up and down on it. For Floridi, an LoA is the set of variables that are attended to. Crucially, these are chosen for the purpose of the interaction, and the success of the interaction depends on well-chosen variables. If Alice is paying attention to the colour of the trampoline she is jumping up and down on, and failing to notice that it is being moved, things might go badly for her. Similarly, if Bob is trying to build a model of a

---

[2] An important exception is C.S. Pierce, who actually introduced the term 'abduction'.

population to predict migration patterns in the coming year, and he fails to include a variable measuring an important influence on migration, such as unemployment rate, his model will not make good predictions.

Floridi emphasizes that we can get into conceptual muddles very quickly by failing to notice that two people are describing the same system at different LoAs. For example, Floridi considers an argument about whether a building once used as a hospital, now converted into a school, is the 'same' building. If Alice is in urgent need of a hospital, the variables of interest to her mainly concern the Accident and Emergency Department, and it is not the same building. If Bob is trying to pick his niece up from school, and knows the way to the old hospital, then for his purposes it is the same building. Alice and Bob getting into an argument over whether it is or is not the same building, treating that question as if it could be answered independently of the LoA each is using, would not be useful.

Modelling is extremely important to science, as we examine in chapter 7 and touch on in many places (see also Morgan and Morrison (1999a)). Science can model well by making LoAs of models explicit. This can be as simple as specifying the variable measured, including the units of measurement used. A physicist and biologist each describing a body of water will use many different variables. Their models do not really disagree, once their varying purposes, and the relevant LoAs, are specified. In chapter 5 we draw attention to a problem in social science methodology. Describing a system using different types of data may lead to significantly different, and even opposite, results. Using data measured at the individual level or at the aggregate level may completely change the landscape. This means, translated into the terminology just introduced, that we have to specify the LoA at which we run the analysis. Simpson's paradox, discussed in chapter 7, can also be seen as a misspecification of LoA. Clarity about LoA is often very important.

In sum, in this chapter we have introduced some different argument forms, names for different broad scientific methods, and ways of thinking explicitly about models, all of which are useful to apply to scientific practice.

PART II

Causality: Accounts, Concepts and Methods

In this part of the book we present a number of accounts, concepts, and methods of causality. These occupy the current debate or are the background against which present-day debates originated.

CHAPTER 4

Necessary and Sufficient Components

4.1 Examples: electrical short-circuit and AIDS

Suppose Alice's house burned down. When the experts examine the site after the fact, they report that the fire was caused by an electrical short-circuit. However, note that the electrical short-circuit, all by itself, was not enough—it was not 'sufficient'—to cause the fire. There are other things that must happen—that are 'necessary'—for the fire, such as the presence of oxygen in the atmosphere. No oxygen; no fire. Also, other things can cause house fires, such as dousing them with gasoline and chucking a Molotov cocktail through the window. Why do the fire investigators report that the cause of the fire was the short-circuit?

The same line of reasoning may be employed in the area of disease causation. For instance, AIDS has several causes: exposure to HIV, repeated risky sexual behaviour, absence of anti-retroviral drugs to debase the virus, and even failure to die of some other cause before HIV infection has the 20 years or so it needs to develop into AIDS. All these factors, plus probably many others, have to go together to constitute the 'sufficient' and complex cause or causes of AIDS. Why do we think of HIV as the cause of AIDS?

This is a non-trivial problem across many sciences, where multiple causes may well be the norm. Sometimes we still want to identify something we think of as *the* cause. Why? And how? Reasoning about causality will be more effective if we can understand these different kinds of causes. We will examine in this chapter several attempts to discriminate among kinds of causes of the same effect—including some useful means to visualize these differences.

4.2 Component causes

We now examine the views. These were presented in three pieces of work developed between the late fifties and mid-seventies. These works have quite distinct scopes and objectives, and yet they share a core idea: causation is understood in terms of some qualified analysis of necessary and sufficient components.

Chronologically, the first piece of work presenting these ideas was Hart and Honoré (1959), who examined legal causality at length, drawing out a number of philosophical theses about causation from this context. The second was Mackie (1974), a philosopher who produced a study in the metaphysics of causation, developing ideas quite close to Hart and Honoré (1959) (a work which is explicitly discussed in Mackie's book). Finally, Rothman (1976) also developed ideas very close to those of Hart and Honoré, and Mackie, but in the context of disease causation in epidemiology. Rothman (1976) seems not to have known of the works of Hart and Honoré, or Mackie. Rothman's line of thinking in epidemiology is still used today, with some refinement, and in his later works the previous thinkers are acknowledged (Rothman et al., 2008). We will first look briefly at these three different contexts for the work, before presenting the core common idea.

Hart and Honoré. Hart and Honoré develop an account of causal *reasoning* in the law and in history. The book analyses causation in legal theory at length, in tort law under different facets, e.g. causing harm, negligence, contract, etc. It discusses issues of moral and legal responsibility and mentions differences between common law and civil law. A key aspect for legal causation is the distinction between causes and conditions. Hart and Honoré still are, in many ways, the main reference in legal causation. Moore (2009) explains that their work is a justification of the law's doctrines of 'intervening causes'. This claim is based on three premises: (i) we need to 'match' legal doctrines about causation and the conditions for moral blameworthiness; (ii) in the conditions for blameworthiness we have to include considerations about a pre-legal and pre-moral notion of causation; (iii) such a notion of causation is by and large provided by the everyday use of the notion.

Mackie. Mackie's book is an investigation of the *nature* of causation. It is a philosophical analysis concerned with ontology, specifically about what causal relations are, while, says Mackie, the sciences find out particular causal relations and causal laws. Mackie thinks that conceptual analysis, and analysis of language, give a rough guide to what is out there in the world (see also chapter 19). The aim of the work is to develop a metaphysical theory of causation, one that is in line with a Millian characterization of causation (for further details on J. S. Mill, see chapter 16).

Rothman. Rothman takes a *practical* approach to causality in epidemiology, aiming to highlight the multifactorial aspect of disease causation (the fact that diseases typically have multiple causes). His work also aims to bridge the gap between metaphysical conceptions of cause and basic issues in epidemiology. Rothman's 'sufficient component model' is a way of thinking about 'effect modification', namely about the 'characteristics' of diseases (the effect or effects) measured at different levels of other variables (the causes). While the *quantification* of effect modification can be different from study to study depending on scale measurement, the sufficient component model provides a qualitative view of the multifactorial aetiology of diseases.

The core of agreement of the three approaches is that causality has to be analysed in terms of necessary and sufficient components of causes, in the way we will now explain. For convenience, we begin with Mackie's and Hart and Honoré's discussion, and then examine Rothman's application to epidemiology in the next section.

We begin with the idea of necessary and sufficient causes:

- Necessary cause: a cause that must happen for the effect to happen.
- Sufficient cause: a cause that is enough all by itself for the effect to happen.

An example of a necessary cause might be infection with tubercle bacilli as a cause of the disease tuberculosis. You can have symptoms a bit like tuberculosis symptoms without being infected with the tubercle bacilli, but without the tubercle bacilli, the disease you have isn't TB. An example of a sufficient cause might be decapitation as a cause of death. Note that infection with the tubercle bacilli is not sufficient for TB. Most people infected by the bacilli do not develop any illness.[3] So there are some sufficient causes, and some necessary causes. There may even be some necessary and sufficient causes although these are uncommon. For instance, a net force acting on a body is a necessary and sufficient cause of acceleration.

The vast majority of the causes we are interested in are neither necessary nor sufficient. Electrical short-circuits are neither necessary nor sufficient for fires. Smoking is neither necessary nor sufficient for lung cancer. Being a member of an ethnic minority is neither necessary nor sufficient for suffering discrimination. Nevertheless, the ideas of necessary and sufficient causes can be very useful in characterizing many causes we are interested in.

According to Mackie, causes are at minimum *INUS conditions*. INUS stands for an Insufficient, but Non-redundant part of an Unnecessary but Sufficient condition (Mackie, 1974, p62). The idea is that when we say that the short-circuit caused the fire, it is insufficient to cause the fire, because it needs other conditions too, such as oxygen. However, it is non-redundant, because in this particular type of set of conditions, it in fact produces the spark. This set of conditions, which includes the short-circuit and oxygen, is unnecessary for fire, because fires can start other ways. However, that whole set of conditions, together, on this kind of occasion, is sufficient to start the fire. In this whole set of conditions, we are pointing to the short-circuit as the INUS condition that is the most salient factor, even though it couldn't have been as effective as it was without the other conditions. Note that the oxygen is also a cause of the fire, although not a particularly salient one. We will discuss this in the following section.

---

[3] It is worth noting that in the nineteenth century an injection with a large number of tubercle bacilli could have been a sufficient cause of tuberculosis, and was considered such at that time. Clearly, medicine evolves in its diagnosis prevention and treatment abilities, and therefore whether causes are necessary or sufficient in medicine may accordingly change.

This view, broadly shared by Mackie and Hart and Honoré, departs from a Humean conception of causation as 'uniform sequence' or regularity (see chapter 15), as if there were pairs of events that are always related. Hart and Honoré say that this is incorrect, and we should follow Mill's approach instead. For Mill, yes, causal generalizations are about regular occurrences of a given kind, but when a complex *set of conditions* is satisfied. Mill (1843, Book III ch5, sec3) says: 'The cause [...] is the sum of the total of the conditions positive and negative'. This definition conceives of causes as 'assemblies' of factors that also include negative factors or absences (on absences, see also chapters 11 and 13). According to Hart and Honoré, Mill's definition gets closer to the causal reasoning of lawyers than Hume's does. Mackie's work shows that our everyday causal reasoning is also permeated by a Millian, rather than Humean, conception.

It is worth noting that, for Mackie, INUS claims are primarily *generic* causal claims. Short-circuits are generally causes of fires. That is one of the reasons why they are dangerous. The factors involved in INUS refer to types of events or situations, and thus express the typical form of a causal regularity: there is an appropriate (generic) set of factors, some of which may be negative, that is both necessary and sufficient for the effect in the field in question. Mackie thinks that what is ordinarily called a cause (and expressed through causal verbs) is an INUS condition or, better, an individual instance of an INUS condition, and it may be a specific state, i.e. the token event 'electrical short-circuit', rather than a generic event. Hart and Honoré instead note that in legal causation the main concern is about single-case causal claims, such as a particular short-circuit causing the fire at Alice's house, rather than generalizations (see also chapter 5).

Note that not all causes are INUS conditions. The example of infection with tubercle bacilli above is not an INUS condition for tuberculosis; it is a necessary condition. If we do find necessary conditions for disease, that is very useful, as by preventing that condition we can be sure we will prevent the disease. Being decapitated is not an INUS condition for death; it is a sufficient condition. Nevertheless, the INUS concept is useful because it allows us to characterize much more precisely the nature of many of the causes that do actually interest us. This can sharpen up our reasoning, and so impact on our practices of inference, prediction and control.

4.3 INUS causes and related concepts

The usefulness of INUS conditions is not exhausted there. The introduction of INUS conditions also led to the specification of related notions in the literature. INUS conditions come along with a 'conceptual package' that also includes the following useful notions and distinctions.

Causal field. This notion was introduced by Mackie, who borrowed it from from Anderson (1938). The idea is simple: causal claims are always made in some kind of context, and in this background there is also the *causal field*. The causal field should not be confused with a part of the cause. It is '[...] a background against which the causing goes on' (Mackie, 1974, p63). It is the specification of the appropriate field that allows a correct distinction between causes and conditions; for instance theories of combustion belong to the causal field and allow us to say that while the short-circuit is a cause, presence of oxygen is a condition (see below).

Causes versus conditions. Hart and Honoré (1959, p36) distinguish between (INUS) causes and conditions. They say that 'conditions are *mere* conditions as distinct from causes because they are *normal* and not because they are contemporaneous with the cause' (emphasis added). Conditions are not the same as the causal field. They illustrate with the example of a person who lights a fire in the open, then a light breeze comes up and spreads the fire to the neighbour's property. The light breeze is a condition, not a cause, of the fire. The breeze is later in time than the cause, (i.e., the person's action to light the fire), and it is 'normal', because there is nothing unusual about a breeze. Recall that Hart and Honoré (1959) situate their discussion in the field of legal causation, where the issue is to decide the legal responsibility for the fire, not the 'physical' cause of the fire. Moore (2009) correctly reiterates that causes in Hart and Honoré's approach are either human voluntary actions or 'abnormal' natural events (see below) that are necessary elements completing a set that are sufficient for the effect, that is the INUS's cause. Mackie (1974, p36) also endorses the distinction between causes and conditions, and notices that conditions are part of the 'total' or 'sufficient' complex cause. In the given circumstances, if the condition (the breeze) had not happened, the effect (the neighbour's house on fire) would not have happened either. Interestingly enough, counterfactual analyses apply to both causes and conditions and, unlike the approaches discussed in chapter 9, they cannot identify the univocal—the sole—cause of an effect.

Normal versus abnormal. Hart and Honoré (1959) notice that the sciences usually want to know about *types* of occurrences in order to establish causal generalizations that describe what 'normally' happens. In ordinary life, and legal contexts, we are often interested in *particular* occurrences, seeking for example an explanation of a particular situation that departs from a 'normal, ordinary, or reasonably expected course of events' (p31). Normal conditions are thus present as part of the usual state or mode of operation. Normal and abnormal conditions are relative to the context, or causal field. In our earlier example of the short-circuit causing fire, the presence of oxygen in the atmosphere is normal. But because the oxygen is among the normal background conditions in which a building stands, it is not considered a cause in the same sense as the short-circuit. A consequence of this is that the distinction between causes and conditions also depends on establishing what are the normal and the abnormal conditions. In a 'normal' situation conditions are not causes, and it is precisely the 'abnormal' situations that make a difference for the production of the effect. Establishing what is normal and what is abnormal—and consequently which are the causes and which the conditions—is driven by practical interests: the reasons behind a causal question also help identify causes and conditions (a point also made by Collingwood (1938)). This is important to scientific practice. For example, in examining biomedical mechanisms in the human body, a temperature of approximately 37° Celsius, and the availability of oxygen and glucose, are normally assumed.

4.4 Rothman's pie charts

As we indicated early on in the chapter, the same approach has been developed in the field of epidemiology in order to provide an account of disease causation. We turn now to that, which will serve to illustrate the usefulness of the account.

Rothman (1976) makes the point that in health situations what we call 'causes' are in fact *components* of sufficient causes, and are not sufficient in themselves. For instance, the measles virus is said to be the cause of measles, but in fact the 'complete sufficient cause' of measles also includes lack of immunity to the virus and exposure to the virus. Likewise, drinking contaminated water is but one component of the complete sufficient cause of cholera, and smoking is but one component of the complete sufficient cause of lung cancer. The account has also been developed in subsequent works (see e.g. Rothman and Greenland (2005) and Rothman et al. (2008)). What Rothman means here is just the same as what is meant by an INUS condition above.

Rothman illustrates his ideas by means of 'causal pies'—see Figure 4.1—still used in epidemiology now. He thinks that a sufficient cause of a disease is generally not one single causal factor, but a complete 'causal mechanism'. Rothman takes a causal mechanism to be a minimal set of conditions and events that are sufficient for the disease to occur. In this perspective, no specific event condition or characteristic is sufficient, by itself, to produce the disease. So the definition of 'cause' does not describe a complete causal mechanism, but only a component of it. It is worth noting that Rothman's use of mechanism differs from the one discussed in chapter 12.

The sufficient component model helps with the concept of strength of an effect and interaction. In a sufficient component model, or pie chart, we display all known factors that play a role in the occurrence of a particular disease. Pie charts, in a sense, visualize contingency tables (see chapter 7), but with an important difference: in a pie-chart aetiological fractions—i.e., the components—do not have to sum up to one. This means that we are not looking for the sum of the components that make the effect necessary. It is a useful heuristic way of thinking about the multifactorial character of most diseases.

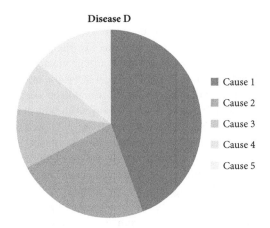

Fig 4.1 Rothman's sufficient component model of disease causation.

A natural question is whether this kind of account clashes with probabilistic approaches that we examine in chapter 8 and that seem to permeate our causal thinking in everyday and scientific contexts. Rothman et al. (2008, ch2) discuss the use of risks and probabilities in epidemiology, but they do not relate these concepts to the sufficient component model. So the first impression one gets is that the model is missing a concept of probabilistic causality, which is well entrenched in epidemiological thought (see for instance Lipton and Ødegaard (2005) and Parascandola and Weed (2001)).

But the authors later clarify that if one wants to introduce a chancy, stochastic component in the model, then it suffices to use a stochastic model to describe a dose-response relationship that relates the dose of the exposure to the *probability* of the effect occurring (see Rothman et al. (2008, p17)). The sufficient component model is a 'deterministic' model, but one that can accommodate 'chance' quite easily: one way is to view chance, or at least some part of the variability that we call chance, as the result of deterministic events that are beyond the current limits of knowledge and observability. In other words, disease causation may well be completely determined by disease mechanisms, but our partial and imperfect knowledge and understanding of disease mechanisms lead us to model these phenomena probabilistically (see also chapter 8).

Finally, Rothman and his collaborators do not claim that pie charts are universally applicable, not even in medicine. They say that there are *some* phenomena that we can explain using this model, but not all. The potential outcome model (see chapter 9), for instance, helps relate individual to population-level causation; graphical causal models (akin to Bayesian nets presented in chapter 7) are useful instead when modelling a causal system. Also, pie chart diagrams of sufficient causes and their components are not well suited to providing a model for conceptualizing the 'induction period', which may be defined as the period of time from causal action until disease initiation (see Rothman et al. (2008, p15ff)). They also do not help with cases where causes have to happen in a particular order for the effects to result. For example, in coinfection with hepatitis B and hepatitis C, the order of infection matters. Having chronic hepatitis B, then acquiring hepatitis C seems to be worse than having hepatitis C first (Crockett and Keeffe, 2005). In general, INUS analyses apply only when components act independently, and not when they interact with one another as hep. B and hep. C infections do. Admittedly, this narrows the scope of INUS causes. So it is specifically the 'multifactorial' aspect of disease causation that pie charts help model.

In sum, in spite of these limitations in application, the idea of component causes, and associated distinctions such as those between causes and conditions, and normal and abnormal conditions, are very useful in sharpening up our causal reasoning.

Core ideas

- An effect often has multiple causes. Epidemiologists call this 'multifactorialism'.
- INUS causes are Insufficient but Non-redundant components of an Unnecessary but Sufficient condition for the effect, such as a short-circuit causing a fire. As it happened, this was the important cause, either in one particular case, or in a type of case, even though there are other ways of starting fires.

continued

Core ideas *Continued*

- INUS causes highlight the difference between causes and background conditions, such as the short-circuit—the cause—and oxygen in the atmosphere—a background condition.

Distinctions and warnings

- The necessary and sufficient components approach does not say that causes are necessary, it says that they are part of a complex set of causes and conditions, or a complex mechanism, that is sufficient for the effect to occur.
- The approach sheds light on the difference between assessing the physical, biomedical, etc. causes of some phenomenon and the 'causes' in the sense of legal and moral responsibility.
- Pie charts illustrate the multifactorial aspect of disease causation, but are by no means the only model for disease causation.

Further reading

Hart and Honoré (1959, chsII and IV); Mackie (1974, chs2, 3, and 8); Moore (2009, ch5); Rothman et al. (2008, ch2).

CHAPTER 5
Levels of Causation

5.1 Examples: personalized medicine and migration behaviours

Suppose during a general health check-up the medical practitioner asks Bob whether he smokes and he answers positively: not very heavily, but still he has smoked on average ten cigarettes a day for fifteen years now. What's the chance Bob develops cancer or cardiovascular disease? After all, Bob has a very healthy life apart from smoking: he exercises regularly, and eats enough fruit and vegetables, although he indulges himself with a glass of wine over dinner. Should the doctor urge Bob to quit smoking? We have extensive medical knowledge, both in terms of risk factors and of mechanisms, about the relation between smoking and various diseases at the population level. Yet this generic knowledge does not immediately inform us about what would, will, or did happen to a particular individual: it does not directly inform us about the single case. This is *one* problem of the levels of causation.

Consider now another example. Demographers study, among other things, migration behaviours of populations. One such study was pioneering in developing new methods for analysing individual and aggregate data. Scientists used data from the Norwegian population registry (from 1964) and from two national censuses (1970 and 1980). They then developed two models, one using aggregate data (e.g., migration rates, percentage of farmers or of other occupations, etc.) and one using individual data (i.e., the same variables as measured at the individual level). In the first instance, the aggregate model and the individual model showed opposite results. According to the aggregate model, the regions with more farmers were those with higher rates of migration. According to the individual model, in the same region migration rates were lower for farmers than for non-farmers. This apparent contradiction in the results was explained away as due to different methodologies for analysing the data. Multi-level models can use aggregate and individual-level variables in the same statistical framework: aggregate characteristics (e.g. the percentage of farmers) explain individual behaviour (e.g. migrants' behaviour). Here, we can run the statistical analysis using individual-level data, aggregate-level data, or a combination or both. The mode of analysis will influence the conclusions to be drawn. And it may happen that the individual-level analysis does not justify inferences about the aggregate level, and vice-versa. This is *another* problem of the levels of causation.

Our causal reasoning, and our ability to infer causes, and predict and control effects, are all influenced by these problems.

5.2 Three parallel literatures

We find three areas where the problem of the levels is discussed. The philosophical literature on causation recognized this issue quite early, but most philosophical discussions were carried out independently of science. The legal literature also acknowledged the distinction of levels of causation, and was informed by the philosophical literature, although the most recent developments have not been integrated yet (in either direction). Finally, the scientific literature had to deal with the problem of levels from a methodological and practical point of view, and again did not cross paths with philosophical discussions. The existence of at least three parallel literatures suggests that this is a fertile area for interdisciplinary collaborations.

5.2.1 The philosophical literature

During the eighties the main accounts of causality under debate were Suppes-like probabilistic approaches (see chapter 8) and Salmon's causal–mechanical account (see chapter 11). A stock example used in that literature has a golf player in several settings. Suppes (1970, p41) reports that the original example is due to Deborah Rosen (see Rosen (1978)). In one example, the golf player, who is just a mediocre player, hits the ball, which flies away from the hole. But a squirrel appears and kicks the ball, which then falls unexpectedly into the hole. We wouldn't expect that the squirrel's kick would be followed by a hole-in-one. But in this particular case, the kick seemed to cause a hole-in-one, even though in general we would judge squirrels' kicks to *prevent*, rather than cause, holes-in-one. There have been several variants of this example, aimed at defending one or the other of Salmon's or Suppes' views. It is a fictional example, a 'thought-experiment', and you can add as many details as you like to make it seem real: speed and direction of the wind, friction of the grass, and so on may enter in the calculation of the probability of a hole-in-one. But this is not the point. The example succeeds, at least, in isolating the following issue for discussion: causal relations in single cases may not meet our generic expectations. This has led philosophers to develop accounts of causation that cope with the fact that squirrels' kicks seem able both to prevent and to cause holes-in-one.

One such account is Elliott Sober's two concepts of cause (Sober, 1984). This is not to be confused with Ned Hall's two concepts (see chapters 19 and 23). Sober suggests that we need two distinct concepts of cause: type and token, alternatively sometimes called population-level and individual-level. The rough idea is that at the type or population-level, squirrels' kicks do lower the chances of holes-in-one, and so tend to prevent holes-in-one; but at the token or individual-level it may happen that a squirrel's kick causes a hole-in-one. The two causal claims do not contradict one another, since they are different claims: a type or population-level claim for one, and a token or individual-level claim for the other.[4]

The philosophical discussion moved to worrying about the relation between the levels, which we will discuss in the next section. But the levels issue was reanimated by Ken Waters (2007), albeit from a slightly different angle. Waters examines the case of 'genome-wide association studies' (GWAS for short) and tries to disentangle different ideas. His main idea is that causes make a difference to their effects, but the difference they make depends on whether they are actual or potential causes.

For example, there have been various GWAS to examine genetic loci possibly associated with Alzheimer's Disease, beyond already-known associations, such as the gene *APOE*. Bertram and Tanzi (2009) survey existing studies, which report over two-dozen new potential loci for susceptibility to Alzheimer's. GWAS typically test several hundreds of thousands of genetic markers, examining single nucleotide polymorphisms (SNPs) chosen to indicate genetic variation among humans. They compare genotype or allele frequencies in individuals affected with Alzheimer's to those in individuals not affected. At the time of Bertram and Tanzi's review, findings for Alzheimer's, barring *APOE*, were not very consistent. In the authors' own study, two genes, $GWA\_14q31.2$ and *CD33*, attained genome-wide significance in the initial study, and in follow-up studies. The review concludes that, in the light of the studies to date, 'the most compelling and genuine non-*APOE*-related published GWAS signals have been observed in *GAB2*, followed by less consistently replicated signals in *GALP*, *PGBD1* and *TNK1*' (Bertram and Tanzi, 2009, pR143). But they note that the list of loci worth investigating is likely to keep growing, and a great deal more work is needed.

Waters' point is that GWAS tell us only that, say, the gene *GAB2* is a *potential* cause of Alzheimer's disease. The biggest trial examined by Bertram and Tanzi had approximately 16,000 cases and controls combined. Studying such large populations is necessary to find genes that may have rather small effects. However, gene *GAB2* in Alice may be an *actual* cause of Alzheimer's disease when she *actually* develops the disease.

One way to understand Waters is that he uses actual and potential for what the levels literature called the individual and population level, or token and type. It is an interesting idea that at least in some cases population studies identify potential causes, and we have then to decide which are actual in particular cases. We will discuss a possible shift in the terminology later.

The actual versus potential causation issue is also rehearsed in the recent literature on causal powers, which in some accounts have an Aristotelian flavour, and may be actual or potential (see chapter 14). There is another related area where the actual–potential distinction arises. Hitchcock and collaborators (see e.g. Hitchcock and Knobe (2009)) use the idea of actual causation in applying probabilistic causal

---

[4] One of the authors (Phyllis) discovered personally that cleaning spilled fuel off the decks—a type or population-level preventer of sailing accidents—can in the token or individual-level case produce a sailing accident.

reasoning and causal modelling to debates on, for example, moral and legal responsibility. They suggest that what counterfactual questions we should ask depends on the *actual* causal story.

5.2.2 The legal literature

In the legal literature there are two, related, problems. One is the scope of different causal claims and how close these are to legal theory. The other is about using one level (typically the general) for causal attribution at the other level (typically the singular). Moore (2009), in particular, is concerned about distinguishing two types of causal statements:

(i) Singular causal relations e.g. 'the spark caused the fire'. This holds between token events or states of affairs.
(ii) Statements of causal generalization e.g. 'sparks cause fires'. This holds between types of events or states of affairs.

The distinction between (i) and (ii) has a bearing on considerations about causal relata, namely on what we allow as legitimate terms that stand in a causal relation. We discuss relata in chapter 21. Moore examines both generalist and singularist theories of causation. Generalist theories aim to reproduce some law-based relation. Moore thinks these theories share the broad problems of counterfactuals (epiphenomena, overdetermination, omissions, etc.—see chapter 9). Singularist theories, argues Moore, survive these problems better. These theories are closer to a common-sense understanding of causation and thereby to the presuppositions of a legal theory of causation. The legal framework considered by Moore is especially of tort and criminal law, so it is understandable that his main focus is to establish responsibility *in the single case*, for facts that have actually occurred.

The distinction between general and singular causation is also discussed by Federico Stella (2003). He addresses problems about freedom and security as they are threatened by scientific and technological progress, especially in modern industrialized countries. One aspect Stella is interested in is how to use scientific evidence in establishing liability. Think of cases such as the thalidomide scandal or diseases caused by occupational exposure. The problem is whether and how to use information of general causation established in science to ascertain *individual* responsibility. This is a typical problem for using epidemiological evidence, which is by definition about health and prevention in *populations*.

A survey of several scholars, ranging from law to epidemiology to philosophy, shows that there is a unanimously recognized problem: the single case cannot automatically be subsumed under the general case. Epidemiological evidence can show us risks and odds at the population level, but not at the individual level. This is explicitly recognized also by agencies such as IARC (International Agency for Research on Cancer) and the EPA (Environmental Protection Agency).

Stella notices that judges and jurists in common law jurisdictions have taken an even stronger stance about the impossibility of general causation informing single-case causal attribution. Specifically, all civil sentences require proof of singularist causation. So what is singularist causation and how can we prove it?

Stella discusses some strongholds in civil and tort law. Some behaviour is considered causal when it constitutes a necessary condition for the event, and yet such behaviour is but one of the many necessary conditions for the occurrence of the event. So a behaviour may not be necessary on its own, but necessary only contingent on some other conditions or broader context. This, notice, is clearly an INUS concept (see chapter 4). Interestingly, next to this INUS idea, we find the idea of subsumption under scientific law, which is basically the D-N model of explanation.[5] This is of course problematic, as philosophy of science has moved beyond D-N and developed alternative models of explanation, for instance based on mechanisms (see chapter 12). Stella also discusses other forms of particularist proof, where 'personalized' evidence is considered. For instance, ballistic examinations can provide information that matches a particular weapon to a particular bullet, in virtue of the unique characteristics of the weapon.

What we can conclude from this brief discussion is that there is a genuine distinction between general and singular causation. A major problem is to ascertain causation in the single-case, and jurisprudence is, overall, aware of difficulties for using general knowledge coming from fields like epidemiology. But when it comes to singularist proofs, the discussion is confined to outdated models of explanation (the D-N model), so there is plenty of space for philosophers of causality to work with jurists to integrate the advancements made in both areas.

5.2.3 The scientific literature

As anticipated above, the problem of the levels of causation is known and discussed in the scientific literature, too—it is just approached from a different angle. The angle is that of measurement of variables, which affects what data is used in the usual corresponding statistical analyses. So in the scientific literature, the problem of the levels of causation is a problem of what *unit of observation* to choose, whilst in the philosophical literature it is about the *scope* of applicability of a causal claim—what the claim applies to.

Traditionally, the social sciences developed two approaches, called the holistic, and atomistic or individualistic, approaches. According to holism, the system as a whole determines the behaviour of the parts in a fundamental way; the properties of a given system cannot be reduced to the mere sum of the properties of its components. According to individualism, social phenomena and behaviours instead have to be explained by appealing to individual decisions and actions, without invoking any factor transcending them. This problem also plays out in other scientific fields. For example, systems biology emphasizes the importance of systemic properties of the whole to the behaviour of the system, and thereby that there are limitations to studying only the system parts.

The above views about social reality 'translate' into the mode of analysis of data. Let us consider quantitative analysis, for instance statistics in social science. A terminological precision about variables is in order:

---

[5] The D-N, or deductive-nomological, model of explanation maintains that to explain a phenomenon is to deduce it from general laws and initial conditions. See Salmon (1989).

- *Individual level variables* measure individual characteristics, which means they take values of each of the lower units in the sample. For instance, we may measure income for *each* individual in the sample and use that individual level variable in the analysis.
- *Aggregate variables* are summaries of the characteristics of individuals composing the group. For instance, from individual level measurements we can construct the *mean* income of state residents in the sample.

The statistical analyses of these two types of variables have counterparts in terms of types of models: individual-level models and aggregate-level models. Aggregate-level models explain aggregate-level outcomes through aggregate-level variables. For instance, in the example at the opening of the chapter, the aggregate model explained the percentage of migrants in a region through the percentage of people in the population having a certain occupational status (e.g. being a farmer). Individual-level models, on the other hand, explain individual-level outcomes using individual-level explanatory variables. In the example, the individual-level model explained the individual probability of migrating through the individual characteristics of being or not being a farmer.

But these models, as we have seen, may lead to opposite results when studying the *same* data set. The reconciliation came from the development of *multilevel models*, in the pioneering works of Daniel Courgeau and Alvin Goldstein (Courgeau, 1994; Courgeau, 2007; Goldstein, 1987; Goldstein, 2003). Simply put, multilevel models are a special type of statistical model used in causal analysis to model hierarchical structures. The first important point is the recognition that society is organized into hierarchies. For instance, individuals form families that form local populations that are aggregated into the national population. Or: firms can be aggregated into regional markets which form national markets which constitute the global market. Or again: pupils form classes that form schools that form school systems. The second important point is the recognition that there is no a priori reason to choose one level of analysis over another. Actually, there are many good reasons to study the *interactions* between the levels. One of these reasons is the contradictory results that may be obtained when studying a phenomenon only at one level. Recall the example of farmers' migration mentioned earlier. To understand what was going on in Norway, scientists had to make claims across the levels, from the aggregate level to the individual level and vice-versa. In fact, what they did was to explain the *individual* probability of migration for farmers through the *percentage* of farmers in the same region. Choosing the model to use at the micro or macro level is not easy—for a discussion see for example Kittel (2006).

Multilevel models are also used in other areas, such as in educational research. Driessen (2002), for instance, studies school composition and primary school achievement, analysing data about Dutch primary schools. The response variable or outcome is 'language and math proficiency'. Driessen is interested in assessing whether and to what extent the following factors help explain differences in proficiency between pupils: parental ethnicity and education, pupil's sex and age, school composition, and ethnic diversity. Some of these factors are individual properties (such as sex and age), but others have a group dimension, such as the composition of the school (the proportion between Dutch and non-Dutch pupils) or the ethnic origins of the pupils. The study shows that there is quite a strong effect of school composition on language, but a weak one on maths; all children, independently of their background, perform worse in schools with high ethnic diversity. These kinds of study are, needless to say, important to ask questions about distribution policy and other measures in education.

Hopefully it is clear that mistakes can be made in causal reasoning by ignoring the levels of causation. It is worth noting that in social science the problem of the levels was identified early by Robinson (1950). He coined the two terms 'atomistic' and 'ecological' fallacies. The *atomistic fallacy* occurs when we wrongly infer a relation between units at a higher level of analysis from units at a lower level of analysis. The *ecological fallacy* occurs when we draw inferences about relations between individual-level variables based on group-level data. Robinson developed these ideas in a paper on illiteracy and immigration. He analysed data from the 1930 census in the US for each of 48 states and the District of Columbia. The statistical analysis of data revealed two apparently contradictory results. On one hand, Robinson calculated the 'individual correlation', among descriptive properties of *individuals*. Here, the correlation was positive: immigrants were more illiterate than native citizens. On the other, he calculated an ecological correlation, between descriptive properties of *groups*. Here, the correlation was negative: the correlation between being foreign-born and illiterate was magnified and in the opposite direction. Here, the correlation was negative: within groups, there was an even greater correlation between being foreign-born and being more literate than native citizens. A finer analysis of the data and of the phenomenon led Robinson to put forward the following explanation: immigrants tend to settle down in states where the native population is more literate, and that's why when analysing data at the individual level and at the group level we find different results. Robinson's moral was indeed that care is needed when handling data and drawing inferences. This problem is a manifestation of Simpson's paradox, discussed in chapter 7.

5.3 Bridging the levels—and the terminology!

5.3.1 The terminological gap

The significant terminological divergence between philosophy and social science is highly likely to impede understanding. In the light of it, we want to suggest a switch in the vocabulary. The 'type/token' and 'population/individual' terms could usefully be replaced with the following terms: generic and single-case. 'Generic' refers to the possibility of replicating a relation, or to the fact that a relation has been replicated. 'Asbestos causes cancer' is generic in that it allows for multiple instantiations. On the contrary, 'Bob's exposure to asbestos fibres caused him to develop cancer' is single-case, as this relation is not repeatable, it is unique. This allows precision in cases that are currently muddled. Some single-case causal relations refer to a population. For instance, in Japan 1966 was the 'Year of the Fire Horse'. There was a popular myth saying it was bad luck to be born that year, especially for women. This folk belief led to a drastic birth drop, of about 25%, in 1966. This is a single-case happening, as it is a one-off case, but it is one that concerns a population.

The 'generic–single-case' terminology also better accommodates another distinction coming from the application of statistics to the social sciences: the statistical individual versus the real individual. The social sciences make extensive use of probabilistic models. One thing we can do in these models is to pick out two 'real' individuals, for which we have measured some characteristics, and ask what happens to them. For instance, for individual A and individual B in the cohort of the migration study in Norway, we may check whether two farmers, who also share other characteristics, end up migrating to a different region or not. Of course, there is no reason why two 'real' observed individuals should follow the same process. It is usually difficult to do this when data come from big surveys (especially because most of the time anonymity has to be preserved). But in other areas of social science this may be possible, for instance in ethnographic research.

However, imagine that you pick out from the cohort two individuals, at random. If you abstract from the 'real' context of these individuals, all you know is that they belong to that cohort, where certain statistical relations between characteristics have been observed. Then, these two individuals are not real, but just 'statistical', and they automatically follow the same process, precisely because they represent, so to speak, the average behaviour observed at the aggregate level. In fact, random processes typically average out on the characteristics of the total number of individuals observed. This is nicely explained by Courgeau:

> In the search for individual random processes, two individuals observed by the survey, possessing identical characteristics, have no reason to follow the same process. By contrast, in the search for a process underlying the population, two statistical individuals—seen as units of a repeated random draw, subject to the same selection conditions and exhibiting the same characteristics—automatically obey the same process. (Courgeau, 2003, p8.)

Table 5.1 recaps the terms used in the literature and where they apply.

5.3.2 The relation between the levels

Clearly, confusion about the levels of causation and concealed switching between the levels, such as when we use a generic claim to infer what will happen in a single case, can lead us to muddled thinking and false conclusions. So it is natural to ask whether there is any systematic relation between the levels of causation.

Sober (1986) hypothesizes the existence of a 'connecting principle', according to which the *likelihood* of the single-case relation is increased by the strength of the corresponding generic relation (substituting our terminology for Sober's). Let's get on to a more scientific example, one that is standard in the philosophical literature: smoking and cancer. Sober's idea is that the more evidence we have about the effects of smoking on lung cancer in general, that is, the more evidence for the generic claim, the more likely (*ceteris paribus*) the single-case relation will be, that is, that smoking causes lung cancer in individual patients. This does not say, of course, why something in accordance with or contrary to our general knowledge actually happens in the single case. It just says, intuitively, how we form expectations about single cases from generic knowledge. This epistemic connecting principle has been 'rediscovered' and reformulated in order to address some conceptual imprecisions in Russo (2009), and also expressed in a more cogent and computational way by Kleinberg (2013). Here, then, the generic level is epistemically prior, and allows us to make inferences about single cases.

TABLE 5.1 Terminology of levels

	Meaning	Example	Used by / context	Applies to
Population / Individual	Refers to a causal claim either at the population level or at the individual level	Smoking causes cancer / Bob's smoking caused his cancer	Philosophical literature, probabilistic causality	Causal claim
Type / Token	Refers to causal relations between characteristics or between instantiated events	Smoking causes cancer / Bob's smoking caused his cancer	Philosophical literature, probabilistic causality	Causal claim
Generic / Single-case	Refers to a causal relation that can be instantiated more than once or to an instantiated relation / Single-case causal relations can also be at the population level	Smoking causes cancer / Bob's smoking caused his cancer / The financial crisis in 2008 caused recession in Europe	Link philosophical and scientific literature	Causal claim
Aggregate / Individual	Summary of characteristics of individuals composing a group / Measure of individual characteristics, taking values of each of the lower units in the sample	Mean income of state residents / Income of each individual in the sample	Scientific literature	Variable

Ellery Eells (1991) takes a different view. For Eells, the generic and the single-case levels are unrelated. In his view there are different 'probability raising' concepts at work at the different levels. In the single case it is a physical, objective probability that we need to track; in the generic case, it is a change in the frequencies that we need to track (for his overall view, see chapter 8). The problem of the levels, including Eells' account, is also discussed by Russo (2009) and Kleinberg (2013). These works discuss how the levels are related in given scientific contexts (the social sciences and computer science respectively). Here, the levels are epistemologically and methodologically related. This means to make inferences at one level, we need (some) knowledge of instances at the other level. Likewise, methods for drawing such inferences at different levels are intertwined.

But there are also different views about the *metaphysical* relation between the levels. Mackie (1974, ch3), for instance, holds the view that the single case is the primary case, and generic causal relations are secondary. A similar view has been defended by Cartwright (1989). She holds that single cases of causings in the world are primary in an ontological sense: what exist are single cases of smoking causing lung cancer in different people at different times. The generic level is in a sense only derived, or resultant, due to repeated instances of single-case causes. This is true, Cartwright thinks, even if our causal epistemology often seeks out generic causes as a means to discovering what happens in the single case.

Daniel Little (2006) tackles the problem of the relation between the levels with reference to the practice of the social sciences. He explains what it means to establish claims at the generic and individual level. While he broadly embraces a 'microfoundations' view (i.e. social 'aggregate' facts are rooted in the action of individuals), he still grants that generic causal claims are real and also theorizes intermediate or, as he calls them, 'meso-level' relations.

Ultimately, the metaphysical and the epistomological relations between the generic and the single-case levels are far from being settled. Effective causal reasoning needs to bear this in mind. It is likely that whether a generic causal claim will justify a single-case inference, or attempt at control, needs to be assessed on a case-by-case basis.

Core ideas

- Causal claims about populations or types are importantly different from causal claims about individuals or tokens—for example, just because medical knowledge tells us that smoking generally causes lung cancer, it doesn't follow that *Bob* smoking caused *Bob's* cancer.
- When the sciences studying hierarchical structures of social systems and other kinds of systems use the language of 'levels', they are talking about levels of modelling, which includes the levels at which variables are measured.
- The generic/single-case distinction helps bridge the gap between terminological differences in the scientific (aggregate vs individual variable) and in the philosophical literature (type vs token; population vs individual).

Distinctions and warnings

- Be careful translating the language between the debates. For example, statistical individuals are not real individuals. 'Individual' and 'aggregate' refer to the measurements of variables, for scientists; while 'individual-level' and 'population-level' refer to the scope of causal claims, for philosophers.
- Mistranslation of this language can cause serious misunderstanding. A claim *about* a population can be a single-case causal claim. For example, the drastic drop in birth rates in Japan in 1966 concerns the whole population, but is single-case.
- There are different types of relations among the levels that we might be interested in: epistemological, metaphysical, or methodological.

Further reading

Eells (1991, chs 1, 3, and 6); Kleinberg (2013, ch2); Russo (2009, ch6); Sober (1986).

CHAPTER 6

Causality and Evidence

6.1 Examples: effects of radiation and smoking causing heart disease

Radiation is a physical phenomenon whereby a 'body' emits energy, for instance electromagnetic, thermic, light energy and so on. An important question in studying radiation is how such emitted energy is then absorbed, either partially or totally, or reflected, or diffused by another physical system. Studying radiation also means studying its effects. We know that radiation (ionizing and non-ionizing) can have harmful effects in organisms. Some are quite well known. For instance, scientists in radiology and diagnostic nuclear medicine measured the sensitivity of different organs and thus established 'effective radiation doses'—the dose of radiation that allows scientists to see something in our bodies without triggering harmful effects (Mettler et al., 2008). Other effects, however, are much less understood and therefore need more *evidence* to be collected and evaluated. These include mutation, cell death, erythema as in sunburn, and heating. One example where we need more evidence concerns the use of mobile phones allegedly causing brain cancer. This is an interesting case, because the evidence itself is under debate. We know that some radiation can have harmful effects on the body, and we know that mobile phones emit radiation. But we don't know whether the radiation mobile phones emit actually has any harmful effects. Another disputed case is whether microwave ovens cause cancer.

The difference between this and the discovery that smoking causes heart disease is striking. Scientists noticed a correlation between smoking and heart disease. This led them to do experiments to try to find a mechanism by which the smoke, entering the lungs, somehow affects the heart. What we are carrying out in such cases is causal inference. We can see that some people suffer from heart disease, and that disproportionately many of them smoke. We are trying to find out whether smoking causes heart disease—and if it does, how. In the different case of mobile phones and cancer, it is difficult to collect clear epidemiological data. For instance, it is difficult—and perhaps not possible—to collect data about exposure to radiation from phones *alone*. We are exposed to many other sources of radiation, e.g. the microwave that we use twice a day to warm up meals; the Wi-Fi repeaters at the university, in the pub, etc.; the x-rays we have had at the dentist's over the years; and natural radiation to which we are all exposed. Moreover, there may be interactions with other factors: does it matter whether, on top of being exposed to radiation sources, we smoke?

Radiation is an interesting case because fundamental science and its applications turn out to be extremely useful for medicine, giving us fantastic diagnostic devices such as CT and MRI scans. At the same time, radiation could also be the cause of an increased burden of disease, especially cancer. So we need to understand more deeply the mechanisms of radiation: how it works, and how we can use it, and within what limits.

Many causal claims we make every day do not require extensive evidence. Suppose Alice usually catches a bus around 8 a.m. and gets to her office by 8.30 a.m. Today, however, Alice missed it. The next bus got stuck in a traffic jam, and Alice arrived at her office at 9.15 a.m. instead. Did missing the bus cause Alice to be late? We would automatically say 'yes'. In everyday reasoning, we may well quickly compare an unusual situation to the usual one we expected—what Peter Menzies (2007) calls 'default worlds'—and so decide what has changed, and caused the unusual effect. While the importance of default reasoning should not be underestimated, finding causes is not usually so simple in science. To find out anything in science, we look for evidence. Causal relations are no exception. Indeed, evidence for causal relations is one thing many different sciences spend a lot of time looking for. Does radio-frequency energy cause brain cancer? What kind or kinds of things do we need to look for in order to establish the existence of such a causal relation?

There are official bodies that collect, evaluate, and explain the evidence for important causal knowledge. If you browse the websites of the World Health Organisation, or of the International Agency for Research on Cancer, you will find pages explaining what relations have been deemed causal, what relations are instead deemed mere associations as yet, and why. Most countries also have a body specifically charged with deciding what medical treatments are allowed or paid for in that country, such as the US FDA (Food and Drug Administration), or NICE (National Institute for Health and Care Excellence) in the UK, or the European Medicines Agency. These bodies give good examples of how we establish the plausibility of causal claims based on the information we have. Thinking about the important roles of regulatory or research bodies such as these shows that there is a further problem of evidence of causality in science: not just the *scientific* problem of how to establish causal claims, but also the problem of *engagement* with science; effective communication between science and both the public and policy makers who are often not specialists in physics, medicine, or any other science.

6.2 What do we want to know?

The first really important thing to realize is that we want to know quite a lot of different things. They are closely related, because of course they are all something to do with whether a causal relationship exists between possible cause C and possible effect E, but they are not all the same. The phenomenon of radiation discussed in the opening of the chapter, and the controversies over its potentially harmful effects, show that we have different reasons for establishing that C causes E, and consequently may require different kinds of evidence. Let us be more specific.

What causes what, and how? We saw above in the smoking and heart disease case that we want to answer: what is the cause of the observed effect of increased incidence of heart disease?—is the cause smoking? We also want to know *how* smoking might cause heart disease. In the mobile phones case, we want to know whether mobile phones cause brain cancer, but we also want to know whether this new cause, mobile phone use, has any other effects. For instance, we can use mobile phones to listen to music, and epidemiological data is reporting more and more cases of hearing loss in young people, who are more likely to listen to loud music through earphones. Is there a link? So when we say that we want to know whether C causes E, this covers a variety of things, like what the cause of E is, or what the effects of C are, or how C causes E. This is important, because what evidence you need can vary slightly depending on what you want to know. This sounds very obvious, but it is easy to forget, especially if you aren't precise about what you want to know.

How much of C causes E? There are complexities even in the canonical cases, where we want to know whether C causes E. Consider for instance the question about the effects of drinking alcohol. Clearly we want to know whether drinking causes, say, liver disease, or not. But that is too simple. We also want to know *how much* drinking causes liver disease. This is because, in small quantities, alcohol (especially red wine) is known to have beneficial effects for health. But we need to establish a threshold between beneficial and harmful effects. We seek, whenever we can, a quantitative answer. It is no use knowing that drinking causes liver disease, but not know whether a glass a day is enough, or whether you need a bottle a day before there is any chance of developing liver disease! Also, you may want to know how much drinking raises the chances of liver disease *for you*. Individual responses to the same exposure may be quite different, so while it may suffice for Alice to drink a glass a day to have problems with her liver, for Bob it may take a bottle. This is of course not an incentive to push your physical resistance to alcohol to the limits, and that's why there are recommended daily intake quantities. Now consider similar questions for radiation: how much do we have to use mobile phones before the radiation has any effect? And does it make a difference whether they are kept close to the ear, in a pocket close to the heart, or in a jeans pocket close to the genitals? Here the question of 'how much' is closely related to the question of 'how'.

Who or what does C cause E in? Even if we can sort that out, we also want to know whether smoking causes heart disease more or less in certain parts of the population—and why. Much of our medical knowledge comes from preliminary studies in animals. If something causes cancer in some animal, especially those that are physiologically most like us, then it might cause cancer in us, too. But *extrapolating* the results from animal studies to human populations is a far from easy task, and the literature has examined under what conditions such inferences are justified (see for instance

LaFollette and Schanks (1996)). In particular, Dan Steel (2008) has put forward an account of extrapolation based on comparison between mechanisms in the 'control' and in the 'target' population. It is also worth mentioning that there is a lively debate about the legitimacy of animal studies, from an ethical point of view. Is it ethical to experiment on animals? Is it ethical to let people suffer dreadful diseases when we might be able to find a cure? The idea that it is unethical to use animals in the lab can be defended, of course, and it would be beneficial to discuss these issues with methodologists. If we stopped research on animals, only observational studies would be left to investigate causation in some diseases, and we discuss the merits and limits of these types of studies in chapters 2 and 7. This is a very difficult issue.

What differences are there between humans? Humans are not all the same, either. Some things affect women more than men; things that are fine for the young can be dangerous for the old. An interesting example comes from 'gender medicine', which concerns understanding how biological sex and gender influence health and disease;[6] this includes reassessing how most drugs are tested. Women experience frequent hormonal changes in the course of the month, and it is not clear how many drugs react to this. Also, women's body mass is on average less than men's, which means that women *may*, by taking a dose prescribed for an average adult, consistently take too much of any drug. So there is a question about what evidence would support a uniform approach in drug testing and in medical treatment across both genders.

There may also be racial differences. Strong sunlight affects the fair of skin far more than those with darker skin, and so those with fair skin may be more susceptible to skin cancer. At the same time, recent research in dermatology suggests that sunlight has beneficial effects for blood pressure and the cardiovascular system. Scientists are studying a mechanism involving nitric oxide (a chemical transmitter in the skin), says researcher Richard Weller in a recent TED talk.[7] This might explain differences in rates of cardiovascular disease across populations that are more or less exposed to sunlight. So, we have evidence that sunlight has adverse effects on our health *and* evidence that it might have beneficial effects on our health. Which one should prevail?

In general, we want to know the distribution of disease across the population, according to specific biomedical or socioeconomic characteristics. Epidemiologists call groups that do significantly better or worse with respect to particular diseases 'sub-populations', and statisticians refer to the statistics of the sub-population when they say 'reference class'. If at all possible, we want to find out whether C causes E in many of these different sub-populations. To do this, we often have to go beyond observing rates of incidence of the disease in the population, and look at laboratory studies, such as using animal or test-tube studies, to understand the biochemistry of diseases and find out what sub-populations matter, and why. This has recently led to breakthroughs

---

[6] Gender medicine can be considered a sub-discipline within medicine, having dedicated publications and scientific associations—see for instance Oertelt-Prigione and Regitz-Zagrosek (2012).

[7] See <http://www.ted.com/talks/richard_weller_could_the_sun_be_good_for_your_heart.html>.

in treating cancer, because we can identify different types of, say, skin cancer by studying the cancerous cells, and so use the most effective treatment for the type individual patients have.

In fact, what population C causes E in turns out to be a far more complex question than it seems. When we do a trial, we try to find out first whether C causes E in the trial population. That is, among the people we actually tried out the drug on. If the trial is well-designed, we can be pretty sure that C causes E in *that* population. But then we have to worry about whether C causes E in *other* populations. If we haven't tried the drug on the elderly, the already sick, the very young, and on pregnant women, it might not work on them. If we did the trial only with British people, it might not work in Italy, or in Canada. This is known as the problem of external validity, or the extrapolation problem, that we mentioned earlier talking about animal studies (see also chapter 2).

Personalized diagnosis and treatment. Finally, what *you* really want to know when you go to your doctor is not any of these things. You want to know whether the treatment will work on *you*. If it works on people in general, but doesn't work on you, then it's of no use to you. There are other possibilities, too. It might work well on most people, but not work as well on you, or even be harmful to you—particularly if you are very young, elderly, pregnant, or already suffering from some chronic disease. This is the problem of inferring from the general to the single case we discuss in detail in chapter 5.

Finally, you want to know what kind of cause C is, in the single or general case. If you get exposed to C, will you definitely get the disease—i.e. is C *sufficient* all by itself for the effect? Or is C a *necessary* cause, so that you have to be exposed to C to get the disease, but you also need other things? Causes can also be neither necessary nor sufficient. We discuss these issues in chapter 4.

It is worth summarizing various related questions about a possible causal relationship between C and E. There are in fact very different aspects of the relation that you may want to find evidence of. For instance:

- What are the effects of C?
- What is the cause or causes of E?
- How could C cause E?
- How much of C causes how much of E?
- In what population or sub-population does C cause E?
- Even if C usually causes E, will C cause E in a particular single case?
- Is C a necessary or a sufficient cause of E, or neither?

This chapter is first about causal inference; about what evidence we use to find out about causes. But we will see the question of what counts as acceptable evidence for inference relates to what we want to do once we have that knowledge. This of course involves prediction and control, and sometimes explanation, as these tend to be the things we want to do! We want to know when C will cause E (prediction), why C causes E (explanation), and how we can make C cause E, or stop C from causing E (control).

6.3 Evidence for causal relations

Arguably, medicine is the science that has most advanced in the project of laying out and trying to standardize explicit strategies for gaining and evaluating evidence for causal relations. This long-standing concern for evidence has led to the development of *evidence-based* approaches, that are also spreading outside medicine, and that we will discuss later in this chapter. So, although the problem of evidence is not confined to medicine, it is the place where many current discussions originate.

6.3.1 Bradford Hill's guidelines to evaluating evidence

Bradford Hill is a well-known figure for medical doctors and epidemiologists. He is particularly well-known because, in the 1950s with Richard Doll, he argued for a causal connection between smoking and lung cancer (Doll and Hill, 1950). Their arguments were based on observational studies, when the mechanisms by which smoking causes lung cancer were still largely unknown. Their positions led to a historically interesting controversy with Ronald Fisher (see for instance Vandenbroucke (2009) for a discussion). But we want to tell another story here.

Another reason why Bradford Hill is well known is that he formulated a set of guidelines for evaluating evidence in epidemiology. Hill's guidelines for what you should consider when evaluating whether C causes E are still considered a valuable tool now. Here is the list (drawn from Bradford Hill, 1965, p295–9):

1 The strength of the association between C and E.
2 Consistency: whether the association is observed in different places, circumstances and times, and observed by different people.
3 Specificity: for example, is the effect specific to particular groups of workers and sites on the body and types of disease? (He notes that some diseases are not specific in this way.)
4 Temporality: take care which of your observed variables is the cause, and which the effect. It might not always be clear which occurs first.
5 Biological gradient: can you see a decrease in the effect with a decrease in the cause, such as with a dose-response curve in medicine?
6 Biologically plausible: does the cause seem reasonable, in the light of the biological knowledge of the day?
7 Coherence with general known facts about the domain. For Bradford Hill, this is about that disease.
8 Can you get experimental evidence if possible? This is evidence where you go beyond observing the cause in action, and act to create it or prevent it, and see whether the effect appears.
9 Analogy: are there other similar known causal relationships? This is important to medicine, as once you find a new mechanism of disease, you may find other diseases that follow a similar mechanism.

It is worth noting that laying out sources of evidence for causal relations achieves only so much. Bradford Hill himself explains why:

What I do not believe—and this has been suggested—is that we can usefully lay down some hard-and-fast rules of evidence that must be obeyed before we accept cause and effect. None of my nine viewpoints can bring *indisputable evidence* for or against the cause-and-effect hypothesis and *none can be required as a sine qua non*. What they can do, with greater or less strength, is to *help us to make up our minds* on the fundamental question—is there any other way of explaining the set of facts before us, is there any other answer equally, or more, likely than cause and effect? (Bradford Hill, 1965, p299, emphases added.)

Bradford Hill emphasizes that he is trying to offer guidelines to help with causal inference, not offering rigid criteria that must be satisfied. We will see the problem of rigid criteria again later in discussing evidence-based approaches.

6.3.2 The Russo-Williamson thesis

Work in philosophy of science—and particularly in philosophy of causality—on methodological aspects of medicine is relatively recent. Unsurprisingly, methodological aspects concern causal inference, specifically what evidence justifies some inferences and not others. The 'Russo-Williamson Thesis' (RWT) arose in this context.

Russo and Williamson (2007a) and Russo and Williamson (2011a) note that Hill's guidelines seem to divide broadly along the lines of the two different approaches to causality we examine in many places in this book: the 'difference-making' approach, and the 'production' approach. Broadly, the difference is between relations between cause and effect that abstract away from how the cause brings about the effect which focus on difference-making, and the relation whereby the cause actually brings about or produces the effect, which focus on production. See chapters 8, 9 and 10 on difference-making approaches, and chapters 11, 12 and 13 on production accounts. There is more on how the distinction arose in the causality literature in chapters 19 and 20. In this chapter we will talk of 'mechanisms' rather than production, as that is the language used in medicine. Hill's guidelines 1, 2, 5 and 8 seem to concern difference-making, while 3, 4, 6, 7 and 9 urge attention to how any kind of difference-making comes about, seeking evidence of an underlying mechanism. So the Russo-Williamson thesis says that *typically* in order to establish causal claims in medicine, we need to establish facts about difference-making and about mechanisms. Or, in other words, that we typically need to evaluate evidence of difference-making and of mechanism.[8]

A few remarks are in order. The Russo-Williamson Thesis is about what kind of *evidence* we have to use to decide what causes what.[9] It doesn't introduce new concepts: the distinction between difference-making and mechanisms already existed in the causality literature, and the ideas of correlations and mechanisms are already used in medical methodology. In the course of Part II of the book, we will see that there are various difference-making concepts in the philosophical literature (probability

---

[8] Interestingly enough, analogous views have also been defended outside medicine. For instance Gerring (2005) emphasizes the role of mechanisms, alongside difference-making, in establishing causal relations in social science research.

[9] For a similar formulation, but with emphasis on ontological, rather than evidential, aspects, see Joffe (2013).

raising, counterfactuals, manipulability) and also various production or linking concepts (processes, mechanisms, information). The RWT says that we use evidence both of difference-making and of mechanisms to establish causal claims. In the light of the discussion of chapters 11, 12, and 13, the RWT can be formulated more broadly, saying that we need evidence of difference-making and of *production*, whether in the form of processes, mechanisms, or other types of linking.

However, it is important to emphasize that the RWT does not claim that there are two entirely distinct types of evidence, as if there were just two types of studies, or instruments for gathering evidence. The RWT concerns the object of evidence; what we need evidence *of*. In fact our methods of getting evidence of a correlation, and of an underlying mechanism, can be highly intertwined. A single study or experiment might provide evidence of difference-making and of mechanisms (see discussion in Illari (2011a)). For example, observations about the mode of transmission of a bacterium, say *vibrio cholerae*, provide evidence about the mechanism underlying such transmission *and* that the bacterium makes a difference to the occurrence of the disease. Note that having some evidence of the underlying mechanism does not imply that we know everything about it. In socio-economic contexts we are most often provided with evidence of difference-making generated by statistical models, and the real challenge is to use or model causal information such that evidence of mechanisms is likewise generated.

It is also important to make clear what is meant by *establish*. In this book we emphasize how useful causal knowledge is, but also how uncertain and tentative it can be. So to 'establish' that C causes E in science never implies certain knowledge. In a sense, causal claims are always tentatively established and the big challenge is to find out enough knowledge to justify action, even in cases of imperfect knowledge. Think again of the mobile phone and cancer example. Do we have enough evidence of difference-making and of mechanisms to establish that mobile phones cause cancer? No. However, if you read the instructions for your mobile phone (have you ever done such a thing?) you might find safety notes saying to hold the phone at some distance from your ear, thus suggesting that available scientific knowledge at least tells us that a small distance may already affect the propagation and absorption of radiation emitted by the phone. In other words, if there is a low-cost, low-risk action, we can act anyway, even if causal claims are far from being established. This, of course, opens another issue, about what evidence supports some precautionary measures rather than others.

This leads us to the question of whether the Russo-Williamson Thesis is normative or descriptive (see e.g. Broadbent (2011)). If normative, the thesis would urge scientists to follow some (perhaps philosophical or conceptual) protocol. If descriptive, the thesis would simply report current or past scientific practices. Neither strategy fits squarely with the RWT, nor with the more general approach that we discuss in chapter 19. The point is not to judge causal conclusions drawn by scientists either in the past or today, but to understand what *evidential elements* science uses in drawing causal conclusions. This surely requires some description of the scientific practice, but also some 'conceptual engineering' that is meant to add to the conceptual toolbox of scientists, and take part in the ongoing debate to improve scientific practice (Floridi, 2011a).

6.3.3 What is evidence of mechanism and why is it helpful?

Saying that evidence of mechanism is helpful in causal inference does not mean we have to have a completely known mechanism to plug in somewhere in the inference process. It means that at least *some* information about the mechanisms, or processes, by which C causes E is useful. This includes 'negative' information. For instance, one reason why it took a long time to establish that the previously bacterium *helicobacter pylori* is a cause of gastric ulcers is that, according to the available biological knowledge, bacteria can't survive in an acid environment such as the stomach. This is evidence of mechanism that suggested that bacteria were not a possible cause. Once a bacterium had been discovered in the stomach, further investigations showed how *helicobacter pylori* could survive in that environment, which led to the revision of our background biological knowledge. Evidence of a difference-making relationship, in this case of the association between *helicobacter pylori* and gastric ulcers, was able to support evidence of mechanism after the revision. The combination of the different pieces of evidence allowed the scientific community to establish the causal relation between *helicobacter pylori* and gastric ulcers at the population level. (For a historical reconstruction of the case, see for example Gillies (2005b) and Thagard (1998).)

Having a plausible or known mechanism, or even knowing just bits and pieces about how a system works, helps with several problems that scientists routinely encounter in their studies based on difference-making evidence, i.e. on evidence of association or correlation. One such problem is confounding (see chapter 2). If you have an association between F and E, F might not be a cause of E if the relationship between F and E is confounded—i.e. F is correlated with both E and the real cause, C. The result is that the 'true' causal relation between C and E cannot be estimated correctly. A stock example in the philosophical literature is that coffee drinking is correlated with both smoking and heart attacks. The result is that we don't know whether smoking causes heart attacks on its own, or via coffee drinking, or both. Evidence of mechanisms should help disentangle the different causal paths in cases like this, exclude implausible paths, and support choices about model building and model testing.[10] The general point to make is that mechanisms supplement correlational knowledge. We see the importance of structuring correlations in data due to Simpson's paradox in chapter 7. Also related to model building and model testing, mechanisms help with the choice of the reference class, to find out in what populations C causes E. Suppose you are designing a randomized trial. You may want to choose to split the population into different sub-groups, for instance men and women, because available biomedical knowledge suggests that the active principle of the new drug to be tested may react differently in different categories of people. And evidence of underlying mechanisms may help with other aspects,

---

[10] In some literature 'mechanistic reasoning' is discussed (see for instance Howick (2011a)). Note that we do not mean mechanistic reasoning when we discuss evidence of mechanism. An episode of reasoning is not a piece of evidence. We can of course reason about evidence gathered, and the reasoning will be part of the process of evaluating evidence. But the reasoning and the evidence are distinct. Solomon (2011) argues that mechanistic reasoning is an important category mistake. In interpretation of evidence, extensive data-processing techniques are also often used, and those are discussed in chapter 7.

for instance how long to run the trial. On this point, the case of the first streptomycin trials discussed in papers from the Mechanisms and the Evidence Hierarchy (MatEH) project funded by the Arts and Humanities Research Council (AHRC) is illuminating (Clarke et al., 2013; Clarke et al., 2014).

In other places in this chapter and in the book, we discuss other uses of mechanisms, such as in causal explanation and external validity. So the idea is not that mechanisms are a panacea to solve all the problems of causal inference, but that they help with scientific practices in several different situations. Correlation is not causation, and evidence of mechanisms helps decide what causal claims our evidence of difference-making relations really supports.

6.3.4 What is evidence of difference-making and why is it helpful?

Evidence of difference-making usually consists of correlations, associations, and dependencies in statistical analyses, but can also comprise counterfactuals (based, for example, on established medical knowledge of case reports (Clarke, 2011, sec4.5)). We have seen that mechanisms help with confounding, which is an ongoing problem for inferring causes from evidence of difference-making. Conversely, however, even if we have evidence of a causal path between C and E—if we know an underlying mechanism connecting C and E—we cannot conclude that C causes E.

Clearly if we can trace a mechanism connecting C and E, then C has some influence on E. However, recall the many different things we want to know. In general, one important thing is whether C has an average effect on E. If we see C go up, or alternatively make C go up, will E also go up? We don't know this yet, because we have traced just one linking mechanism. The problem is we don't know whether there are also other unknown paths that link C and E, which may *mask* the known mechanism. A stock example in the philosophical literature, also discussed in Steel (2008, p68), concerns the causal path from exercise to weight loss. When you exercise you burn more calories—this mechanism is fairly well understood. However, exercise also increases appetite. So difference-making here helps in finding and quantifying—via statistical analyses—the overall effect of exercise on fatness or thinness. The quantification is important. Currently, our usual way to estimate effect-size is in looking for correlations.

Confounding and masking are two sides of the same coin. In the confounding case, mechanisms help us decide what variables to control (or even to exclude) in the causal structure. In the masking case, difference-making helps us discriminate the different effects of multiple or complex causal paths.

Evidence of difference-making, whether gathered from statistical analyses or single-case observations, can also provide an entirely novel epistemic access to an unknown domain. Philosophers of science know how difficult it is to reconstruct the process of hypothesis generation. Sometimes it is an accidental observation that triggers an idea about an unknown cause of disease, and sometimes it is persistent, sustained and disciplined investigation. But sometimes the mechanisms of disease causation are unknown, and in this case often a correlation in a well-conducted, controlled study is our first solid piece of evidence of a previously unknown cause. Just such a finding by

Richard Doll and Richard Peto was influential in making scientists think that smoking causes lung cancer ((Doll and Peto, 1976), see also discussion in Gillies (2011)). Evidence of mechanisms, however good, cannot provide the same information as evidence of difference-making. So the point here is to understand how they complement each other in several scientific tasks, rather than to argue that one or the other is more important.

6.4 Evidence-based approaches

This leads us to evidence-based approaches, which can be seen as methods for assessing evidence. We lack space here to trace the history and development of evidence-based medicine and of other approaches. Good resources to follow up on this are, for instance, Howick (2011*b*) or Killoran and Kelly (2010). Evidence-based medicine in particular has attracted criticism, for instance by Worrall (2002) or La Caze (2011). We will not examine all these quarrels in detail. The literature has examined the merits and limits of randomized trials, or of 'old' medical theories, sufficiently. We will instead highlight some aspects of evidence-based approaches (EBAs) that are relevant or useful in reading other chapters of the book.

The first aspect is that EBAs include a variety of methods of gathering evidence (many of them discussed in chapter 2). The better-known methods are of course Randomized Controlled Trials (RCTs), but they also use meta-analyses, and numerous observational studies, as well as qualitative methods (see for instance the various procedures used by NICE to come up with guidelines). EBAs, and especially evidence-based medicine (EBM), produced *hierarchies* of such methods of gathering evidence (NICE (2006), OCEBM Levels of Evidence Working Group (2011)). These are usually called hierarchies *of evidence*, but we need to be careful to distinguish between what we count as evidence and what we count as a method for gathering evidence. They are not quite the same.

In chapter 2 we provide a survey of kinds of evidence-gathering studies. These are particular trial designs; protocols for gathering evidence. Different trials can be used to gather evidence of the same thing; and the same kind of trial can be used to gather evidence of different things. For example, we can get evidence of mechanism from many places, such as breakthrough technology in biochemistry, which lets us see microstructure that we could not see before. This was the case in Rosalind Franklin's famous X-ray crystallography photographs of DNA. But evidence of mechanism can also be obtained by extensive and repetitive experimental interventions. This was the case with Crick and Brenner's work in 1961 cracking the genetic code using chemical mutagens (see discussion in Bell (2008)). In assessing this evidence, we need to evaluate both the method by which the evidence was generated, and what exactly the evidence is evidence of. Similar caution is needed in evaluating technologies for evidence, such as medical diagnosis. For instance, Campaner (2011) and Campaner and Galavotti (2012) discuss 'biopsy evidence' or 'spectroscopy evidence' as manipulationist or interventionist evidence, which they take to be another type of evidence. But biopsy and spectroscopy are techniques or methods for gathering evidence, which can be used to gain evidence of different kinds of things. For instance, using the distinction in objects of evidence above, a skin biopsy will reveal whether the mole is malignant, therefore raising the probability of developing skin cancer, or it could be used to inform doctors as to how the mole developed (if the patient was not born with it). So, the biopsy is not itself evidence, but a method to allow doctors to generate evidence and ultimately establish causal claims. This is not to downplay the importance of evidence coming from interventions. Such evidence is indeed crucial in many cases, yet we should carefully distinguish what we gather evidence *of* and the *methods* for gathering evidence.

Increased care in distinguishing evidence from the method used to generate it changes the landscape of the discussion entirely. For instance, the issue is no longer to debate whether RCTs are better than any other method of gathering evidence, but under what conditions, and for what purposes, one method is *more suitable* than another method. Solomon (2011) discusses EBM along these lines. She evaluates EBM with respect to its goals: to establish efficacy of treatments, not to discover new medical knowledge. But we want to know many different things, so EBM cannot provide a complete epistemology of medicine. But within its scope of application, EBM has achieved a lot, specifically about about how to design both observational and experimental studies, and about how to compare different studies on the same topic. (See e.g. Cochrane Collaboration, Levels of Evidence table issued by the Oxford Centre for Evidence Based Medicine). This doesn't mean that EBM methods are foolproof, and problems with the hierarchy, especially the upper part, have been discussed in the literature (see e.g. Stegenga (2011), Osimani (2014)).

6.4.1 A procedural approach to evidence?

EBAs are spreading well beyond medicine. They are routinely used in public health, and also increasingly in social science. There must be a reason why they look so promising in settling scientific issues. One reason has to do with the need for a more procedural, structured protocol for a scientific method. In a sense, it is reassuring to think that so long as we follow *the* method, we will reach reliable conclusions; ones that are also easily translatable into clear and unequivocal answers for the decision-makers. Is this new teaching method effective? Randomize a couple of classes in your school, run the trial, and you will find the answer. Statistics won't lie, and you will be able to tell the ministry whether it is a good idea or not to invest in this new teaching method. But isn't there important information that we miss out this way? Can we always decide about causal relations by using statistics? Or is there some information that qualitative methods will instead be able to provide? This chapter has provided reasons to think that we need more than evidence of difference-making to establish causal relations; the whole book, moreover, tries to demonstrate the usefulness of a *plurality* of methods and of concepts for causal analysis.

It is possible that the rigidity of early evidence hierarchies was off-putting for people working in less statistics-prone, more qualitative or context-oriented domains, who see the challenge as being to integrate observational methods, qualitative methods, and evidence of medicine, in such a way that all can be assessed in establishing causal knowledge. This concern was anticipated by Bradford Hill, who called what he drafted *guidelines*, not criteria. He saw the danger of putting evidence in a hierarchy that is then treated rather like a straightjacket, underestimating the complexity of evaluating evidence, especially evidence that is of different things, and derived via different methods.

An example of how evidence-based approaches can trip up is discussed by Cartwright in the context of evidence-based policy. Cartwright (2011) illustrates with the example of the 'Bangladesh Integrated Nutrition Policy' (BINP) that even good policies, established according to the most rigorous evidence-based principle, may fail. Cartwright explains how a policy which worked in Tamil Nadu failed when applied in Bangladesh, a different population than its initial trial population, in a context with a different social structure. So one may object that this is a point about external validity. It is. But it is also about what is missing in the procedural and protocol EB approach: sensitivity to context.

One suggestion coming from these debates on EBM is that we shouldn't look at evidence from the perspective of a rigidly hierarchical approach, but from the perspective of aiming for *integration*, and analogy is made to reinforced concrete. Reinforced concrete is a composite material. It is made of concrete, which withstands compressive stress, and of steel, which withstands tensile stress. The composition of the two types of material makes reinforced concrete good for construction. The analogy, we suggest, works for evidence. If various independent sources of evidence converge on something, that's much better evidence. But this goes beyond a form of 'triangulation' of evidence; it encourages integration of different information coming from different sources. We have seen earlier that we need evidence of difference-making for some reasons, and evidence of mechanisms or production for other reasons. Evidence of difference-making and production are *complementary*. On balance, integrating evidence lets you do much more with what is actually available.

Core ideas

- There are many different causal things we want to know, such as 'Does C cause E?', 'What is the cause of E?', and 'How much of C causes that much of E?'. It is worth being careful what you are trying to find out, and what evidence might be needed.
- Integration: Given that all evidence is fallible, using multiple sources of fallible evidence that converge on a causal structure is better than relying on a single source of evidence, because however good it may be, it can occasionally fail.
- Evidence of difference-making and of production (in this chapter, often of mechanisms) are complementary because each addresses the primary weakness of the other.
- Evidence of difference-making:
  - Advantage is eluding masking, that is being outweighed by an opposite mechanism or combination of mechanisms.
  - Disadvantage is possibility of confounding (and chance).
- Evidence of mechanism:
  - Advantage is eluding confounding (and chance).
  - Disadvantage is possibility of masking.

Distinctions and warnings

- Everybody agrees that evidence supports causal claims. The controversial part is what counts as evidence. Everyone agrees that evidence of difference-making is important to support causal claims, but not everyone agrees we also often use evidence of mechanisms. There is variation across scientific fields on this issue.
- The evidence required to establish a causal claim is not necessarily the same as the evidence needed to take action. For example, a low-cost, low-risk precaution might well be taken on the basis of quite modest evidence of possible harm.
- Evidence and evidence-gathering methods are not the same thing. For example, a Randomized Controlled Trial (RCT) is a systematized method for gathering evidence, which can be used to gather evidence of many different things.

Further reading

Bradford Hill (1965); Howick (2011b); Clarke et al. (2014).

CHAPTER 7

Causal Methods: Probing the Data

7.1 Examples: apoptosis and self-rated health

Consider the functioning of the apoptosis mechanism, which is a mechanism of controlled cell death. In apoptosis, the cell dies in a programmed order, and so causes less inflammation or damage to other cells than necrosis, which is cell death in which the cell contents usually escape into the surrounding fluid. Apoptosis is a mechanism helpful to the survival of multicellular life, as part of the mechanism to keep organs to their normal size by balancing out the growth of new cells. It is also the mechanism by which a tadpole's tail disappears! There is quite a lot of data and some background knowledge for this mechanism. One question we may want to ask is whether we can find a *causal structure* that underlies the data, and we might try to build a model of the system, using our background knowledge and our data. Suppose, in particular, that the following variables are available: agents that can damage DNA, such as levels of radiation; *p53* gene status and p53 protein level, which are both signals of the cell trying to repair itself; and expression levels of caspases. This is highly simplified, but the levels of these things fluctuate in the normal cell. Various mechanisms try to repair any damaged DNA. If DNA repair is successful, the *p53* gene will cease to be highly expressed, and p53 protein will fall back to normal levels, being controlled by a negative feedback loop. But p53 levels can keep rising, with the protein failing to be broken down, and triggering a caspase cascade that disintegrates the cell. What happens depends on relative levels, and so quantitative modelling is needed to understand the causal structure of this mechanism. This can be done using background knowledge of these relationships, and extensive data. This then allows us to make more fine-grained quantitative predictions about what will happen under various conditions.

Alternatively, consider a different case from social science. Suppose you are a demographer or a sociologist, and you receive data collected after some surveys conducted in the Baltic countries in the nineties. You are asked to study 'health' in these populations and for this period. Data, however, are not about 'health' but rather about how people *perceive* their health. You want to establish the determinants of 'self-rated health', including social factors—such as social support or stress—and other factors such as age or physical health. Suppose you have, in your data set, the following variables: education, social support, physical health, locus of control, psychological distress and alcohol consumption. In this case, you want to 'dig out' a causal structure—the structure of the network formed by these variables—from the data received. One way we can do that is using causal methods such as structural models, described later.

So although some of our reasoning about causality, our attempts to predict and control the world, are qualitative, modern science is increasingly quantitative. Causal modelling methods have been developed to meet the increasing need to represent our causal knowledge quantitatively. They have also been turned to use in inferring causal structure from large data sets.

7.2 The need for causal methods

To use causal methodology effectively, and interpret its conclusions properly, it is important to reflect upon its procedures. The forefathers of experimental method, such as Francis Bacon or Isaac Newton, did that very carefully. They described in detail what an experiment is, how to conduct it, its strengths and weaknesses in giving us epistemic access to causal relations, and so on. The topic is still alive in modern times, such as in Fisher's *The design of experiments* (Fisher, 1935), where he developed in detail a methodology for agricultural studies. We introduce experimental methods in chapter 2, and further discuss them in chapter 16. However, experiments are only the beginning. Bacon thought that once we gathered the data, generalizations would be relatively obvious. Sadly, this is not so. We frequently have to *probe* the data further to find causes. This is why the field of statistics has grown a lot since the nineteenth century, and developed applications in many scientific disciplines, from demography and sociology through to physics or population genetics. In this chapter we aim to provide the reader with a toolbox to understand causal methods that follow this 'statistical' tradition: what we do after we get the data.

For the purpose of exposition, we simplify the presentation of causal methods as sharing two common features:

(i) Some kind of (sometimes simplified) representation of the structure of causal relationships among the crucial variables in the dataset, such as in a graph, a set of equations or a table.
(ii) A probability distribution attached to the variables in the dataset.[11]

To provide a *comprehensive* presentation of causal methods exceeds by far the ambition of a short chapter like this, so we will not attempt to do so. Instead, we offer first a very minimal understanding of formal causal methods (one that Bayesian net

---

[11] A probability distribution is a function that assigns a probability to each of the possible values of a variable. For an accessible introduction to the basic notions in probability and statistics, see Bandyopadhyay and Cherry (2011).

advocates, graph theorists and structural modellers would accept), then explain the differences between a few of the most widespread approaches, and finally convey some understanding about key notions involved and their difficulties. We will have succeeded if, next time you have to deal with formal methods for causality, you will at least be able to recognize what is at stake. If you want to know more about each method, you should of course read the authors mentioned, and ultimately undergo the messy and frustrating experience of playing with any formal tool you intend to use.

7.2.1 Two aims for causal methods

Causal methods can be used either to model *available* knowledge of causal relations in order to make predictions or explanations, or to model observed phenomena in order to *discover* the causal relations. The two examples mentioned in the opening of the chapter illustrate these two usages:

- Representing available causal knowledge. Casini et al. (2011), for instance, discuss recursive Bayesian networks (RBNs) as a possible modelling strategy for the apoptosis mechanism (see also Figure 7.1, which is borrowed from that paper). The goal is to represent existing knowledge through formal tools so that it is easier to handle, and handle quantitatively, for instance to make inferences.
- Finding out what causes what. Gaumé and Wunsch (2010), for instance, analyse data from two surveys conducted in the Baltic countries to find out the mechanisms explaining the variable 'self-rated health'. In other words, they use a modelling strategy to figure out what causal relations hold in the data set.[12]

In some cases, modellers will be doing both tasks, but even so, these two separate aspects are worth bearing in mind.

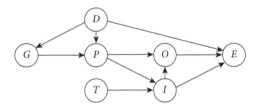

Fig 7.1 RBN representation of the functioning of the mechanism of apoptosis. *D*: Damaging agents; *G*: p53 gene status; *P*: p53 protein level; *O*: Apaf1 level; *E*: caspase 3 level; *T*: TNF level; *I*: caspase 8 level.

Casini et al. (2011).

---

[12] There is also another class of models for discovering causal relations: automated reasoning or algorithmic search. These models search for causal relations in the data using algorithms. Given a data set, they impose a number of restrictions (see e.g. Markov conditions or the faithfulness condition discussed later), then run an algorithm on the data set, and 'bootstrap' causal relations. See for instance Spirtes et al. (1993) and, to some extent, Pearl (2000). Another way of describing the difference between this and the other class of models is between human learning of a causal model and machine learning of a causal model.

Some further remarks are in order. First, it is important to bear in mind that every method in this chapter has its strengths and weaknesses. There is no perfect method. So there is no a priori reason to prefer one method over another. The choice of method depends on the data at hand, the type of phenomenon under study, the kind of inferences we are interested in and so on. What is important is whether the method can represent the data available, whether its assumptions are at least approximately satisfied and whether it can give the answer desired.

Second, formal models are always partial representations of a (real, worldly) system. We typically make a selection of which variables to include in the model, and make several simplifying assumptions about the structure of the phenomenon and other properties of the system. This is not just for 'causal modelling', though. Any modelling enterprise implies some degree of simplification and abstraction. There is a long tradition in philosophy of science that discusses what information we gain and what information we lose during modelling (for an overview of problems about modelling, see Frigg and Hartmann (2012)). One important consequence is that models are fallible. We can build the best model ever and get things wrong. We can build very different models to explain the same phenomenon, and face the problem of understanding which one best approximates reality.

Third, science may aim at tracking the truth, whether the truth is about the Higgs boson, the causes of cancer or the effects of solar storms. But *truth* is often too big a question to answer through methods that have their inherent limitations. Instead, we can usefully concentrate on the *validity* of our models, namely to what extent we can trust them and their support of our scientific conclusions. Validity, however, is not the same as truth. Models can be false, strictly speaking, and yet be useful. For example, you might explain to small children that mummy is getting fat because she has a baby in her tummy. Mummy of course is not getting fat, and the baby is in fact in her womb, not her stomach. But in giving the children a strictly false explanation, you deliberately draw on vaguer concepts familiar to them to allow them a greater understanding of what is happening—you give them a model of the world they can grasp.

The validity of models has to be evaluated with respect to their set or sets of assumptions and to the background context used. Validity has also been discussed in relation to *statistical* properties of a model, or the possibility of replicating the study. But there are many other considerations that play an important role in deciding about validity. One is what we want the model to be valid for: to establish causal relations in the population of reference? Or to export the results to different populations? See also chapter 2, where the *locus classicus* of Campbell and Stanley (1963) and Cook and Campbell (1979), as well as more recent discussions such as Jimenez-Buedo and Miller (2009) and Jimenez-Buedo and Miller (2010), are mentioned.

It is important to bear in mind that we need to resist the myth that the automation of reasoning will solve the problem of validity. It doesn't. Automation of reasoning helps a great deal with making difficult situations more tractable from a computational point of view. But, to put things plainly: we should forget 'naked statistics' in the sense of uninterpreted statistical results. There is always a modeller behind a model, and so it is important to specify the starting point of the model (the set or sets of assumptions) and of the modeller (her or his background knowledge).

With these preliminary warnings in place, we can now begin drawing the road map for causal modelling. The aim is not to provide you with comprehensive knowledge about causal modelling, but to give you hints about where to get the information you need for your study and research. For instance: What are the key concepts, and why are they important and/or debated? Who are the Bayes' Nets advocates, who are the structural modelling people, who has worked in econometrics? The crucial idea we want to convey is that there is great diversity in research methods and there is no a priori reason why one method is better than others.

7.3 The most widespread causal methods

Causal methods are widely used across the sciences. Here we guide you through the ones that are most often used or that have raised debate. It is important to note that the models below come in two varieties: purely associational, and causal. That is because models, in the first instance, denote a statistical technique used for various inferential practices, and what makes them causal is an interpretation, which requires an extra set of assumptions and/or of tests performed. When it comes to the task of prediction, the non-causal version can be more powerful. A job of the causal interpretation is to help us with control, the aim being to improve our prediction of what will happen when we deliberately intervene to disturb the system modelled.

Bayesian networks (BNs). BNs are graphical models that represent the probabilistic dependencies and independencies between variables. Graphs are *directed*, meaning that the arrows go from one variable to another, where the variables are represented by nodes. Graphs are also *acyclic*, meaning that even going through all the paths in the graph you can't run into a cycle. Every variable (the nodes in the graph) has a marginal probability distribution attached, and the arrows in the graph represent *conditional* probability distributions. The most important feature of BNs is the (Causal) Markov Condition, which we explain below. It is important to note that this condition is built in to the definition of a BN. BNs were first developed in the 1980s, gained popularity in the 1990s, and are still a flourishing area of research, thanks to the work of Korb and Nicholson (2003), Neapolitan (1990), Pearl (1988), Pearl (2000) and Pearl (2009), Spirtes et al. (1993) and Williamson (2005).

Structural models (SMs). Structural models belong to the broader family of 'regression models' (see Freedman (2005a) for an accessible presentation and a critical assessment). These models can be expressed in the form of lists of equations or a recursive decomposition describing a system. Both formats include variables, containing information about the observations and parameters often to be estimated, and providing key characteristics of the relations between the variables. In econometrics, the term 'structural modelling' was used by the pioneers in the Cowles Commission to indicate that the equations in the model represent an economic structure, which is given by economic theory (about e.g. the behaviour of a rational, ideal, economic agent) rather than inferred from empirical data (about e.g. the behaviour of real economic agents)—see for instance Haavelmo (1944), Koopmans (1950), Hood and Koopmans (1953).

It is largely in this sense that Hoover (2001) or Heckman (2008), for example, use the term 'structural'. Mouchart et al. (2010) instead use the term to convey the idea that the scientist *models* the structure, based on background knowledge, data analysis and interpretation of results against the background (so that the structure is not given a priori by theory). Technically speaking, SMs are defined as a family of probability distributions. The goal is to 'give structure' to these distributions, which are indicated by economic theory or suggested by preliminary analyses of data and background knowledge, ordering conditional and marginal components that represent causes and effects. (See also discussion of the recursive decomposition later in this chapter.) Pearl (2000) and Pearl (2011) use the term 'structural' in a way that generalizes BNs.

A structural equation is often written as $Y = \beta X + \epsilon$, where Y denotes the effect, or outcome, X the cause, β quantifies 'how much' X causes Y, and ϵ introduces an error component, or a chancy element, in the equation. The peculiarity of such an equation is that it is interpreted as expressing a *causal* relation between X and Y, and so it is not symmetrical with respect to the equality sign. This means that rewriting the structural equation $X = \frac{Y-\epsilon}{\beta}$ has no *causal* meaning, although it has algebraic meaning. The other important aspect is the interpretation of the parameter β, which has to represent the 'structural', rather than accidental or spurious, aspect of the relation between X and Y. In this form, a structural equation suggests that the relation between X and Y is linear. An even *more* general form of a structural equation is $Y = f(X, \epsilon)$, which instead does not specify the way in which X and Y are related; part of the structural modeller's task will involve precisely the specification of a particular form. One last note: structural models usually describe a phenomenon using *systems* of structural equations rather than one single equation.

Multilevel models (or hierarchical models). These are statistical models having the peculiarity of analyzing, in a *single* modelling framework (to be expressed in equations), two distinct types of variables: individual, and aggregate. In a sense, multilevel models are variants of structural models. Individual-level variables measure characteristics of individual units of observation, such as income of a person or of a firm. Aggregate variables measure characteristics of groups, such as the *average* income per capita in a population, or the percentage of people with income above a certain level. The advantage of putting both variables in the same framework is to model interactions between levels: in multilevel models we can see how aggregate variables have an effect on individual-level behaviour. These models are widely used in the social sciences, especially demography and educational research. The main developers of multilevel models are Goldstein (2003) and Courgeau (2007), but see also Greenland (2000). In chapter 5 we discuss the issue of the levels in more detail, along with some case studies.

Contingency tables. Contingency tables are a basic form of statistical modelling that orders the variables in the data in rows and columns and analyses the relations between them. This method is also called 'categorical data analysis' (CDA) because the variables involved are categorical, in the sense that the measurement scale consists of categories. For instance, religious affiliation can be categorized as 'Catholic, Protestant, Jewish, other' (here, the order does not matter); or attitudes towards abortion

TABLE 7.1 **Trivariate table with sentence as the dependent variable and social status and record as independent variables**

	Absolute Frequency				Proportions			
Social status	High		Low		High		Low	
Record	Criminal	Clean	Criminal	Clean	Criminal	Clean	Criminal	Clean
Sentence								
Severe	25	2	95	17	0.63	0.08	0.81	0.49
Lenient	15	22	22	18	0.37	0.92	0.09	0.51
Sum	40	24	117	35	1.00	1.00	1.00	1.00

can be categorized as 'disapprove in all cases, approve only in certain cases, approve in all cases' (for which the order matters). In causal analysis, *multivariate* crosstabulations are used, meaning that more than two variables are included and we need to choose explicitly which variables are the causes. See for instance Table 7.1, taken from Hellevik (1984). In social science, CDA has been used and developed since the original work of Boudon (1967). For recent accounts and improvements see also Hellevik (1984) or Agresti (1996).

This list is not nearly exhaustive. In reading the scientific literature you will find many other models. The terminology and techniques vary a great deal from discipline to discipline. But there is something they have in common, which is their 'inner' statistical skeleton, upon which we need to put flesh to get to some causal interpretation. Recall the distinction we began with between models through which we can establish only associations (correlations, dependencies), and those through which we can establish causal relations. This distinction applies to nearly every type of model mentioned above. Which kind of model we have—associational or causal—depends on how much information is filled in for each of the three categories in Table 7.2: background knowledge, assumptions, and methodology. (Moneta and Russo (2014) discuss a similar table in the context of econometric modelling.) Line one in Table 7.2 points out that to establish associations we need much less background knowledge than to establish

TABLE 7.2 **Associational models versus causal models**

	Associational models	Causal models
Background knowledge needed	Choice of variables	Causal context; theoretical knowledge; institutional knowledge, etc.
Assumptions made	Statistical	Statistical; extra-statistical; causal
Methodology used	Model-based statistical induction	Hypothetico-deductive methodology

causal relations. Line two indicates the kind of assumptions made in associational and in causal models: clearly, causal models need more assumptions. Line three highlights different broad methodologies at play in associational and in causal models.

7.4 Key notions in causal methods

Now we can review some key notions in causal models. The list is not exhaustive but hopefully will help the reader grasp what is at stake, what controversies are discussed, and so on. Note that the following are features of *models*, not of the world. There is a great deal of confusion over this point. Strictly speaking the following key notions can only be formally defined on a model. Of course, the system the scientist wishes to model may be more or less susceptible to successful modelling using a model that must satisfy these formal definitions, but strictly speaking the *system* does not satisfy the Causal Markov or Faithfulness Conditions, and so on.

Simpson's paradox. Simpson's paradox refers to a situation where, given the correlation of two variables in a population, once the original population is further partitioned, the correlation is reversed. This paradox is actually a case of 'confounding', where the correlation between two variables is due to a common cause rather than a causal relation between them (see also Wunsch (2007) and chapter 2). The paradox takes its name from the statistician who drew the scientific community's attention to this problem (Simpson, 1951). A stock example to illustrate is the case of a presumed gender bias in the admissions at the University of California, Berkeley. In 1973, admission figures showed that of a total of 8442 men applying, 44% were accepted, while of a total of 4321 women, only 35% were accepted. However, when individual departments were taken into account, figures showed that in no individual department was there a significant gender imbalance.[13] The interpretation of such results, according to Bickel et al. (1975), is that women tended to apply to more competitive departments and for this reason the overall admission rate for women comes out as lower than for men. Cases of Simpson's paradox can happen in social contexts, as we have just seen, but also in biology and sociobiology, for instance in explaining the evolution of selfish and altruistic behaviours (see Sober and Wilson (1998)). Multilevel models can also be seen as an analytical tool to deal with a 'Simpson's paradox type of problem' (see chapter 5). Bandyoapdhyay et al. (2011) examine Simpson's paradox closely and challenge the dominant causal interpretation of Pearl (2000) by distinguishing three types of questions: (i) why Simpson's paradox is a paradox, (ii) what the conditions for its emergence are, and (iii) what to do when confronted with a case of Simpson's paradox. They argue that whether Simpson's paradox has anything to do with causality depends on which question one is asking. It is their contention that causality plays a significant role *only* with regard to the 'what to do' question.

Simpson's paradox is extremely important, because it is not rare. Any data set will give you different statistical results, depending on how you partition it—so you had

---

[13] For the interested reader, the tables are reported in the Stanford Encyclopedia (Malinas and Bigelow, 2012).

better get that partition right! Data sets can be partitioned many ways, so we could discover, for the Berkeley case, that more blonde women get in than blonde men, but fewer dark-haired women than dark-haired men. In this case, though, the idea is that the partition by department is the right partition to detect gender bias in admission, as all the admissions procedures take place at the department level. In the following discussion, we look at various ways we structure and constrain data sets. It is worth bearing in mind that the structures we impose on data really—*really*—matter.

(Causal) Markov condition. In Bayes' Nets (BNs), the Markov condition states that, for any node in the graph, this node becomes independent of its non-descendants, once it is conditioned upon its parents. In other words, given a node in the graph, its ancestors influence it purely by influencing its immediate parents, with no separate influence like a gene whose expression has skipped a generation transmitted in a hidden form. In the BN in Figure 7.2, according to the Markov Condition, whether the pavement is slippery (X_5) only depends on whether it is wet (X_4), and it is probabilistically independent of any other variable once we conditionalize on its parent (X_4) by holding its value fixed. The Markov Condition is the pillar of Bayesian networks. It becomes *causal* when nodes in the graph are explicitly interpreted as causes and effects and arrows as causal relations. According to the Causal Markov Condition, whether the pavement is slippery (X_5) only depends on whether it is wet (X_4), and it is *causally* independent of any other variable once we conditionalize on its parent (X_4) by holding its value fixed. In this case, the Causal Markov Condition says that *direct* causes are the important ones. So, in a sense, it fixes the structure of causal relations. It is important to bear in mind that the Markov Condition is *built in* to BNs, so if a graph or structure does not satisfy the condition, it is not a BN. The CMC is the object of lively discussions in both the philosophical and scientific literature.

We direct you here to some examples of useful discussions of BNs. For a defence of the CMC see Hausman and Woodward (1999) and Hausman and Woodward (2002), and for a criticism see Cartwright (2002). See Williamson (2005, ch.4.2) for argument that, for the CMC to hold, it is crucial to specify what interpretation of causality

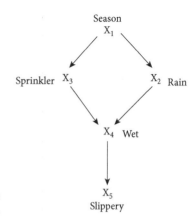

Fig 7.2 A simple Bayesian Network (example borrowed from Pearl).

and of probability are adopted. The CMC has also been discussed in relation to indeterminism—see for instance Drouet (2009). See Suárez and San Pedro (2007) for discussion of whether the CMC is refuted by quantum mechanics, or whether quantum mechanics constitutes an interesting case to assess whether the CMC holds. The hotly disputed question of whether causal models should always satisfy the Markov Condition is also discussed by Gillies and Sudbury (2013), who draw on examples from medicine, where multi-causal forks, i.e. causal forks where several factors cause the same effect, are widely used.

Faithfulness condition. The faithfulness condition has been thought to be the converse of the CMC. In a sense, while the CMC states that *if* a graph satisfies the Markov condition, *then* it is a BN, the faithfulness condition states that *if* a graph is a BN, *then* it has to satisfy the Markov condition. According to the faithfulness condition, all probabilistic dependencies and independencies have to be represented in the graph, otherwise the graph is unfaithful. The graph is unfaithful when there are hidden dependencies; things happening relevant to what is shown, but that we cannot see in the graph. This condition applies to BNs. It is important for causality because it provides a methodological principle to double-check whether graphs are BNs or not (for a discussion, see for instance Hitchcock (2011)). Bear in mind, though, that this condition must be interpreted in the light of what we have said about models in general: they all abstract from a more complex reality. There will always be something unrepresented by any model. In practice it is often the case that minimal probabilistic dependencies are not represented, and the model is still useful for its intended purpose. In this case, the faithfulness condition is, more literally, that all *relevant* probabilistic dependencies and independencies are represented in the graph.

A stock example discussed in the literature concerns the relations between the use of contraceptive pills, pregnancy and thrombosis (see Hesslow (1976)). Contraceptive pills lower the probability of pregnancy, and both contraceptive pills and pregnancy increase the chances of thrombosis, as shown in Figure 7.3. Suppose now that the two paths *pill* →⁻ *pregnancy* and *pregnancy* →⁺ *thrombosis* balance, i.e. they cancel out, and any probabilistic dependence between the two vanishes. Yet, the graph shows a path, so the model is unfaithful. This is of course a highly simplified situation. In more realistic situations, appropriate partitioning of the population would help avoid such problems.

Recursive decomposition. Earlier, we defined a structural model as a family of probability distributions. However, we need to say more in order to have some useful

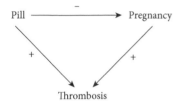

Fig 7.3 Hesslow's example: contraceptive pills, pregnancy and thrombosis.

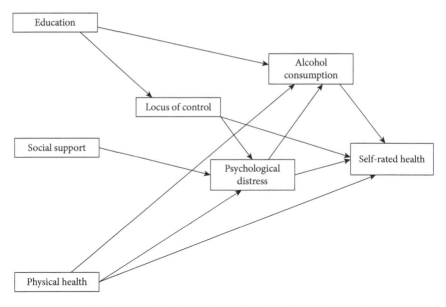

Fig 7.4 Causal graph for the study on self-rated health in Baltic countries.

information about causality. The recursive decomposition is, simply put, the structure of the distributions. Intuitively, the recursive decomposition tells us what causal role (direct cause, indirect cause, or interacting cause or effect) each variable plays in this family of distributions. In more technical terms, we decompose an initial joint probability distribution into marginal and conditional components. The notation we will now use is quite shallow, but if we can grasp the basic idea behind probabilistic causality (see chapter 8), then at least we will get a sense of what a recursive decomposition is. Considering the Baltic study, we'll now reason over the corresponding causal graph (see Figure 7.4).

We begin with all the variables together: education (*Ed*), social support (*Soc*), physical health (*Phy*), locus of control (*Loc*), psychological distress (*Psy*), alcohol consumption (*Alc*) and self-rated health (*Self*). This is the initial joint probability distribution:

$$P(Ed, Soc, Phy, Loc, Psy, Alc, Self)$$

Here, we can't distinguish causes and effects. So we need to decompose this joint distribution. Background knowledge and preliminary analyses of data loom large here. The recursive decomposition is the causal structure represented in the graph, where several paths to self-rated health are identified. First, we list all the nodes that have an arrow going into self-rated health (*Self*). Then, for each of these nodes, we track their immediate parents. We do this for alcohol consumption (*Alc*), psychological distress (*Psy*), and locus of control (*Loc*). Physical health (*Phy*) has a direct effect on self-rated health but does not have immediate parents, which also means that it is exogenous. Last, we list the nodes that have no incoming arrow (and that influence other nodes), namely education (*Ed*), social support (*Soc*) and physical health (*Phy*). This is the sequence of conditional and marginal components:

$$P(Self|Alc, Psy, Loc, Phy)$$
$$\cdot P(Alc|Ed, Psy, Phy) \cdot P(Psy|Loc, Soc, Phy) \cdot P(Loc|Ed)$$
$$\cdot P(Ed) \cdot P(Soc) \cdot P(Phy)$$

So the recursive decomposition 'translates' into probabilistic relations the various paths (sub-mechanisms) acting in the global mechanism depicted in the graph.

Two things are worth bearing in mind:

(i) We can of course interpret the recursive decomposition just as sequences of marginal and conditional *probabilities*, and therefore change the order of the Xs and the Ys as we like—this won't make much difference in *probabilistic* terms. But to get a *causal* perspective, it does matter how we choose the order. In a causal analysis, we have to establish whether alcohol consumption is a *cause* (direct, or indirect) of self-rated health. In causal terms, it matters a lot what order we give to the variables. And of course, a lot will depend on the available background knowledge. The question of whether conditional probabilities preserve meaning when read in either direction is also known as 'Humphreys' paradox'. Simply put, it is intuitive to grasp the causal meaning of a conditional probability: $P(A|B)$ says that the probability of A occurring depends on B, i.e., B causes A (see also chapter 8). But, according to probability theory, $P(B|A)$ is also meaningful. Should we interpret $P(B|A)$ as saying that, likewise, A causes B? Humphreys (1985) originally drew attention to this issue. More recent discussions include Humphreys (2004), and Drouet (2011).

(ii) A causal interpretation of the recursive decomposition goes hand in hand with a *mechanistic* interpretation. The recursive decomposition can in fact be seen as (statistically or probabilistically) representing a mechanism, that can be broken down into smaller parts (the marginal and conditional components). This is of course one view, defended by e.g. Mouchart and Russo (2011). Another view, and perhaps the most widespread one, holds that mechanisms, especially in social contexts, are ultimately what scientists often call 'black boxes' which is to claim that we cannot really know the mechanisms. See for instance the discussion in Chao (2009). It is of course possible that we will never fully understand the mechanisms of some phenomena. Yet, the message of the mechanistic interpretation is that structural modelling (in the sense of explanatory and causal modelling) aims at *opening* black boxes.

Exogeneity. The concept of exogeneity is closely related to the recursive decomposition in structural models. It has a long history, as it has been developed since very early work in econometrics. The recent literature in causality and economics sees in Engle et al. (1983) the locus classicus. They discuss a number of different concepts in the econometric literature and display their connections with exogeneity through the introduction of supplementary conditions. However, they owe a great deal to other works, especially to the idea that exogeneity is a condition of *separation of inference* (see e.g. Barndorff-Nielsen (1978), Florens and Mouchart (1985), Florens et al. (1993)).

This means that the parameters of the marginal and conditional components in the recursive decomposition are independent, and this allows us to distinguish the process generating the causes and the process generating the effect. In this way, we can make more precise the expression often used that 'exogenous means generated outside the model'. The exogenous causes are those for which we do not seek a generating process, or mechanism, *within* the model, and that are instead what we consider the causes of the outcome variable under consideration.

Figure 7.4 reproduces a simplified version of the causal structure of the Baltic study mentioned earlier. In this graph, education, social support, and physical health are exogenous variables; this is indicated by the fact that these variables have no 'inbound' arrows, just 'outbound' arrows. In this model, there is no generating mechanism *for these variables*, while these variables are in the generating mechanism of other variables. Locus of control is influenced by education and therefore is not exogenous (in this model). Likewise, psychological distress is generated by locus of control, social support and physical health. It is important to bear in mind that whether variables are exogenous is an *empirical* matter. In other words, *if* exogeneity tests are successful, *then* we can interpret these variables as causes in the model.

Invariance. The concept of invariance is also closely related to structural models and to recursive decomposition. Invariance was originally developed in early econometrics (see for instance Frisch (1995), Marschak (1942), Haavelmo (1944)) and it has been further discussed in recent times by Woodward (2003), as we also discuss in chapter 10. Invariance means that the relations under analysis, and expressed in the form of a recursive decomposition, are 'stable', or undisrupted. This is something we can check in different ways. In experimental contexts, stability can be tested by intervening on the cause-variable and observing whether the causal structure is disrupted and to what extent. In observational contexts, stability can be tested by checking whether the causal structure holds across partitions of the data set (subpopulations). We require invariance because in structural models we want to establish causal relations that are generic, i.e. repeatable (see also chapter 5); invariance also helps rule out accidental relations (such as the correlation between the increased presence of storks and of birth rates in Alsace) and spurious relations (such as the correlation between yellow fingers and lung cancer).

It is important to bear in mind that in causal modelling invariance is an *empirical* matter: if the relations pass tests of invariance, then the relations are *deemed* causal. It is worth recalling, at this point, two aspects of causal models that we emphasised earlier: (i) no matter how good our model is, we may get wrong results, and (ii) the notions hereby presented are properties of models, not of real, worldly systems. That is to say that the modelling enterprise aims precisely at establishing whether, and to what extent, relations between variables are invariant. Whether it is 'true' that the relations between variables of interest are invariant is precisely what we try to find out in science, and this is why we replicate studies, or re-analyse data, or apply a model to different data, etc. We won't know a priori, but only by doing more and more research. Another strategy, that we discuss in more detail in chapter 10, is to *define* causality through invariance: if the relations are causal, then they must show invariance properties. The approach described in chapter 10 is different from the one just sketched. The main difference is that manipulationist approaches take invariance as a characteristic that defines causality, whereas invariance can also be considered a test for causal relations.

Modularity. Modularity is mainly discussed in the context of Bayes' Nets and of structural models (in Pearl's sense). Modularity is a property of a causal system, according to which we can modify the system 'bit by bit', which means we can modify only one part of the system, holding fixed the rest, and without disrupting the whole thing. In a sense the idea is not very different from the marginal-conditional decomposition discussed above, but with one important exception. Modularity is generally discussed in terms of whether it is possible to *modify*, or *manipulate*, the system in such-and-such a way. Instead, in the recursive decomposition we do not modify anything, but consider dependencies and independencies between variables, that may have been originated by manipulations of the system, or may just have been observed as in the case of a survey.

The reason why this notion goes in tandem with manipulationist (or Woodwardian) invariance is apparent: invariance says that the causal relation between variables X and Y is such that by wiggling X—keeping everything else fixed—Y will change and the whole system won't be disrupted. The relation between X and Y has to be modular, then, otherwise we won't be able to intervene in such a 'surgical' way. However, the marginal-conditional decomposition does not necessarily describe the behaviour of a system being *manipulated*, as data can be purely observational. The recursive decomposition identifies mechanisms and sub-mechanisms in a purely observational way, and tries to describe the different parts, modules or components in a system with fewer stringent restrictions.

One question concerns the status of the modularity condition. In manipulationist approaches, which we discuss in chapter 10, modularity is an essential feature of causal relations. But we could see it from a different angle: modularity is an empirical matter, so we are interested in finding out *whether* and even to what extent a system is modular because this will give us valuable knowledge about where to intervene, for instance for policy purposes or in lab experiments.

The key notions just discussed are all crucial to understanding what the models themselves mean, and also for understanding the various fierce debates that take place over a modelling approach, or between modelling approaches. But they have also taken on an interesting position in the philosophical literature, because they are also interesting causal concepts that have been sharpened up by the causal modelling literature.

Do not forget that formally these are features of models, not of populations or of worldly systems. However, populations can be more or less susceptible to a particular kind of modelling, by *approximately* fulfilling the criteria. And knowing these key notions can be helpful in thinking about what we want from causes, to guide what we are looking for. For example, the Causal Markov Condition allows us to see how useful it is to find a proximate cause of a target effect of interest, which makes the action of more distal causes irrelevant. This is very handy, because using that cause alone gives us a great deal of control of the target effect. Alternatively if we find a system that is at least approximately modular, that is great, because we know there are ways we can disturb parts of the system, and the overall system will not be affected.

In this way, these concepts have become useful in sharpening our causal reasoning, affecting our epistemic practices, and the use we can make of the causes we find.

Core ideas

- In much of science we do a lot of processing of data to find correlations, using these to find causes.
- There is no gold standard. The 'best' causal method depends on the case at hand.
- Correlation is not causation, and models that help us establish correlations (or associations) have a different apparatus to models that help us establish causal relations.

Distinctions and warnings

- Automated procedures for causal inference help, but do not solve the problem of getting causal relations out of rough data: background knowledge is vital!
- Assumptions and background knowledge are key for establishing the validity of models.
- Technical terms like 'exogeneity', 'structural' or 'modularity' have specific meanings, and need to be used carefully.

Further reading

Pearl (2000, chs 1, 5); Freedman (2005b, chs 1, 2, 5); Russo (2009, ch3).

CHAPTER 8

Difference-making: Probabilistic Causality

8.1 Example: smoking and lung cancer

It is difficult to overestimate the importance of probability and statistics to science. They have proven to be an enormously useful and powerful tool (see also chapter 7). Consider for instance studies on the causes of cancer. It is nowadays undisputed that smoking causes cancer. Evidence is abundant, having been accumulated for decades. (For a brief history of smoking and cancer see White (1990) or Witschi (2001).) But up until the fifties, smoking was even considered a good medication for a sore throat or for lung infections. The scientific community became suspicious because epidemiological studies revealed a sharp rise in lung cancer rates, and a strong correlation between smoking and cancer. More precisely, studies revealed that smoking tended to be associated with cancer irrespective of gender, socio-economic status or other factors; that is in each sub-population smokers had a higher risk of developing cancer. In other words, smoking increases the *probability* of developing lung cancer. Care is needed, however, in spelling out what probability means.

The *International Agency for Research on Cancer* (IARC) seeks to classify agents that are carcinogenic or possibly carcinogenic (which means they cause or possibly cause cancer). Tobacco smoke (active and passive) is classified in group 1, which means that it is 'carcinogenic to humans'. Note that 'carcinogenic to humans' is a *type*, or *population level*, or *generic* claim (see chapter 5). While there is an established causal link, according to IARC, between smoking and cancer, there isn't an established causal link between smoking and cancer for any particular individual person. The individual response to exposure is personal and we all know smokers who never developed cancer (again, see chapter 5).

To see further why probability is relevant, consider IARC's classifications of carcinogens. Groups 2A and 2B of the IARC classification indicate that there is evidence of carcinogenicity, but evidence that is not conclusive in establishing causal links. In the *Preamble to the IARC Monographs* we read:

This category includes agents for which, at one extreme, the degree of evidence of carcinogenicity in humans is almost *sufficient*, as well as those for which, at the other extreme, there are no human data but for which there is evidence of carcinogenicity in experimental animals. Agents are assigned to either Group 2A (*probably carcinogenic to humans*) or Group 2B (*possibly carcinogenic to humans*) on the basis of epidemiological and experimental evidence of carcinogenicity and mechanistic and other relevant data. The terms *probably carcinogenic* and *possibly carcinogenic* have no quantitative significance and are used simply as descriptors of different levels of evidence of human carcinogenicity, with *probably carcinogenic* signifying a higher level of evidence than *possibly carcinogenic*. (IARC, 2006, emphases in the original.)

Human papillomavirus type 68 is, for instance, in group 2A as a probable cause of cervical cancer. There are a number of screening programmes in place to detect cervical cancer as early as possible, and also vaccination programmes. A review on radiofrequency electromagnetic fields is currently in preparation, as an example of group 2B.[14]

Probability and statistics are not the only tools we use to establish carcinogenicity and we address the problem of evidence for causal relations in chapter 6. Here, though, we have illustrated the motivation for thinking that causal relations should be probabilistically analysed. Assessment of probabilistic relationships has come to be one of the most important tools for finding causes, and it is also one of the most publicly visible tools of science, with the media constantly reporting some correlation or risk factor or other. So thinking probabilistically affects causal inference, and prediction, and thereby also causal reasoning.

8.2 Is causality probability-altering?

In view of the importance of statistical tools in the sciences, an account of causality in terms of probabilities is a very natural starting place for causality in the sciences. Due to this, the idea of a probabilistic characterization of causality has been discussed in the philosophical literature at least since the sixties. In the early works of Good and Suppes the intention was to define 'cause' in terms of 'probability'. But later accounts did not pursue this strong view, instead trying to establish bridges between causality and probability by showing how, for instance, one could derive probabilistic conclusions from causal premises (for a discussion, see Gillies and Sudbury (2013)).

What probabilistic accounts share is that they exploit the formalism of probability theory in order to define 'cause', and also other related notions we shall meet shortly, such as 'spurious cause' or 'negative cause'. The core idea is that a cause makes its effect more (or less) probable. This means that probabilistic accounts are 'difference-making' accounts, where the relevant difference-making is that causes make a difference to the probability of occurrence of the effect. Various probabilistic theories use different formalisms to make the idea of 'causes making the effect more or less probable' more precise, and that is how these accounts differ from each other. Pioneer theorizers of probabilistic causality include Hans Reichenbach, I. J. Good, Patrick Suppes and Ellery

---

[14] Website of IARC, <www.iarc.fr>, accessed on March 19th, 2012.

Eells. We will not present the complete formalisms developed by these scholars, concentrating instead on explaining some elements of them. Also, we will not compare the various measures of causal strength offered by these and other authors, as this becomes very technical, and an informative discussion is provided by Fitelson and Hitchcock (2011).

I. J. Good was a mathematician who offered a quantitative explication of probabilistic causality in terms of physical probabilities in the early sixties (see especially Good (1961a) and Good (1961b)). Good wished to offer a concept of causality that was not sharply deterministic, as causes don't always produce the effect, the same cause may produce a different effect in some cases, and an effect may be produced by different causes in different cases. Good wanted to capture this idea using the formalism of probability theory, which allows him to capture it *quantitatively*.

We will use the work of Suppes (1970) to make this clearer by providing a basic formalism. Suppes' formalism, at least the one presented in the first half of his book, is quite accessible, and it is the starting point of many other contributions in the field. Suppes' theory is more sophisticated and complex than what we present here, but we don't need to present technicalities. To grasp the heart of probabilistic theory, we will first consider the typical and most widespread definitions of positive and negative causes. We use formulae and definitions, as probabilistic theories do, to help the reader grasp and make more precise basic ideas of causality.

The first idea to make more precise using Suppes' formalism is that causes make a difference to the probability of the effect.

Definition 8.1. *The event C_t is a prima facie positive cause of the event $E_{t'}$ if and only if*

(i) $t < t'$
(ii) $P(C_t) > 0$
(iii) $P(E_{t'}|C_t) > P(E_{t'})$

This says simply that (i) the cause, such as smoking, occurs before the effect, such as lung cancer; that (ii) the probability of the cause is greater than zero; and that (iii) the probability of the effect (lung cancer) given the cause (smoking) is greater than the probability of the effect all by itself. In other words, developing lung cancer is more probable if we smoke.

Definition 8.2. *The event C_t is a prima facie negative cause of the event $E_{t'}$ if and only if*

(i) $t < t'$
(ii) $P(C_t) > 0$
(iii) $P(E_{t'}|C_t) < P(E_{t'})$

This says simply that (i) the cause, such as quitting smoking, occurs before the effect, such as lung cancer; that (ii) the probability of the cause (quitting smoking) is greater than zero; and that (iii) the probability of the effect given the cause is *less* than the probability of the effect all by itself. In other words, developing lung cancer becomes less probable if we quit smoking.

Suppes offers two definitions to capture both the more intuitive case where causes make effects more likely, and negative causes where causes make their effects less likely. While many accounts focus on the *probability-raising* requirement, in fact what is really at stake is a *probability-changing* requirement.[15] Prima facie, what we are interested in is whether the putative cause makes a difference to the probability of the putative effect. This is important because in many contexts, such as in medicine, we are interested rather in causes that lower the probability of the effect. For example, does physical activity prevent cardiovascular problems? Does a healthy diet prevent cancer? These are negative causes of the effect. These and cases like them can be captured if we have a broader concept of probability changing rather than just probability raising.

But prima facie causes are just ... prima facie. It is not difficult to imagine that, with the available data, we get the probabilities, and therefore the causes, wrong. Here is a stock toy example of how this may happen (we also use this example in chapter 2). Here we make its probabilistic structure more precise. Having yellow fingers is correlated with having lung cancer, so yellow fingers are a prima facie positive cause of lung cancer. But of course, medical knowledge tells us that the genuine cause of lung cancer must be something else, for instance cigarette smoking. Cigarette smoking fully accounts for the correlation between yellow fingers and lung cancer and in the literature we say that cigarette smoking 'screens off' yellow fingers from lung cancer. More formally:

Definition 8.3. $C_{t'}$ *is a prima facie cause of* $E_{t''}$ *and then there is an earlier event* F_t *that screens off C from E:*

(i) $t < t' < t''$
(ii) $P(C_{t'} \wedge F_t) > 0$
(iii) $P(E_{t''}|C_{t'} \wedge F_t) > P(E_{t''}|C_{t'})$

This just says (i) that F is earlier than C, which is earlier than E (in our example, smoking is earlier than yellow fingers, which is earlier than lung cancer); (ii) that the probability that both C (yellow fingers) and F (smoking) happen is greater than zero; and (iii) that the probability of C (yellow fingers) given that F (smoking) has happened is greater than zero, and, although C (yellow fingers) and E (lung cancer) are correlated, that correlation disappears if we introduce the factor F (smoking). Then we say that F screens off C from E. Smoking screens off yellow fingers from lung cancer. This idea was developed by Reichenbach (1956), who was motivated by problems in statistical mechanics (see chapter 2). In particular, Reichenbach was analysing the arrow of time and he tried to get the direction of time through the direction of causation of physical phenomena.

In real science, finding cases of screening off is far from trivial. In medicine, this is very important to correctly identifying symptoms and causes of diseases. High blood pressure causes strokes, but high blood pressure is also a symptom of an earlier cause,

---

[15] For a different view see, however, Gerring (2005), who argues that probability raising provides a 'unitary conception of causation' in the social sciences.

and it is vital to identify the right chain of cause–effect events in order to prescribe the most effective treatment.

Here is another example from social science. Correlations between migration behaviour and the establishment or dissolution of households are well known. The problem is of course to establish the direction of the causal relation, namely whether family problems increase the probability of migrating, or whether the intention to migrate, for example for professional reasons, instead makes separation more probable. The complication in examples like this is that the probabilistic relation may flow in either direction, and the direction suggested by the analysis will greatly depend on the data gathered. For instance, in some contexts or cases household dissolution is the root cause of migration. However, changes in households are recorded much later (as divorce takes time), so migration is de facto observed *before* household dissolution. This would give the impression that migration causes household dissolution, although the temporal order is just an artefact of how data are collected (for an example, see Flowerdew and Al-Hamad (2004)). To decide about such cases, we usually need more than probabilities: we need articulated background knowledge (see chapter 7).

We will now introduce graphs as a useful heuristic tool to visualize what the previous definitions mean. The simplest situation is the one in which we want to establish whether C causes E:

$$C \longrightarrow E$$

We have seen that probability changing may be positive or negative depending on whether the putative cause C increases or decreases the probability of E. Once we introduce a third event F, there are a number of possibilities. One is that we are dealing with a causal chain:

$$F \longrightarrow C \longrightarrow E$$

But we may also be in another situation, one where F is a common cause of C and E. This is the most typical case of screening off and the one that has generated a long and still-lively debate on the nature and methodology of the principle of common cause, especially in the 'EPR correlations' (see e.g. San Pedro and Suárez (2009) and Hofer-Szabó et al. (2013)).

Identifying the correct causal structure is precisely the problem that causal modellers in physics, social science, medicine and so on constantly strive to address—see chapter 7.

We now introduce a further nuance that affects the subtleties of many probabilistic theories. First, note that in the discussion so far we have equivocated between two senses of 'causes'. One is to talk in general terms; the other is to talk about specific causes. So, for example, we are interested in two different causal claims about smoking and lung cancer: 'smoking increases the chances of developing lung cancer in the population' and 'Alice's smoking makes Alice more likely to develop lung cancer'. We discuss this difference thoroughly in chapter 5.

This problem has interested many proponents of probabilistic theories. I. J. Good, for instance, provides precise probabilistic formulations for these two ideas: (i) tendency of an event to cause another event; (ii) the degree to which one event actually causes another. These need different probabilistic measures. His *Calculus* is rather technical, so, again, we shall not present his mathematical formalism. But we wanted to mention Good's work because these two ideas—tendency of an event to cause another event and the degree to which one event actually causes another—influence other probabilistic theories, and come up elsewhere in the book (see chapter 5 about the levels of causation, and chapter 14 about capacities and powers—which are like tendencies).

Eells (1991) develops his probabilistic theory in a similar direction, thoroughly discussing ideas of type causation, such as smoking causes lung cancer, and token causation, such as Alice's smoking causes her lung cancer. We will now examine the different ways of specifying these two varieties of causes probabilistically, using Eells's formalism. First, a note about language. The literature within causality often uses the language of type versus token causes, and population-level versus individual-level causes, to mark the same distinction. We explain these in detail and also discuss an alternative vocabulary in chapter 5.

Eells characterizes probability raising for type-level causality:

$$P(E|C \wedge K_i) > P(E|\neg C \wedge K_i),$$

where C and E stand for cause and effect, and K_i represents the background context. This says that smoking causes lung cancer if it raises the probability of lung cancer, taking into account all the relevant background possible causes, confounders, etc. It clearly concerns a population or type. This is quite a strong requirement, and in a sense it is unrealistic as we are seldom in a position to control the totality of the environment around the putative cause. In practice, if we make the probability-raising principle above *relative* to a model which specifies relevant background factors, then we can avoid considering all possible causal factors, precisely because the model incorporates the boundaries of and possible conditions for the causal relation (see Gillies and Sudbury (2013)).

Once we have this idea of type-level probability raising, we can reason about token probability raising. Basically we need to identify key points at which the probability of token events is likely to change, and we can draw a probability trajectory which is tailored to the case at hand. Eells illustrates the token probability trajectory using a stock toy example that we also discuss in chapter 5: the squirrel and the golf ball. Imagine a golf player who hits the ball straight in the direction of the hole, which gives a high probability of a hole-in-one; then suddenly a squirrel appears and kicks the ball. The probability of a hole-in-one drastically decreases at the squirrel's kick, but then again suppose the ball hits a branch and falls into the hole nonetheless. We could draw the probability trajectory of the ball and see that after hitting the branch again the probability gradually rises to 1, when the ball falls into the hole (see Figure 8.1).

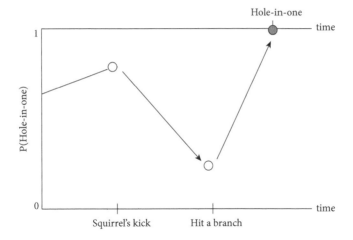

Fig 8.1 Probability trajectory.

Eells' account is not without problems. For instance, it is very hard to determine the probability trajectory of an event. We can a posteriori reconstruct it, by re-calculating the probability of the event in question, given the available information. But is this good enough? For example, could this line of reasoning be used in court? A historical sentence in Italy in 2012 judged the bosses of the company Eternit guilty in the 'Eternit trial' because exposure to asbestos led many workers to develop lung cancer and die. It is very difficult to assess the probability trajectory of developing lung cancer of any individual deceased worker. But epidemiology can help establish how likely it is that salient facts in the exposure history did in fact raise the probability of developing lung cancer. So we can fill in some details of the 'trajectory'.

This also illustrates the limits, not only of Eells' probability trajectory, but of the probabilistic framework altogether. On one hand, it was an immensely important move because it liberated causation from deterministic connotations and tied it to actual scientific practice that employs probability and statistics. On the other hand, the kinds of formalisms laid out here are merely the bones of methods that need to be much more complex and sophisticated to be useful in science. Probabilistic accounts have not gone far enough to account for complex causal scenarios not formalizable solely in terms of probabilistic relations between events. For this, see chapter 7. Another important limitation is that time is presupposed rather than properly analysed. Suppes even omits the time indexes we have included above precisely because he *presupposes* that causes precede effects in time. The problem with this presupposition is that most of the time, in actual causal inference in the sciences, we just don't know the correct temporal ordering, or we don't know the time lag between exposure and outcome.

So probabilistic reasoning is important to reasoning about causality in the sciences. The ideas of negative and positive causes, screening off and levels of causation are all important to understanding methodological practices in the sciences, and to interpreting their results so as to be able to use them effectively in prediction and control.

But there is a good deal more to understand than this. We have barely begun, but the 'probabilistic turn' initiated by Good, Suppes and others puts us on a good track to explore the relations between causality and probability.

8.3 Beyond probabilistic causes

8.3.1 Determinism/indeterminism and predictability

The 'probabilistic turn' in the philosophy of causality had to fight against an entrenched concept of deterministic causation. Causes had traditionally been conceived of as necessary, and often also sufficient, elements for the occurrence of their effects (see chapter 4) and causality had long been thought to be part of deterministic systems only. In a deterministic system, given the initial conditions and the laws, we can in principle predict future states of the system exactly.[16] Systems described in classical physics, such as, for example, Newton's laws of motion, are paradigmatically deterministic. But not all systems are like that. Quantum physics, for instance, describes indeterministic systems, with examples such as radioactive decay and Brownian motion of particles in a fluid. Here, events can be predicted only with certain probabilities and this may give the impression that events are not caused, or at least are not deterministically caused. Einstein-Podolsky-Rosen (EPR) correlations are also an example of an indeterministic system. Quantum mechanics (QM) is one of the most controversial topics for both physicists and philosophers. John von Neumann (1955) held the view that QM is a-causal in the sense that it is non-deterministic. But if we accept the idea that causes may be probabilistic, then there is no reason why QM should be a-causal. Part of Suárez's work, for instance, is devoted to defending a causal interpretation of EPR correlations (see e.g. Suárez (2007)).

All this prompts two remarks. First, we should not confuse predictability with determinism. Second, we should not confuse determinism with causality. Much work in the philosophy of causality has been devoted to spelling out and defending a *probabilistic* concept of causality. Besides the scholars mentioned above, Wesley Salmon also tried to provide a causal account of primarily physical phenomena that are not deterministic (Salmon, 1980b; Salmon, 1984). The distinction between determinism/indeterminism and predictability is also related to the next issue.

Paul Humphreys (1990, sec17) distinguishes between 'theories of probabilistic causation' and 'probabilistic theories of causation'. In theories of *probabilistic causation* we aim to provide a conceptualization of causal relations that are inherently probabilistic, stochastic, chancy or indeterministic. So we may well talk about 'indeterministic causation'. In *probabilistic theories* of causation we aim to provide a probabilistic characterization or modelling of causal relations that may or may not be probabilistic in themselves. In this second case we may either admit that there exist cases of deterministic causation and indeterministic causation, or we could hold that causation is all

---

[16] There are difficult cases, most notably chaotic systems, which can be described by deterministic equations, but future states of the system are so highly sensitive to initial conditions that future states cannot be exactly predicted in practice.

deterministic, but our modelling of causal relations is probabilistic because we don't have complete knowledge or because of measurement error.

8.3.2 Interpretation of probability in probabilistic causality

We have been talking about causes that make a difference to the probability of the effect. We have also been talking about two levels at which we can describe causal relations: the type and the token level. The philosophy of probability asks how the idea of 'probability' that we used to specify these ideas ought to be interpreted. What do we mean when we say that smoking increases the probability of lung cancer?

There are a variety of views in the literature—for an introduction, see Gillies (2000). The most important distinction is between views that hold that probabilities are some kind of feature of the world, and views that hold probabilities to be features of the strengths of the beliefs of agents (e.g. scientists). The strength or degree of belief is a classic Bayesian interpretation: the probability of smoking causing lung cancer is just a representation of how strongly the agent, say a doctor, believes it. More objective interpretations of probabilities, in terms of some feature of the world, tend to be either frequency or propensity interpretations. Frequency interpretations say that probabilities represent frequency of incidence in a population. So the probability that smoking causes lung cancer represents the frequency of appearance of lung cancer in populations of smokers. But this does not work in the token case. One person does not display a frequency of getting lung cancer. Instead, this kind of case can be given a propensity interpretation: each person has a propensity to get lung cancer, which will be influenced by their other risk factors for lung cancer.

In discussions of probabilistic causality, most philosophers have also offered an account of the interpretation of probability, although Suppes does not take an explicit stance. Good says that he is interested in physical probabilities (frequencies or propensities) but that his formalism is about epistemic probabilities (namely, degrees of belief) because degrees of belief are what we use to 'measure' objective probabilities. Eells holds a 'mixture' of a frequency view for type-level causation and a propensity view for token-level causation (Eells, 1991, ch1.2).

This debate is ongoing. An interesting point is that it looks like the question of the interpretation of probability cannot be answered independently of the context in which the question arises. This concerns not only the question of whether we are dealing with type or token causal relations, but also the specific scientific domain. Russo (2007) and Russo and Williamson (2007b) argue that generic or type-level causal relations need a frequency interpretation and that single-case causal relations need an objective Bayesian interpretation—their arguments apply to the social sciences and to causal models for cancer, respectively. Daniel Courgeau discusses probability in social science and holds a form of objective Bayesian interpretation (see Courgeau (2004) and Courgeau (2012)). In Suárez (2011) several contributions discuss the propensity view of probability in physics. Gillies (2000) holds the view that the social and physical sciences require different interpretations. The question of whether these domains need different interpretations of probability is related to the issue of whether probabilities are objective features of the world that we can more or less directly access, or whether they represent agents' beliefs. Of course, this opens up other directions of research, notably about the relations between probabilistic modelling, knowledge and reality.

8.3.3 A development: probabilistic temporal logic

Kleinberg (2013) enriches probabilistic theories of causality with temporal logic and with a data-mining approach. Kleinberg makes a simple idea—that positive causes raise the probability of their effects—methodologically sophisticated by incorporating additional tools, namely temporal logic, into the probabilistic framework. This allows her to give a fine-grained analysis of probabilistic relations that happen in time. Thus her theory is not subject to typical counterexamples that abound in the philosophical literature such as barometer readings being correlated with storms, or pregnancy with contraceptive pills and thrombosis.

The probabilistic relations between the mentioned pregnancy, the use of contraceptive pills and thrombosis is a stock example. The problem arises because contraceptive pills raise the probability of thrombosis, but pregnancy also raises the probability of thrombosis. Since contraceptive pills lower the probability of pregnancy, they therefore lower the probability of thrombosis, via this different path. They both lower and raise the probability of thrombosis (see also chapter 7). The framework developed by Kleinberg (2013) can address these prima facie counterexamples which have plagued philosophical accounts because it makes temporal relations explicit and because other important aspects related to the modelling (for example, the partitioning of the population) can be incorporated into the framework.

Kleinberg also articulates the relation between type and token causal relations. The interest here is (i) to discover causal relations at the generic level and, on the basis of such knowledge, (ii) to be able to say something about the single-case (particularized diagnosis, for example). Kleinberg represents causal relationships by means of formulae in probabilistic temporal logic and defines a new measure for the significance of such putative causal relationships. The type-level relations identified in this way are then used to evaluate token-level explanation. To this end, Kleinberg reworks the connecting principle originally proposed by Sober (1986), which has been largely overlooked in previous literature, in a way that is more sophisticated and computationally tractable than the discussion of Russo (2009). As we also discuss in chapter 5, the relations between type and token causes are quite complex. Kleinberg's work has the merit of constraining the issue to a particular domain and of developing computational tools that make the intuitions of pioneer theorizers of probabilistic causality computationally tractable. This work has already been applied to some data sets containing electronic data in hospitals and in finance, and has great potential for wider uses in areas where time series are analysed.

In summary, probabilistic approaches to causality are one of the earliest approaches of philosophy of science. They allowed the clarification of different kinds of causes, such as negative and positive causes, and connected with the formal tools of the sciences. They are still generating useful work, in connection with developing computational methods, increasingly applied in the sciences.

Core ideas

- Since causality does not seem to require determinism, and probabilities are such a crucial tool in the sciences, it seems causality might be related to probabilities.
- We can use probabilities to make precise some important causal ideas, such as negative and positive causation. Overall it seems that causation, if probabilistic at all, concerns probability *changing*.
- There is an important difference between type- or population-level causes, such as 'smoking causes lung cancer', and token- or individual-level causes, such as 'Alice's smoking caused her lung cancer'.
- The important idea of 'screening off' can be described probabilistically: F screens off E from C if the correlation between E and C disappears when we introduce F.
- Favoured interpretations of probability in probabilistic causality include objective interpretations, particularly frequency in a population as the population tends to infinity, or propensity views; and subjective degrees of belief.

Distinctions and warnings

- Remember the distinction between positive and negative causes, as the importance of negative causes is easy to forget.
- The difference between type or population causes and token or individual causes is also important: it is not always simple to infer between the two. That smoking causes lung cancer in the population with a certain probability does not mean that Alice has that same probability of getting lung cancer if she smokes.
- Determinism, indeterminism and predictability are distinct concepts. Some probabilistic phenomena are indeterministic (e.g. quantum mechanics) and this affects (probabilistic) predictions. But for other phenomena, for instance cancer, it has not been established that they are indeterministic. Even so, probabilistic modelling is a useful and successful method for eliciting relations between variables.
- The distinction between 'theories of probabilistic causation' and 'probabilistic theories of causation' is related to the previous point: it is one thing to have a theory about causation as *inherently* probabilistic, and it is another thing to have a theory that models causal relations using probabilities.
- Interpretations of probability in probabilistic causality are all problematic. Can we really consign so much of science just to degrees of belief? What does it mean to say 'frequency in a population as the size of the population tends to infinity'? No populations are infinitely large! How do we understand Alice's individual propensity to get lung cancer? How do we estimate that?

Further reading

Hitchcock (2011); Kleinberg (2013, ch2); Williamson (2009).

CHAPTER 9

Difference-making: Counterfactuals

9.1 Example: mesothelioma and safety at work

Progress in medicine is noteworthy for advancing our understanding of disease aetiology. We now know that mesothelioma is a tumour that develops in the lungs due to severe and prolonged exposure to asbestos. Before this causal relation had been established by the scientific community, few, if any, safety measures concerning asbestos were taken in work environments such as factories, or in the construction of buildings. Some patients affected by mesothelioma, or their relatives, have sued their employers because—they claim—*had they been protected* against asbestos exposure in the work environment, cancer wouldn't have developed. So, the argument goes, negligence, in this case failure to protect workers, is the cause of some states of affairs, in this case mesothelioma. On 13 February 2012 the court in Turin, Italy, issued a historical verdict against the managers of the asbestos factory 'Eternit'. They were held liable for not having adopted proper safety measures in the work environment. The charge concerned some 2,200 asbestos-related deaths in the past thirty years.

The Eternit case exemplifies what philosophers call 'counterfactual reasoning'. This is important in many places in science. It is particularly used for reasoning to the consequences of known causal structure. But such reasoning is also used as part of the process of discovery of causal structure, as counterfactual reasoning is also used to generate hypotheses. For example, we might reason, 'if asbestos had not caused this incidence of mesothelioma, we ought to be able to find some other cause'. In ways like this, we reason from a possible causal structure to observable consequences, which we can test. This case also illustrates how understanding causality in biomedical sciences, and other areas, is very important to the law. We need valid causal reasoning in the law to hold someone legally responsible for negligence. And we need valid causal reasoning in biomedicine and epidemiology to establish whether proper protection would have avoided cancer development.

9.2 The unbearable imprecision of counterfactual reasoning

This way of reasoning is ubiquitous in everyday and in scientific situations where we try to reason about what causes what. For the time being we will abandon complex scientific and legal cases, and illustrate the core ideas through everyday examples. For instance, we might say, 'Had I heard my alarm clock, I wouldn't have missed the bus', or, 'If we were to take the sun away, the earth would fly off into space'. We seem to be able to make sense of such claims.

In a way, though, our ability to make sense of the kinds of claims above is puzzling. We call these kinds of claims 'counterfactuals' because they are contrary-to-fact conditionals. A conditional is just an if–then claim: *if* one thing, *then* another thing. If you hear someone say, 'Had I heard my alarm clock, I wouldn't have missed the bus', this is an if–then claim whether or not the words 'if' and 'then' actually appear in the sentence. In this case you also realize that the person speaking did not hear the alarm clock and did miss the bus. So the claim is contrary-to-fact. When we make and understand such claims, we are reasoning about things that didn't happen.

Counterfactual reasoning is particularly easy to use in everyday life, and might be one of the most common ways for people to think about causes. As we have mentioned above, it is also important in science and in the law. So it's probably a good idea to see if we can think with counterfactuals as precisely as possible. Unfortunately, reasoning with counterfactuals can be terribly imprecise. This is because lots of counterfactuals are meaningless or at least very difficult to interpret. Think about this one for a couple of minutes:

If Julius Caesar had been running the US during the Cuban missile crisis, he would have been lining up ballistae on the Florida coast, pointing at Cuba.

The problem is that there is no standard way to assess a counterfactual like that, and different people will try to assess it in different ways. One person will think:

'Well, Caesar's only ballistic weapons were ballistae; he didn't know about any of ours, so that's what he'd try to use.'

Another person might think:

'If Caesar were kidnapped and transported to the US of the Cuban crisis in something like Dr Who's Tardis, he'd be so shocked and confused he wouldn't be able to do anything.'

Another might think:

'If Caesar, with all his tactical genius and leadership ability, had been born in the US, and had risen to be in charge of the army in the Cuban missile crisis, he'd know all about our tactics, and he would probably . . .'

Different people can think different ways about this, and there's no standard for serious discussion, because there's nothing to force you to think one way rather than another. You can generate lots of counterfactuals like this; just pick your favourite character and famous event or historical period:

'If Catherine de Medici had tried to assassinate JFK, she would have . . .', or

'If Steve Jobs had been in charge of technological development for Gengis Khan, he would have . . .'.

This kind of discussion can be quite fun,[17] but if we want to reason with counterfactuals seriously in science, the law or in history, we need to do better.

9.3 Philosophical views of counterfactuals

9.3.1 Lewis: possible-world semantics for counterfactuals

One attempt to do better is that of David Lewis, who is the main philosophical theorizer of counterfactuals. One of his aims was to provide a precise, logic-based and common standard for assessing counterfactuals.

Lewis gives us the counterfactual account of causation: A causes B if and only if had A not been, B would not have been either. This account falls under 'difference-making' because the occurrence or non-occurrence of the cause makes a difference to the occurrence or non-occurrence of the effect. That is to say, using a counterfactual, we reason about how things would have been different, if the putative causes had not been there.

First, Lewis developed what is called 'possible-world semantics', as, in his view, the meaning of counterfactual claims depends on what happens in other *possible* worlds. The idea is that, to figure out whether a counterfactual conditional is true, we compare what happened in this world with what we think would happen in another world that is very similar to this world. The main difference between this world and that other possible world is that the event we are considering happened in this world and not in the other world. So we think things like: in this world, I missed my alarm clock, and I missed the bus. But in another possible world where I heard my alarm clock, I got up and was ready in time, and I caught the bus. In this world, the sun is there at the heart of the solar system, and round and round the planets go. But in another possible world where the sun was taken away, the planets stopped going round and round, and instead all flew off into space, in whatever direction they were moving when they lost the sun.

So far, this is just the start of Lewis's idea, as it is still imprecise. What Lewis developed in his possible-world semantics is a more precise semantics to evaluate the truth or falsity of a counterfactual. Classical propositional logic is not very helpful for counterfactuals, as any counterfactual is formally true, in virtue of the fact that its antecedent is false. That is, classical propositional logic treats 'If I had heard my alarm clock, I would have caught the bus' as true just because 'I heard my alarm clock' is false. This is a well-known paradox, known as the paradox of material implication.

---

[17] We particularly recommend <http://www.youtube.com/watch?v=n6MNciSb85E>. If this link has gone dead, Google "counterfactuals big bang theory" and see what videos you get! Alternatively, watch *Big Bang Theory*, series 4, episode 3.

Lewis's alternative possible-world semantics rests, first, on the assumption of the existence of a plurality of worlds, amongst which there is also our actual world. This position is known as modal realism. Second, worlds are compared with each other on the basis of their *similarity* or closeness. Worlds are thus ranged according to their similarity to our actual world: the closer the world is, the more similar it is to the actual world. To order worlds, we use a relation of comparative overall similarity, which is taken as primitive. Briefly, and very informally:

A world w_1 is closer to our actual world w_a than another world w_2, if w_1 resembles w_a more than w_2 does.

The truth of the counterfactual we are trying to assess is then ascertained by an 'inspection' of what happens in other possible worlds. Given any two propositions A and B, the counterfactual reads: 'if A were true, then B would also be true'. The counterfactual operator is defined by the following rule of truth:

The counterfactual $A \mathbin{\Box\!\!\rightarrow} B$ is true (at a world w_i) if, and only if:

  (i) there are no possible A-worlds, or
  (ii) some A-world where B holds is closer to w_i than is any A-world where B does not hold.

The second case is the interesting one, since in the first case the counterfactual is just vacuous. Notice, moreover, that in the case where A is true, the A-world is just our actual world, and the counterfactual is true if, and only if, B is also true. This is enough to give a sense of the semantics of counterfactuals. Lewis's theory has a complex logical basis that we will not discuss in detail (Lewis, 1973; Stalnaker, 1968; Stalnaker, 1975). In a nutshell, he suggests that there are two separate elements relevant to comparing the similarity of a possible world with this world, the actual world:

  (i) How similar are the laws of nature in the possible world to those in our world?
  (ii) How extensive is the spatio-temporal region of the possible world *perfectly* matching our world?

Lewis thinks the laws of nature are more important to similarity between worlds than exact match of particular fact, while approximate matches of particular facts don't matter very much at all. This is because, for Lewis, the laws of nature are more important than small local happenings. Two worlds that vary in their laws of nature could be so wildly different from each other that they were very dissimilar. We may be utterly unable to function in a world with different laws of nature from ours, as the things that happened would be completely unexpected. Two worlds that vary in the exact happenings in a spatiotemporal region, however, but which then unfold according to the same or similar laws of nature, are not so different. If we found ourselves in a world with fewer planets, or stars, but where they behaved very like our planets and stars, at least we could rapidly come to expect and predict what would happen in the future.

There's also a difference between two different kinds of violations of the laws of nature, for Lewis. In a deterministic world that exactly matches our world, to get anything at all to happen differently will require some violation of the laws of nature. That's just what it means to say the world is deterministic. So if my alarm clock doesn't ring in the actual world, we need a violation of the laws of nature to get it to ring in a world with the same laws as our world, that exactly matches our world up to the point where the alarm clock rings. Lewis calls the minor violation to change the possible cause a small miracle; a 'divergence miracle'. This is different, Lewis thinks, from the kind of larger violation that would be required to get me to wake up anyway in the other possible world, even if my alarm clock didn't go off. This is a 'convergence miracle'. Basically Lewis thinks it takes less of a violation of the laws of nature to get perfect-match worlds to diverge from each other than to get not-quite-matching worlds to converge to become a perfect match of each other.

With this in mind, Lewis thinks claims like 'if I had heard my alarm clock, I wouldn't have been late' are true—in our world—if some world where my alarm clock goes off and I'm on time is closer to our world than any world where my alarm clock goes off and I'm late anyway. This claim turns out to be true in Lewis's view because the world which exactly matches this, but has a small local miracle to make my alarm clock ring, and then has untouched laws of nature, so I wake up on time, is closer to actuality than a world where there's a small local miracle to make my alarm ring, but then a further convergence miracle to get me to sleep on anyway, undisturbed.

Worlds may not be deterministic, and if so these relationships can be thought of probabilistically, but the general idea is here. If everyone assessing a counterfactual claim uses the same semantics, they will agree in their assessment, and can have meaningful and useful conversations about counterfactuals. Of course, there may be some vagueness in that assessment, so that there is some controversy. But the controversy will be resolvable. This is a big advance over the mess described above!

9.3.2 Problems for Lewis's account

This is great as a reasoning strategy, and it certainly gives us some useful insights into causality. But Lewis wants a fully reductive account of causality in counterfactual terms. He wants to be able to replace the idea of causality with his counterfactual story. There are two main problems: objections to modal realism, and the circularity objection. They are interrelated.

Lewis claimed that all these possible worlds are *real*, which is why his view is called modal realism. There really is a world where the sun is taken away, and the earth flies off into space. That world is just not our world, the actual world. Lewis needs modal realism for two reasons. First, because he analyses causation in terms of counterfactuals, and to ascertain whether C causes E we need to say what makes the appropriate counterfactual true. If there are no real possible worlds, the story goes, there is nothing that makes statements about these possible worlds *true*. This presupposes that there is always something that makes a particular causal claim true (see chapter 21).

Some people don't like what they see as a wild metaphysical claim, and this is the most frequent objection to Lewis. Lewis claims that his possible-world realism is actually parsimonious. After all, we know that at least one world exists, and we are just saying there are more. This is like seeing some ducks and thinking that more ducks exist, not like postulating new kinds of things, like seeing ducks and swans and postulating the existence of swucks! Lewis is right that we can have good reason to posit more of the same kind of thing, but few people have been persuaded that postulating more possible worlds is similar to postulating more ducks.

The second reason Lewis needs modal realism is to give a non-circular, a *reductive*, theory of causation. Lewis tries to reduce causality to counterfactual dependence grounded in similarity relations between possible worlds. For this to work, the similarity relations between possible worlds have to be independent of our perspective. But this seems unlikely, given the highly theoretical nature of their construction. The idea of 'distance' between possible worlds is wholly metaphorical. Even if the worlds were real, we still categorize them as closer or more distant using judgements of similarity. Many have thought that our judgements of similarity between possible worlds actually follow what we have found out about causality in *this* world. Why would we think that differences in laws mattered more than differences in particular facts, if it weren't that we think those kinds of laws matter more to causality? If this is so, then our causal views are structuring the similarity judgments that we use to ground the counterfactual dependence that was supposed to explain causality. There is no reductive theory of causality in terms of counterfactuals.

However, it is still possible to find the idea of possible worlds useful to clarify thinking about exactly the sorts of claims above—claims everyone makes sometimes. These possible worlds can be treated as *fictions* we use to structure our reasoning, not as real worlds. As we will see in other chapters, especially chapter 21, there are other accounts of the truthmakers of counterfactual claims—best system laws, capacities, invariance, validity. We don't have to have infinitely many real possible worlds to legitimize counterfactual reasoning.

One view of possible worlds has been to treat them as collections of propositions. Instead of a concrete world, the world where my alarm clock went off is thought of as a collection of propositions stating what is 'true' at that world. These could be written down as a list of sentences:

- My alarm clock went off.
- I woke up.
- I caught the bus.
- The curtains were blue.
- I brushed my hair at 8.07 a.m. precisely.
- etc.

Note the triviality of the propositions expressed in the later sentences. Original suggestions of possible worlds as collections of propositions specified that the set of propositions had to be maximal and consistent, which can be thought of quickly as entirely specifying the whole world. To assess the counterfactual about my alarm clock, we think about a set of propositions that also specifies the geological state of the rocks in that world, a billion years before my birth. Happily, this seems unnecessary to the reasoning task we began with. Fictional possible worlds only need to be specified with enough detail and precision to sharpen up the reasoning. These details can be agreed in debate with other people you are arguing or reasoning with. For a discussion of fictionalism about possible worlds, and the relation to treating possible worlds as collections of propositions, see Nolan (2011).

9.3.3 Rescher's method of validation

In different contexts such as psychology, legal causation or statistics, other forms of evaluation for counterfactual *reasoning* have been developed. We give here a very short primer on the method developed by Nicholas Rescher (2007). This method is a real alternative to Lewis's modal realism for the causality literature.

Rescher's motivation is to escape modal realism. Can we evaluate conditionals, including counterfactuals, avoiding possible-world semantics entirely? Rescher thinks we can. First, we have to define counterfactuals in terms of beliefs: a counterfactual 'purport[s] to elicit a consequence from an antecedent that is a belief-contradicting supposition, on evidence that it conflicts with the totality of what we take ourselves to know' (Rescher, 2007, p74). This way we can avoid reference to events and to the truth or falsity of the antecedent. It is all phrased in epistemic terms, appealing to beliefs and evidence.

Rescher's formalism is quite simple. We represent a counterfactual thus:

$$p\{B\} \rightarrow q.$$

This reads:

If p were true, which we take not to be so—not-p being a member of the set of our pertinent beliefs B (so that $\neg p \in B$)—then q would be true. (Rescher, 2007, p81.)

We then need to construct a deductively valid argument where the belief-contradicting supposition at stake entails the consequent. In the demonstration, we have to add suitable supposition-compatible beliefs. In other words, for the counterfactual to be true, the consequent q has to be derivable from the combination of q plus some appropriate subsection of the set of background beliefs. It will matter a lot what 'belief-compatible assumptions' we include in the proof. The choice of these assumptions is guided by a principle of conservation of information, namely 'prioritising our beliefs in point of generality of scope and fundamentality of bearing' (Rescher, 2007, p105).

Counterfactual validation then means restoring consistency in an optimal way by prioritizing information. For instance, conceptual relations have priority over mere facts, norms of practice will advantage some facts over mere matters of brute contingency. The acronym 'MELF' indicates the order of the priorities: Meaning, Existence, Lawfulness, Fact. MELF indicates how precedence and prioritization work in the absence of case-specific specifications to the contrary and guides us in choosing what background beliefs to use and what beliefs to discard in the derivation.

Consider the following example borrowed from Rescher (2007, p106):

If this rubber band were made of copper, it would conduct electricity.

A counterfactual like this arises in an epistemic context where the following beliefs are relevant:

1. This band is made of rubber (factual belief).
2. This band is not made of copper (factual belief).
3. This band does not conduct electricity (factual belief).

4. Things made of rubber do not conduct electricity (lawful belief).
5. Things made of copper do conduct electricity (lawful belief).

Let us negate the second belief: 'This band is made of copper'. The goal is to restore consistency between the initial set of beliefs. So we will have to choose between keeping 3 or 5. That is to say, we have to choose between a particular feature of this band and a general characteristic of things made with copper. But 5 is more informative than 3, as it is a lawful belief, not just a factual belief, and thus has higher priority. Once we accept 5, we can also validate the counterfactual: it is true that if this rubber were made of copper, it would conduct electricity.

MELF gives a priority order for keeping and eliminating beliefs. It is important to notice that in factual contexts we give priority to *evidence*—i.e. the most supported proposition is the one that is most strongly evidenced; in *counter*factual contexts, instead, we give priority to 'fundamentality', which means priority is given to propositions expressing meaning, existence and lawfulness, rather than propositions expressing factual evidence. Rescher (2007, ch11) provides more technical details on the logics to be used for MELF considerations in derivability. It is an open area of research whether Rescher's method can also illuminate aspects of causal reasoning and be more widely applied.

9.4 Counterfactuals in other fields

9.4.1 Counterfactuals in the law

Counterfactuals are also widely used in the legal domain. Moore (2009, chs16–18) provides a thorough discussion of their meaning and use in this field. In particular, counterfactuals are dominant in tort and criminal law, where they are usually called 'but for', 'sine qua non' or 'necessary condition' tests. This indicates that legal theory identifies the natural (pre-legal) relation of causation with counterfactual dependence. This is what 'received' legal theory *says*.

Moore explains that counterfactual theory lies in a view of morality according to which 'our actions must make a difference for us to violate our basic obligations' (Moore, 2009, p371). Moral responsibility is captured by some kind of counterfactual *test*, comparing 'how the world is after our actions with how the world would have been if, contrary to fact, we had not done the actions in question' (Moore, 2009, p371). Moore then discusses two approaches to evaluating counterfactuals: the covering-law model and possible-worlds semantics. Moore expresses reservations about the Lewisian solution, preferring a covering-law model, where laws can be mere regularities of nature, or relations between universals, or relations between conditional probabilities; we then use such laws to explain facts, notably by subsuming them under laws of nature (Hempel, 1965; Chisholm, 1946; Chisholm, 1955).

Causation by omissions, or absences, deserves special attention as the idea is important to legal theory. Causation by omission involves causing some outcome by *failing* to do something, such as by not taking due care, perhaps not fencing off a dangerous drop. It can be difficult to say why not doing something can cause some kind of positive outcome. Moore thinks that Lewis's construal of counterfactuals does not solve the problem, as Lewis's standard analysis involves positive event-tokens. For Moore, we should think of counterfactuals as relating *ranges* of events in groups functionally related to each other. Therefore, in the case of omissions, the counterfactual does not relate to the particular, token *non*-event, such as not fencing off the drop, but a whole range of particulars, that is all instances of the kind of act omitted, such as different ways of barring the drop. We discuss alternative solutions to the problem of omissions in chapters 11 and 13.

Moore thinks that counterfactual dependence is not sufficient or necessary for the existence of causation, and draws on arguments similar to the ones found in the philosophical literature, pointing to problems about epiphenomena, overdetermination and of course omissions. Still, counterfactual dependence is useful to help decide moral and legal *responsibility*. Moore (2009, ch18) discusses the use of counterfactual dependence in several cases. It emerges, from Moore's presentation, that counterfactuals are highly pertinent to legal theory. He points to an important distinction between causation and responsibility (whether moral or legal). Although we can cast doubt on the plausibility of counterfactual dependence as a *reductive account* of causation, counterfactuals help determine responsibility, and this is what makes them extremely useful in legal reasoning. For further discussion, see also Rosen (2011), Ferzan (2011), Moore's response in Moore (2011) and Schaffer (2012).

9.4.2 Counterfactuals in psychology

The psychological literature on the development of children's reasoning, and on causal learning, has also taken an interest in counterfactual reasoning, and its relation to causal reasoning. It seems agreed that people reason about causes, and also reason counterfactually. But the relation of causal reasoning to counterfactual reasoning remains disputed. Perhaps our causal understanding supports our counterfactual reasoning; perhaps vice versa. Perhaps reasoning causally and reasoning counterfactually come to much the same thing. The psychological literature is investigating these possibilites. Hoerl et al. (2011) is a very useful collection of recent literature.

Counterfactual reasoning and causal reasoning seem to be related in some way or other. People often generate counterfactual claims unprompted when they are asked to think their way through complicated causal scenarios, so they seem to find thinking in terms of counterfactuals useful. Note that this doesn't yet mean either that causal reasoning explains counterfactual reasoning, or that counterfactual reasoning explains causal reasoning.

Nevertheless, some psychologists have offered a counterfactual process view of causal reasoning, which holds broadly that thinking counterfactually is an essential part of causal reasoning, at least in a core set of cases. For example McCormack et al. (2011) note that even if counterfactuals do give an account of truth conditions for causal claims, which is Lewis's claim, that does not entail that people actually engage in counterfactual reasoning, as the process of causal reasoning. Technically, the two views are independent. Nevertheless, the idea is an interesting one, and if the process of causal reasoning essentially involves counterfactuals, that would seem to lend some support to Lewis's view.

However, there are some empirical reasons to be cautious about the counterfactual process view. For example, very young children, aged 5–7, can reason causally, discriminating causal-chain structures from common-cause structures using temporal cues, for example. But when they are asked counterfactual questions about the structure, they don't give answers that are consistent—reliably consistent—with their choice of structure. Adults do. This suggests that children's counterfactual reasoning is not as secure as their causal reasoning at this age (McCormack et al., 2011). Alternatively, when asked to assess a cause with a negative effect, adults will generate counterfactuals, but they tend to be counterfactuals that focus not on altering the actual cause of the outcome, but on supposing that possible preventers of the outcome occurred (Mandel, 2011). This seems to be a case of using counterfactual reasoning that doesn't follow causal reasoning.

Alternatively, Roessler (2011) examines whether counterfactual reasoning is so often associated with causal reasoning because it is needed to grasp some of the commitments of causal claims. Adults only have to think explicitly about causal scenarios that are more complicated than those we navigate intuitively on a daily basis. So the idea is that adults use counterfactuals to help them understand such scenarios by working out the commitments of the causal claims in the scenarios.

When it comes to complex causal scenarios, psychologists also investigate the possibility that we posit causal mental models (Tenenbaum and Griffiths, 2001; Gopnik et al., 2004; Sloman, 2005; Gopnik and Schulz, 2007). These involve having an idea of a causal structure in our minds, a structure that extends beyond a small number of causal claims. So for example, Alice might know quite a lot of causal claims about her car. A causal mental model of these claims, which might represent the mechanism of operation of the car (see chapter 12), helps Alice imagine counterfactuals, to think about multiple possible effects she might want to know about. This makes her causal reasoning much easier, even for complex scenarios.

Finally, psychologists investigate the relation between causal claims and responsibility judgements (Lagnado and Channon, 2008; Gerstenberg and Lagnado, 2010). It seems that norms or practices in place will affect people's judgements of responsibility, as you might expect, but they also affect their judgements of causation itself. Sytsma et al. (2012) describe experimental work where pens displayed publicly on a desk run out. If a particular group of people are allowed to take pens, and a particular group are not supposed to take pens, people in the experiment tended to think those not supposed to take pens were both more responsible than the other group, and more of a cause, for the lack of pens. This kind of result is in accord with Menzies (2011), and is particularly interesting given Moore's account described above.

In short, the psychological literature on counterfactual reasoning and its relation to causal reasoning is currently unsettled, but there is very interesting work there, worth attending to as it unfolds further.

9.4.3 Statistical counterfactuals

Counterfactual reasoning does not permeate only natural language but also scientific methodology. There is something natural in using counterfactual reasoning, because most of the time we want to know what *made the difference* for the occurrence or non-occurrence of an effect. I have a headache. Had I taken an aspirin half an hour ago, it would have gone. We mentioned earlier how a conditional like this can be multiply ambiguous. What if I went for a walk instead? What if I had a shiatsu massage? Would I still be suffering from a headache? Maybe not, but that's not the question that interests scientists, and particularly statisticians.

Let's focus on aspirin. How do I know that aspirin is an effective treatment? I should compare two versions of myself, at the same time, one taking the aspirin, one taking nothing. The causal effect would be the 'difference' between the result of taking and not taking aspirin. We know we cannot actually do that. Holland (1986) has called this the 'fundamental problem of causal inference', namely that we cannot observe the causal effect and non-effect *for the same individual at the same time*. Moreover, the effectiveness of treatments has to be established at the population level (see chapter 5), so we need to compare groups of individuals. Ideally, we would have to compare the same group taking and not taking aspirin at the same time, which is of course impossible. So the fundamental problem of causal inference applies to groups too. Yet, nothing prevents us from imagining what would happen if the same individuals did and did not take aspirin. How can we exploit this intuition and formalize it in a statistical framework?

Donald Rubin and Paul Holland are the main theorizers of the 'potential outcome model' or 'counterfactual model' in statistics (Rubin, 1974; Rubin, 1986; Holland, 1986; Holland, 1988). They provide a formal apparatus to estimate the *average causal effect*, namely the difference between the average response if all individuals were assigned to the treatment and the average response if all individuals were assigned to the control.

There are some important assumptions behind the approach. First, the potential outcome model searches for *effects of causes*, not causes of effects. We search for the effects of aspirin, not for the causes of headache. This is important, because most of the models presented in chapter 7 are interested instead in the causes of effects. Second, the formal apparatus will make sense only if the causal factors are *manipulable*. The approach is meant to study the potential effects of treatments, so in this context this is a reasonable assumption. Problems arise, as we shall see later, if we try to extend the approach beyond its initial scope. Third, individuals have to be equally 'exposable' to the treatment, called the 'stable unit treatment value assumption' (SUTVA). If this assumption is violated, the estimation of the causal effect is even more difficult.

Notice that this is not a full-blown experimental approach. It is only 'quasi-experimental', in the sense that it aims to reproduce the pillars of experimentation, namely comparisons between different instances, cases or groups (see chapter 2 and Mill's methods in chapter 16). The 'experimental equivalent' to the potential outcome model is randomized controlled trials (RCTs) that we discuss in chapter 2. RCTs compare different test and control groups, one that takes the aspirin, the other taking nothing, or a placebo, or another drug. Brand and Xie (2007, p394) explain this idea clearly:

The potential outcome approach to causal inference extends the conceptual apparatus of randomised experiments to the analysis of nonexperimental data, with the goal of explicitly estimating causal effects of particular 'treatments' of interest.

Aspirin, or any other drug, is routinely tested using RCTs. But what happens with other types of 'treatments', or factors, especially in the social domain? How do we test the effects (e.g. future earnings) of going to a public or private college? An RCT would take a population and split it in two, randomly assigning some people to a private college and some to a public college. Obviously, we cannot do that. So, where do we get the data about the cases and controls?

The answer is: from 'matching'. The objective of matching is to find in the population individuals that are similar in all possible respects, the only difference being the administration of the 'treatment'. Statisticians have formally shown that in the case of *perfect* matching, the results are equivalent to randomization. In practice, however, perfect matching is not possible, and several techniques have been developed to find *optimal* matching cases (for instance, propensity score matching or nearest neighbour matching). Once we have the cases and the matched controls (rather than experimental controls), we can compute the average response for the two groups, and then the difference, thus obtaining the average causal effects.

This approach has become quite popular in social science (see for instance Morgan and Winship (2007)) but it has also been severely criticized. Dawid expresses a serious concern about the fact that the controls are never observed, but just estimated from matched cases (Dawid, 2001; Dawid, 2007). Russo et al. (2011) examine the counterfactual approach and compare it with 'structural models' (see chapter 7); they are particularly concerned with the fact that the approach has limited applicability, as the sciences are also interested in finding the causes of effects, for which other methods are more suitable. There are also concerns about the requirement that causes be manipulable factors, which rules out attributes such as ethnicity or gender from the range of possible causes (Rubin, 1986; Holland, 1986). Bukodi and Goldthorpe (2012) have similar worries, arguing that the potential outcome model does not suit many sociological studies because (i) the potential outcome model cannot deal with causality when individuals' goals, beliefs or desires are involved, and (ii) the potential outcome model explicitly deals with effects of causes, whilst they are interested in causes of effects.

We presented here the conceptual underpinning of an approach that attempts to *formalize* counterfactual reasoning in quasi-experimental situations. It goes without saying that this only gives you a taste of the potentialities and limits of the approach. Just as with the approaches discussed in chapter 7, you will need to delve into the writings of these authors and properly engage with them to use them any further.

It is very clear that counterfactuals, whether Lewisian, legal, psychological or statistical, are important to causal reasoning, and may well be vital to working out the implications of known causal structures. They may be used in this way to generate further hypotheses for testing. Counterfactuals provide hints that an event did or did not occur because of another event's occurrence; however, some scholars contend that counterfactuals cannot produce an explanation of the occurrence of a phenomenon, because they do not give any hints about how the phenomenon was produced or could be produced. We discuss these views in, for example, chapters 10 and 12. Reasoning, whether about everyday cases or scientific cases, helps us decide what evidence is needed, or admissible, or compelling. But reasoning itself is not evidence (see chapter 6). Nevertheless, counterfactuals have a vital *heuristic* value for reasoning about causal relations. In view of this, greater precision in counterfactual reasoning can be very valuable. We have shown how various fields achieve this.

Core ideas

- A counterfactual is a conditional with a false antecedent, such as 'If I had heard my alarm clock, I wouldn't have been late'.
- Reasoning counterfactually seems to be ubiquitous in everyday life, and also very useful in science.
- Counterfactual reasoning is potentially disastrous if you don't assess counterfactuals consistently, so a consistent way of assessing them is useful.
- 'Philosophical' counterfactuals aim to pick out the cause of some states of affairs by analysing the truth values of the conditional statement that expresses the corresponding 'contrary-to-fact' situation. Their scope is therefore single-case.
- 'Statistical' counterfactuals aim instead to establish whether, on average, a treatment or intervention proved effective across the population. Their scope is therefore generic. The distinction between single-case and generic is also addressed in chapter 5.
- Counterfactuals have been of particular interest in legal theory. Legal theory often uses a counterfactual test to decide issues of legal responsibility, including responsibility for *lack* of action, such as in negligence.
- Psychologists have discovered that causal reasoning and counterfactual reasoning often happen together. They are investigating this relationship, and also the relationship of both styles of reasoning to responsibility judgements.

Distinctions and warnings

- Don't confuse reasoning with evidence. Precise reasoning is useful, but it's still reasoning, not evidence.
- 'Philosophical' and 'statistical' counterfactuals share the same idea but not the same scope and methods.
- Most people find Lewis's modal realism—the existence of an indefinite number of other possible worlds—a bit too much. Possible worlds can instead be thought of as fictions constructed to make counterfactual reasoning precise.
- In particular, characterizing similarity relations between possible worlds without drawing on our understanding of the causal structure of the actual world may well be impossible. If it is, then a reductive account of causality in terms of counterfactuals defined over other possible worlds is, in turn, impossible.
- It is worth remembering that counterfactuals might well be assessed in different ways by different fields.

Further reading

Menzies (2014); Rescher (2007, ch. 1, 9); Reiss (2012b); McCormack et al. (2011).

CHAPTER 10

Difference-making: Manipulation and Invariance

10.1 Example: gene knock-out experiments

The promise of what is now called the 'post-genomic era' is to shed light on the deepest mechanisms of disease or behaviour, illuminating human nature in unprecedented ways. Of course we do lots of big observational studies, looking for statistical associations between genes in populations, seeking out the secrets hidden in our genes. But another crucial way scientists try to fulfil the promise is by performing 'gene knock-out' experiments and gene 'overexpression' experiments. These kinds of experiments go beyond observing the gene to *manipulate* it, inhibiting or promoting its supposed action, to allow us to decide what its role is, what it causes.

It is easy to grasp the core idea of a gene knock-out experiment. If a particular piece of coding DNA has been sequenced, then we know what polypeptide chain it will code for, but we can be unaware of the function played by the resulting functional protein (which is, crudely, a folded polypeptide chain) in the functioning of the cell, or the organism, of interest. One way to investigate further is to 'knock out' that gene. This is usually done by introducing something to the organism, or cell, that binds specifically with that strand of DNA, preventing it from being transcribed to mRNA, so preventing the cell from making whatever protein that DNA strand is normally used to make. We then observe the cell, or the organism, to see what changes. The manipulation of the gene here is key, because it suggests that the changes we observe are effects of knocking out or stimulating the gene. If we observe changes, we have discovered the gene's role. We return to a notorious problem with gene knock-out experiments shortly.

What can we conclude from this kind of scientific practice? Well, it seems manipulations play a vital role in science in establishing causal relations. Some philosophers have developed this idea into a fully-fledged manipulationist theory of causality.

This theory is clearly related to causal inference, having been developed out of one of our methods for causal inference, but we will see that it is also related to our aims of controlling the world, and explaining it. Ultimately, it will also impact on our causal reasoning.

10.2 The manipulationists: wiggle the cause, and the effect wiggles too

In the late 1990s and early 2000s some philosophers, headed by Daniel Hausman and Jim Woodward, made the notion of manipulation the cornerstone of their account of causation (Hausman and Woodward, 1999; Hausman and Woodward, 2002). The account was developed in great detail in Woodward (2003), and many philosophers have adopted this popular account—see for instance Waters (2007), Baumgartner (2009), Glennan (2010b). Simply put, the account is that C causes E if and only if, were we to manipulate C, E would also change. For example, if we were to knock out the gene for brown eyes, we would no longer get brown eyes; if we were to give a patient with a strep infection penicillin, he or she would recover, and so on. This is a difference-making account because, by manipulating the putative cause, we endeavour to *make a difference* to the putative effect. By making such manipuations we can establish *that* C is causally related to E, but not yet or directly *how* they are related (see chapter 12).

The manipulationist view seems to apply straightforwardly to experimental contexts. In such contexts manipulations are performed in order to find out what causes what. Experiments are perhaps the paradigmatic method for gaining causal knowledge, as we hope they grant us direct epistemic access to the effects of our actions, and so excellent evidence of causal relations. (See however chapter 2 for some critical reflections on the role of experiments in the scientific enterprise.)

Note that although the account develops the concept of causality from the method of experimental interventions, which are made by people, it does not *require* interventions to be made by people. The theory is thus not 'anthropocentric' (see chapter 17), although Hausman and Woodward say they are inspired by early agency theorists—namely Collingwood (1940), Gasking (1955) and von Wright (1971). Manipulationism is not intended to be an account of causation in terms of agency. The goal is to provide an account of causation as it is used in scientific contexts. As we will see, manipulationists wish to apply the account to a wide range of experimental contexts, including the physical and life sciences, and they also hold that it is applicable in contexts where experiments are typically not performed, such as the social sciences or astronomy.

To give more detail, the heart of manipulationism is the notion of 'invariance under intervention'. What is 'invariant' is an 'empirical generalization'. In the jargon of philosophers, an empirical generalization is a relation—typically empirically observed in a data set—between, say, variables X and Y, where that relation is change-relating or variation-relating, i.e. it relates changes in variable X to changes in variable Y. X and Y are the putative causal factors; thus an empirical generalization relates changes in the putative cause-variable X to changes in the putative effect-variable Y. So for example we see an association between smoking and lung cancer, and also between smoking and yellow fingers. These associations are interesting, but a change-relating relation might of course be spurious or accidental. For instance, yellow fingers may be statistically associated with lung cancer but this is the case because both yellow fingers and lung cancer are effects of a common cause, cigarette smoking. Alternatively, an increased number of storks may be statistically associated with an increased number of births, but we know that such a correlation does not correspond to a real causal relation: storks don't really bring babies!

So we need to discriminate between causal and non-causal change-relating empirical generalizations. For manipulationist theorists, change-relating relations are causal just in case they are 'invariant', or stable, under a sufficiently large class of interventions. Interventions are manipulations that modify the cause-variable figuring in the change-relating relation. For Woodward, an intervention I on X must satisfy three conditions (Woodward, 2003, ch3):

(i) the change in the value of X is totally due to the intervention;
(ii) the intervention will affect the value of Y, if at all, just through the change in the value of X;
(iii) the intervention is not correlated with other possible causes of Y (other than X).

Simply put, conditions (i)–(iii) mean that if Y is affected, this is *solely* through the action of X, triggered by the intervention I. Any possible influence of I on Y, or of other factors Z_i on X or Y, is ruled out if the action of changing X is truly an intervention. Interventions are thus the means of establishing whether changes in the cause-variable do or would bring about changes in the effect-variable, without disrupting the relation between the cause and the effect. If they do, or would, then on the manipulationist account the invariant empirical generalization is causal. So for example, if we intervene on yellow fingers, we will not see change in incidence of lung cancer—as long as we obey the crucial restriction and do not reduce yellow fingers by reducing smoking! This would not be an intervention as the manipulationists define an intervention, as smoking is itself associated with lung cancer. Alternatively, if we intervene on smoking, fulfilling all the criteria, we will see change in rates of lung cancer. This would allow us to conclude that smoking causes lung cancer, whereas yellow fingers do not. We can also use such knowledge to inform our efforts to control the world. So long as the criteria are fulfilled, we can draw the conclusion whether it is a person making the intervention, or a freak accident in the lab that creates the new circumstances. Note that reasoning on yellow fingers and lung cancer illuminates the main theses of manipulationism because we already have strong causal theories about lung cancer development. We will see later that this is not always the case. A crucial problem in science is to design the right intervention, or to design alternative methods to test invariance if interventions are not feasible.

For Woodward, invariance under intervention also gives an account of explanation in science. For Woodward, we explain a phenomenon like lung cancer when we can answer questions about what would have happened under a range of different conditions. We can say what would have happened if people smoked more: there would have been a higher incidence of lung cancer. What would have happened if smokers cleaned their fingers: those who were susceptible would still have got lung cancer. And so on.

Woodward calls these questions we can answer 'what-if-things-had-been-different' questions, and writes:

> In this way, [invariant generalizations] ... give us a sense of the range of conditions under which their explananda hold and of how, if at all, those explananda would have been different if the conditions ... had been different. It is this sort of information that enables us to see that (and how) the conditions ... are explanatorily relevant to these explananda. It is also information that is relevant to the manipulation and control of the phenomena described by these explananda. (Woodward, 2003, p191.)

So Woodward's core idea is that we can explain phenomena when we can see how altering the initial conditions would change the outcome. It is important to Woodward that we can say not only what conditions actually affected a phenomenon, but those that could have affected it. This is the range of conditions explanatorily relevant to that phenomenon. For Woodward, we have a better explanation if we understand more of these conditions, if our generalization is invariant over a range of interventions. He says:

> This framework ties the depth of the explanations in which a generalization figures to the range of invariance of that generalization and the range of what-if-things-had-been-different questions that it can be used to answer, and generalizations and theories that score well along this dimension will (at least often) be relatively general and unifying. (Woodward, 2003, p373.)

To make this clearer, we can also consider a stock example used by manipulationist theorists, drawn from physics (see e.g. Woodward (2003)). The ideal gas law states that the state of a gas is determined by its pressure P, volume V, and temperature T: $PV = nRT$, where n is the number of moles of gas and R the universal gas constant. This empirical generalization is invariant under a whole range of interventions on the temperature. Therefore the generalization correctly describes how manipulations or interventions on the temperature of the gas would affect the gas pressure, holding fixed the gas volume. Because this empirical generalization is invariant, it is *potentially* exploitable for manipulation and control. We do not actually need to perform such interventions, but if we did, then we could see how the effect-variable would change upon interventions on the cause-variable, and the relation between the cause- and effect-variables would remain stable. So we have understood the relevant causes, and we have explanations of the various effects. Supporters of Russell-inspired arguments against causality in physics (see chapter 17) will raise the question of which variable, in the ideal gas law example, is the cause: pressure, volume, or temperature? Russell would have argued that this is a functional, rather than a causal, law. But manipulationist theorists can reply that what confers explanatory power on the ideal gas 'law' is not that the gas law is a *law*, but rather the fact that it can express invariant relations between the variables. Different variables can be either the cause, the effect or held fixed depending on the behaviour we want to explain—changes in pressure, or volume or temperature.

Manipulationism can be a very intuitive way of reasoning about causality, capturing how we move beyond observing patterns in the world—patterns we describe using correlations. The account certainly helps direct attention very strongly to exactly the right features of experimental design; the right factors to isolate your experimental intervention from. You really cannot do an effective experiment to see whether yellow fingers cause lung cancer if you use smoking cessation as the means to get rid of yellow fingers. So it is useful in causal inference. It is also useful in reasoning about causes, including control. It is set up to link to the purposes in coming to know causes—so we can use our knowledge in successful actions. Do we recommend cleaning fingers to avoid lung cancer, or quitting smoking? We have also seen that it connects nicely with some of the explanatory questions we want to ask about causes. We will consider in the next section what limitations there might be on thinking about causality from the point of view of the manipulationist account.

10.3 What causes can't we wiggle?

10.3.1 What is an intervention?

It is true that the manipulationist account applies nicely to experimental work. But however crucial experimental interventions are in science, a lot of important work in science is not experimental, but observational. For example, most of our evidence for smoking causing lung cancer is observational (see for instance Cornfield et al. (2009)). The reason for this should be quite clear—we can't force people to smoke! There are three distinct excellent reasons for not performing experimental interventions:

(i) Ethical and legal constraints prevent you, such as in the smoking case.
(ii) It is physically impossible, as in trying to double the orbit of the moon to observe the effect on tides.
(iii) The intervention wouldn't tell you what you want to know, because it is likely to be 'structure-altering'.

We will return to structure-altering interventions below. Ethical and legal constraints and physical impossibility prevent interventions in many cases where we remain interested in causes—in medicine and astronomy, and many other domains. These domains do not seem to be domains that are less important, or where we have less knowledge. As an example, let's consider an observational model in social science. To push the argument as far as possible, we'll consider factors that cannot be manipulated, for instance gender. Suppose you are an economist and you wish to assess the impact of gender (being a woman or a man) on salary level or first employment. Of course, if we only could, it would be great—great *epistemologically*!—to get a group of women involved, make them men, send them to the same employers and see how their salary and careers evolve. That's not what economists or sociologists do. Instead they analyse data sets containing observations about gender, about salary levels, and other factors like age, ethnicity, education, and so on. To see whether being a woman is causally related to a lower salary, scientists try to see whether there is a relation between gender and salary that is invariant across variations, for instance being a woman and young, being a woman with a degree, being a woman and British-Chinese, etc. And these relations for women are then compared with the corresponding categories for men: being a man and young, a man with a degree, a man and British-Chinese. If these relations turn out to be invariant, across age, and educational and racial groups, we have good reason to think that being a woman causes you to have a lower salary (see also chapters 2 and 9). This does not mean that there are no interventions of any kind in social contexts. In social science, causal relations are often established without interventions, but these causal relations nevertheless ground socio-economic policy actions, such as anti-discrimination legislation tailored to the specific socio-economic context in which it applies.

Considerations about what we can or cannot wiggle have led some scholars to hold the view that causes are just those factors that can be manipulated in an experiment and that all individuals are potentially exposable to. According to this view, attributes such as gender shouldn't be considered as causes, which of course raises an important debate for causal analysis in the social sciences—see the discussion of statistical counterfactuals in chapter 9 and references there.

But the problem of deciding what can be wiggled and therefore what counts as a test manipulation arises in natural science contexts too. For instance, Samaniego (2013) considers spin-echo experiments and discusses four possible manipulations for controlling the variable 'environmental perturbations': (i) reducing the viscosity of the sample, (ii) changing the temperature of the sample, (iii) introducing magnetic inhomogeneities and (iv) producing a perfectly homogeneous field. It turns out that these interventions violate at least one of Woodward's criteria. Samaniego discusses possible alternatives to overcome this difficulty and nonetheless provide a causal explanation for spin-echo experiments. One alternative is to weaken the constraints for interventions.

Manipulationists also want to apply their account to all causes in any domain. In light of this problem, Woodward says that his account works with real interventions but does not *require* them, instead saying we can approach such problematic domains by considering hypothetical or ideal interventions. So there are three importantly different types of 'interventions' in the manipulationist account:

(i) Real, practical, feasible interventions.
(ii) Hypothetical interventions, which are feasible, but in practice not done.
(iii) Ideal interventions, where we can imagine possible interventions, but it is not possible to perform them.

Hypothetical interventions include making people smoke, while examples of ideal interventions include moving the moon. Maybe one day we will be able to move the moon, but there will always be some interventions that are impossible. Consider altering conditions at the Big Bang to see how that affects the universe. However much science advances, we won't be able to do that! Nor can we change someone's race or gender—we cannot rewind time so that a 43-year-old black female doctor is born a white boy, and undergoes the long sociological influence of that change.

For this very large class of cases, the manipulationist account is no longer aligned with some causal methodologies that are widely used, notably in social science research. One might even worry that manipulationist thinking will actually damage a large class of knowledge, by leading us to downgrade such very important results—results which are currently structuring what many people consider important problems to tackle in our society. We look at a popular manipulationist response to this and other problems in a later section in this chapter, once we have considered another problem.

10.3.2 Modularity

'Modularity' is a word that needs to be treated with care, as it means different things in the philosophical and scientific literatures. In manipulability theory, it has a very particular meaning. For manipulationists, in the model describing the system, causal relations are modular if the causal structure of the underlying system isn't altered when one makes interventions on the putative cause. So for example, a system that satisfies this is: the causal relationship between smoking and lung cancer isn't altered by intervening to reduce smoking. We can alter cancer rates by intervening on smoking behaviour, precisely because the relation between smoking and cancer still holds.

It may sound counterintuitive, but many systems do not behave in such a nice, simple way. This has been well known to social scientists since Lucas' work on macroeconomic policymaking. Lucas was concerned with the validity of predictions about the effects of economic policies, on the basis of the 'known history', especially if available data is at the aggregate level. The problem is that sometimes economic policies alter the causal structure of the population, which means that models used to design the policy, which were based on the previous causal structure, become obsolete. In Lucas' words:

> Given that the structure of an econometric model consists of optimal decision rules of economic agents, and that optimal decision rules vary systematically with changes in the structure of series relevant to the decision maker, it follows that any change in policy will systematically alter the structure of econometric models. (Lucas, 1976, p41.)

Lucas' critique shares some similarities with the views of Soros (1994). In particular, Soros develops his theory of 'reflexivity' according to which the social sciences face the problem of studying phenomena that involve individuals thinking and acting, and reacting to the results of studies concerning them. A consequence of this is that results of studies may be biased, precisely because social actors will alter the social process, while they are under investigation.

The 'North Karelia Project' is a good example of both the Lucas' critique and of Soros' reflexivity (Puska et al., 2009). A massive public health action was launched in the 1970s in North Karelia (Finland) to reduce coronary heart disease mortality rates. Numerous activities were carried out, from health services in contact with individuals, to massive media campaigns about healthy dietary habits. The results were good, but at the beginning net changes due to the interventions could not be identified very clearly. The problem lay in the comparison with data coming from the neighbouring province of Kuopio, which was chosen as the 'control' area. Positive changes in coronary heart disease mortality rates were observed in both the target and the control area. The reason is that people in Kuopio were influenced by the programmes carried out in North Karelia. The intervention modified not only the causal structure in North Karelia (intended) but also in the province of Kuopio (unintended). It is worth mentioning that the North Karelia Project raised important issues about 'exporting' policy actions to different populations (see e.g. Wagner (1982) and McLaren et al. (2006)).

Steel (2008) calls these types of intervention 'structure-altering', because they change the causal structure of the system you are trying to intervene on. Lucas suggests that, if we want to predict the effects of policies, we need to study parameters at another level, trying to understand the preferences or resources that govern behaviour at the *individual level*. In other words, Lucas advocates a 'microfoundations' approach to policy evaluation. Soros, instead, points out that we should develop economic models that resemble reality more than the widespread models of neo-classical economics. Accurate prediction of responses to policy decisions remains a tricky issue. This issue is also relevant to chapter 7.

The problem of structure-altering interventions is also a serious challenge in biochemistry. Here, experimental changes to proteins and so on are generally legal and possible, and of course biochemical molecules don't have knowledge of experimental interventions. But systems are complex and interacting, so that tracing the effects of changes is difficult. This is a serious problem with finding the roles of genes. A notorious outcome of gene knock-out experiments is: no change! This does not mean that the knocked-out gene had no effects on the system prior to being knocked out. It is a consequence of the fact that the human body has multiple back-ups. If you stop one system that allows you to, say, metabolize a sugar to release energy, one of the other systems will kick in, perhaps metabolizing a different sugar. This is excellent news for your survival, but it makes experiments a bit of a nightmare. Such tests involve structure-altering interventions, because the gene knock-out makes the system switch from one mechanism for the effect, to another mechanism for exactly the same (or a closely similar) effect. You cannot find out what the gene does simply by blocking it.

Recall that the manipulationist account is motivated by an attempt to connect an account of what causality is or means to our causal methodology, and also to the uses we make of causal knowledge—our need to control the world. On one hand, in cases where we see structure-altering interventions, the manipulationist account is again divorced from the actual methodology used to discover these important causes, where we find causes in the absence of invariant change-relating generalizations. On the other hand, an interesting feature of these cases is that we are seeking control. We want to be able to make effective policy interventions, and intervene on genes, proteins and so on to cure disease. However, in these cases we do not achieve control by finding and using invariant change-relating generalizations. We are forced to do something more complex. We can alter the system in many many ways, and try to trace what happens. These alterations do not fulfil the three criteria required to count as an intervention in the manipulationist account. Frequently, the system is so complex that we mathematically model the entire system, and run computer simulations in an attempt to get a grasp of what the system will do as a whole. This is the increasingly important work of systems biology, and there have been rather striking successes.

For example, the action of HIV in individual cells turns them into HIV factories; and its action in the human body is to disable the immune system to allow HIV to run wild; all this is immensely complex and interactive. HIV is notoriously resistant to therapies, because the HIV virus is extremely simple and evolves so quickly that individual treatments do it little long-term damage. (See Hutchinson (2001) for a discussion.) However, we have finally got effective therapies because what is needed is a cocktail of drugs that targets HIV at multiple points in its efforts to replicate itself and damage the body's immune system. We hit it everywhere we can, trying to stop it entering cells, leaving cells, copying itself, inserting itself in cellular DNA, transcribing that DNA, and attacking white blood cells. We alter the system, while simultaneously trying to block the alterations to structure that we expect to result from every intervention we make. Our actions are much less the carefully defined surgical intervention of manipulationist theory, and more a desperate bludgeoning everywhere we can reach! We cannot yet stop HIV completely, but we can finally shut down the system it requires well enough to inhibit its action for many years. In this important exercise of control, invariant change-relating generalizations are few and far between.

10.3.3 Conceptual versus methodological manipulationism

There is a popular manipulationist response that accepts that in these important kinds of cases, the manipulationist account of causation is indeed divorced from the relevant methodology for causal inference. However, it is claimed, the 'wiggle the cause, the effect wiggles' idea still captures something that lies at the heart of the *concept* of causality, in all cases. Strevens characterises this as distinguishing explanatory or methodological manipulationism on one hand, from conceptual manipulationism on the other (Strevens, 2007; Strevens, 2008). Strevens detects both strands in Woodward's work. On methodological manipulationism, the aim of the manipulationist account is to connect our philosophical account of causality to the methodology and use of causality. On the conceptual account, though, the aim is only to capture the concept of causality itself.

We have already said enough about the usefulness and limits of methodological manipulationism. Manipulationism is an excellent way to connect causality with use, methods and causal explanation, in experimental domains. But those connections all weaken in non-experimental domains. It is not a good idea to downgrade results in those domains, or engage in a desperate emulation of a manipulationist methodology where that is impossible.

The usefulness of conceptual manipulationism depends on the point of giving an account of a concept of causality. We discuss this in more detail in chapter 19. In brief, our view in this book is that philosophical accounts of concepts of causality are useful in so far as they help guide our thinking and our reasoning, helping us frame problems of control, prediction, explanation and causal inference. If they can do this, then they can help identify evidence, design experiments and work out what we still don't know; and these are immensely useful tasks.

How manipulationist thinking can still help. There are ways in which the manipulationist account can still be useful to think with even when it doesn't apply methodologically. We can do more than observe the world, and we should intervene if we can. For example, in many cases where we cannot intervene on the variable of interest, we may be able to intervene on what scientists call a 'proxy variable'. For example, we can't change someone's gender, but scientists have run trials where they take women's CVs, attach a photo of a man, and change the name on the CV to a masculine name, and track whether the incidence of invitation to interview rises. This can be very illuminating about our original question of whether gender affects job prospects. It is also perfectly true that, on the basis of our causal understanding of this problem, we can think 'If we could change that person's gender, their salary would alter.' So hypothetical and even ideal interventions might help us in our thinking. One thing it might do is focus our thinking on the control of the world we are so often seeking in our studies, whether they are experimental or observational.

How manipulationist thinking might hinder. However, there are also ways in which thinking solely in manipulationist ways about a non-experimental domain might impede sensible thinking, and this is a good reason for caution about thinking manipulationism captures a concept that is *always* at the heart of causality, in any domain. It is true that we are often seeking control, but also true that control cannot always come from intervening on a single variable, as shown by the HIV example above. Further, sometimes prediction of what will happen to an undisturbed system is quite enough of an epistemological challenge. Consider the multiple models of financial systems that have recently been developed in an effort to understand what caused the financial crisis, in the hope that we will notice warning signs in the future. (Interestingly enough, models announcing and anticipating the current economic crisis existed well before it occurred but went largely, or perhaps deliberately, unnoticed. One example is Soros (1994), the first edition of which was published in 1987, and also later Soros (1998) and Soros (2008). Another example is Pettifor (2003).)

Manipulationist thinking can skew our assessment of evidence when it is merely observational—and this is particularly odd in cases where our aim is merely prediction. Finally, we can imagine ideal or hypothetical interventions, either in our attempts to design experiments, or in putting our causal knowledge into practice. But these imaginings are only imaginings, especially when they are divorced from actual practice. We would urge extreme caution in unconstrained counterfactual reasoning—it can be illuminating, but it cannot be mistaken for evidence. (See the section on the unbearable imprecision of counterfactual reasoning in chapter 9!)

In response to these problems, Russo (2011a) and Russo (2014) sketch the main features of a revised notion of invariance as invariance across variations, of which invariance under intervention is a special case. For Russo, invariance is the really important concept at the heart of causality—manipulationists are exactly right about that. However, Russo says that invariance is so important for causal reasoning that it needs to be wide enough a notion to be applicable to experimental and observational studies alike. If we adopt her 'invariance across changes' version, 'invariance under intervention' is automatically subsumed: invariance across changes covers variations that are due to interventions on variables, alongside variations that are observed in the data set, subject to context-appropriate checks for invariance. In other words, invariance need not be defined in terms of manipulations, as after all what we try to track are *variations* in the putative effects due to *variations* in the putative cause. Changes in the effect- and cause-variables can be either observational or due to manipulations—see also chapter 16.

10.3.4 Explanation

Finally, while the manipulationist account of explanation has good features, as we have said, allowing us to answer a range of explanatory questions we often do want to answer, it might be thought that there are other things we also seek in scientific explanations.

First, invariant empirical change-relating generalizations are *generalizations*. This means that we can offer explanations of general tendencies, such as the incidence of smoking in a population; but we cannot obviously give explanations of single-case events, like a particular person getting cancer. It is harder to make sense of the relation between Alice's smoking and her cancer having invariant properties.

Second, many philosophers have thought that explanations involve pragmatic and contextual aspects (see e.g. van Fraassen (1980) on this). We don't seek explanations for many things. For example, background knowledge tells us that according to our most accredited biomedical theories, men do not get pregnant. As a result, we don't seek explanations for men using birth control failing to get pregnant.

Finally, the manipulationist account of explanation makes no attempt whatever to address *why* invariant change-relating empirical generalizations hold, while it seems that much of science is interested in explaining why these and other kinds of generalizations hold. Our explanation for why men don't get pregnant seems to be our understanding of the mechanisms that do generate pregnancy, and our knowledge that men lack essential elements of those mechanisms, such as a womb. Woodward suggests that distinct mechanisms underlie different invariant change-relating generalizations. This might be true, but he clearly thinks we can get *explanation* without reference to any particular mechanism. Many disagree, with a large literature now claiming that in biomedical and social sciences, one explains by giving a mechanism (see chapter 12). This problem might be thought particularly important in systems where one cannot usually find invariant change-relating generalizations of the kind that the manipulationists seek at all, such as social systems, genomics and HIV therapy. Here, a serious explanation requires a description of what's happening to the whole system, with many structure-altering interventions included.

Ultimately, there is very good reason for the popularity of the manipulationist account. Many many scientists have found it useful. It is very intuitive, and gives a lovely connection between an account of causality, an important methodology, and our needs in using causal knowledge. Its main problem is that it applies so fluidly in some domains, it can trip you up in observational contexts. But once you have a firm grasp of its limits, it can be extremely useful.

> **Core ideas**
>
> - Manipulation is an essential notion for causal analysis. In particular, the manipulationist theory sets out criteria for experimental interventions that can be used to test causal relations.
>
> *continued*

Core ideas Continued

- An intervention *I*, in Woodward's account, has to have three characteristics: (i) *I* modifies only the cause *X* (but not other factors), (ii) modifications in the effect *Y* are due only to *I*, (iii) *I* is not correlated with other causes of *Y* (other than *X*).
- The manipulationist theory also links the account of causality to one of our purposes in finding causal relations—control.
- Modularity is a property of causal relations which holds when we can intervene on the putative cause without disrupting the causal relation itself. If a causal relation is modular we can more easily test it, and also use it.
- Woodward's account of causal explanation is that we can explain an effect when we can answer a range of what-if-things-had-been-different questions concerning that effect.

Distinctions and warnings

- Some causes do not seem to be manipulable, such as race or ethnicity. We can *imagine* manipulating whatever we want. The problem is, if we are doing empirical research, really working with empirical data, we cannot directly manipulate factors such as race or ethnicity—particularly not without disturbing other causes of, say, discrimination. We may not always have a clear grasp, even in imagination, of what such manipulations would involve.
- Nevertheless manipulationism, even using imagined manipulations, still has heuristic value, as it can help with generation of hypotheses and reasoning.
- Not all causal relations are modular. Some interventions are 'structure-altering' in Steel's terms, which means that the attempted intervention on the cause alters the previously existing causal relation we are attempting to test, or to use in policy.

Further reading

Woodward (2013); Steel (2008, ch8); Strevens (2008).

CHAPTER 11

Production Accounts: Processes

11.1 Examples: billiard balls colliding and aeroplanes crossing

Causality is not always difficult to understand and establish. We have all had some experience of easy-to-grasp causal relations. Think of billiards or pool. There are several balls on the green table, and the goal is to get your own in the holes using what is really just a stick. Whenever you hit the ball strongly enough with your cue, it will move. Whenever a ball hits another ball strongly enough, it will move or change direction. It seems very easy to trace the lines, or processes, that these balls follow, and detect the causes, the collisions, that make these processes start, change or end.

But we see similar things, and don't interpret them in the same way. Suppose you are looking at the sky on a bright day and tracing the tracks of aeroplanes that cross over. A collision between aeroplanes would result in a crash, not just a change of direction, of course. Planes moving are real causal processes. But consider now the shadows created by these aeroplanes on the ground. They cross, and yet nothing happens to them. They move on undisturbed. But the shadows, and their movements, are physical processes too. We can see and track the shadows as easily as we see and track the planes. So how do we distinguish processes that are genuinely causal from those that aren't?

There is also a highly theoretical question that Salmon (1984) thought of interest to physics. We know from physics that real causal processes are subject to lots of constraints of physical law, such as an absolute restriction on nothing moving faster than the speed of light. (This is why the faster-than-light-neutrinos debate was so very important.) But perhaps pseudo-processes, such as moving shadows, are not subject to the same constraints. Consider a laser rotating very fast at the centre of an enormous dome. The laser rotating seems to be a causal process, so it is restricted by law. So is the speed of the light travelling from the laser to the wall of the dome. But the moving point of light on the wall of the dome is a pseudo-process. In theory, if the dome is large enough and the laser rotates fast enough, the point of light could appear to travel faster than the speed of light. Salmon thought this was another interesting reason to try to distinguish between real causal processes and pseudo-processes.

11.2 Tracing processes

Recall the distinction between accounts of difference-making and accounts of production (see chapter 6). Accounts of difference-making focus on detecting *that* there is a causal relation, through appropriate difference-making relations, e.g. probability raising or counterfactuals. Accounts of production, instead, focus on the *connection* between cause and effect, rather than on causes making some kind of difference to their effects. Let us clarify. For a correlation to hold, it doesn't matter *how* it holds. The link between cause and effect can be left as a black box, as long as some kind of relationship between cause and effect can be detected. Accounts of production focus on *how* cause and effect are connected. This chapter looks at processes, chapter 12 looks at mechanisms and chapter 13 looks at information transfer.

Process-tracing has a long tradition in philosophy of science. We begin by offering a chronological presentation of the main views developed. The technical details aren't essential here, so we will concentrate on the core ideas (for a thorough discussion, see Dowe (2008)). What is interesting about sweeping through the accounts in this way is that it shows how different philosophers have tried to identify different aspects of physical reality—trying to home in on the aspect of reality they take to be key to understanding what a causal process is.

Causal lines. Bertrand Russell is mainly known for his attack on the notion of cause (Russell, 1913). He was concerned that the 'philosophical' notion of cause, mainly tied to necessity, is not at work in science, especially in physics. Later, Russell (1948) argued that causality can be explicated using the notion of a 'causal line'. Causal lines are space-time trajectories that persist in isolation from other things. A trajectory through time of a ball on the green billiards table is a causal line. When it hits another ball and changes direction, this is another causal line. A person's trajectory through time is also a causal line. You at birth, and at five years old, and so on, are all part of your causal line. Such a notion is enough to describe the link between two points in spacetime. It allows us to infer features of different points in your causal line, such as to expect you to look similar tomorrow to how you look today.

Mark method. The 'mark method' was first given by Hans Reichenbach (1956), and then also picked up by Wesley Salmon. Reichenbach wanted to explain the asymmetry of time through the asymmetry of causality (see also chapter 17). His idea was that a causal process that is marked at the beginning will transmit or carry this mark until its end, being manifested at all stages in between. But a causal process marked near the end will not transmit the mark to the beginning. A causal process is one where a mark at the beginning is found again at the end. So for example if we mark a moving car, we will find the mark still on the car when it stops at its destination. This idea is being picked up again in modern accounts of production (see chapter 13).

Transference. Aronson (1971) gives the first transference account. He is explicitly concerned with giving a physical account of causality, an account independent of the actions of people. He says that the physical sciences are not primarily concerned with manipulating nature. He thinks we can find a physical relation that gives the direction of causality—a direction we find in our causal language, when we say things like 'Alice knocked over the cup'. Aronson is concerned only with changes that result from two objects interacting, not internal changes within one object, which he calls 'natural changes'. Most current theories of causality do not make this distinction. Aronson thinks that if bodies are undisturbed, they take on a pattern of behaviour such as constant motion in a straight line, alterations to which require causality. His core idea is that if C causes E, then before E occurs, the two bodies are in contact, and the cause object transfers a quantity such as momentum, energy, force and such like to the effect object during the contact. The direction of transference is the direction of causation. So Alice knocked over the cup because Alice came into contact with the cup, and transferred momentum to the cup. Similarly for billiard balls interacting, and so on.

Fair (1979) specifically aims for a physicalistic reduction of cause, and to elucidate the relation between causation and physical science. He says he is more specific than Aronson about what gets transferred, claiming that physics has told us that causation is really a transference of energy and momentum. That is how causes and effects are related. He writes:

We have come upon a very plausible candidate for what causes are as described by physics: they are the sources of physical quantities, energy and momentum, that flow from the objects comprising cause to those comprising effect. The causal connection is a physical relation of energy-momentum transference. (Fair, 1979, p229.)

Fair does say that it is not easy to reduce our everyday language of causes to the language of physics (see also chapter 19). But, broadly, when a baseball shatters a window, it transfers energy to the flying pieces of glass. The direction of energy transfer is what determines that the baseball shatters the window, and not that the shards of glass cause something in the baseball. For Fair, the account is dependent on what the physics of the day tells us, and may change if physical theory moves on.

Processes. The next big process view of causality drew on all the preceding process views. Wesley Salmon developed an account, later modified to accommodate criticisms by Phil Dowe (Salmon, 1984; Salmon, 1997; Dowe, 1992; Dowe, 2000a). For Salmon and Dowe, processes are conceived of as world lines of objects, and *causal* processes are those that possess conserved quantities, and, for Salmon, also transmit them. Conserved quantities are any quantities that are universally conserved (e.g., mass-energy, linear momentum, or charge), as described by physical theory. When causal processes intersect, they exchange conserved quantities, changing each other, and this is a causal interaction. If the processes are instead *pseudo*-processes—think about the aeroplane shadows—no quantity is transmitted, and nothing is changed by apparent 'interactions' such as shadows crossing. Salmon has solved his original problem: pseudo-processes such as the rotating spot of light on the dome can appear to travel faster than the speed of light, because they are not transmitting conserved quantities.

Boniolo et al. (2011) is the latest process account developed just recently as a follow-up of the Salmon-Dowe account. They notice that if we follow the Salmon-Dowe criterion of exchange of conserved quantities, there will be cases where such exchanges won't be detectable and, consequently, we should conclude that there is no causal process or causal interaction going on. This new approach allows us to detect causal relations by tracking transmission of extensive quantities. Some are conserved, some others are not, and in these cases we'll have a more powerful criterion.

An important aspect of the extensive quantities account is the role of the observer and the choice of the level of abstraction (see chapter 2) at which we ask a causal question. Consider the following example. Let us put a resistor (electric heating) in a room and turn it on. Is there a causal interaction between the heating and the outside world? If we try to track exchange of conserved quantities between the resistor and the room, we'll find none. But we buy electric heating precisely because we expect the electric heating to warm up the room, so there must be some causal interaction going on. Instead of tracking conserved quantities, we can successfully track the exchange of *extensive* quantities—in this case entropy. One could reply that in that system there are molecules colliding and transferring energy and momentum. The point, however, is that this means changing the *level of abstraction*. We find exchange of conserved quantities by stopping talking about the resistor and the outside world any more, to talk about atoms. This solves the problem only by changing the question!

Another set of examples are called 'stationary cases'. These are those in which some quantities (e.g. temperature) are time-invariant. Some quantities in a given system may have zero variation because of complex compensations going on. In such cases, it will suffice to find a point where quantities, even if they are not conserved, are exchanged. In these ways Boniolo et al. (2011) try to provide a *tool* to detect causal relations that is broader in scope than Salmon-Dowe's.

In a variety of ways these theories attempt to say what causation in the world is, independently of how we perceive it. In various ways they all draw on physical theory in their accounts of causation, and clearly some think that looking at physics is the way to tell what causality really is. There is certainly something to this idea of a causal link, but we can see that is it not at all easy to capture! Earlier in his career, Russell (1913) famously contended that causality is not used in physics, and the variety of scientific contexts considered in this book is an encouragement to look beyond the realm of physics to understand causality.

11.3 How widely does the approach apply?

11.3.1 Motivation for the account

So the view that has gradually emerged from the literature on processes is the idea of a causal link as a continuous line in spacetime that transmits or propagates some kind of physical quantity or quantities. The account is an attempt to answer the question of what there is between a putative cause and a putative effect, while accounts of difference-making discussed in, for example, chapter 8, help us establish that there is a causal relation between a putative cause and a putative effect because some appropriate difference-making relation holds. As we have said, sometimes causal inference consists of tracing causal processes, as when we track the movement of colliding billiard balls. What is in between cause and effect matters.

An important motivation of Salmon for his account was that the 'cause is put back in to the because', to paraphrase his words. Salmon meant that the physical process leading up to an effect itself explains the effect, which is what Salmon calls an 'ontic' explanation, rather than some kind of *description* of that process explaining the effect. But, in some way or other, all of the thinkers in this approach wished to describe what causality 'really' is, independent of people, and they all turned to physical theory to look for the answer. The question whether explanation is ontic—i.e. *processes* (or mechanisms) do the explaining—or epistemic—i.e. the *narrative* of processes (or mechanisms) does the explaining—is contentious. Illari (2013) argues that mechanistic explanation requires both ontic and epistemic constraints.

11.3.2 Applicability

Probably the most immediate challenge for the process-tracing approach is its applicability. The views are framed in terms of physical theory, using examples from physics, and physical quantities that make sense in the context of physics. But these quantities do not seem to be relevant to understanding causality in other sciences. Machamer et al. (2000, p7), referring to biology, say:

> Although we acknowledge the possibility that Salmon's analysis may be all there is to certain fundamental types of interactions in physics, his analysis is silent as to the character of the productivity in the activities investigated by many other sciences. Mere talk of transmission of a mark or exchange of a conserved quantity does not exhaust what these scientists know about productive activities and about how activities effect regular changes in mechanisms.

Russo (2009, p18), referring to social science research, says:

> These two concepts [causal interaction and causal process]—and not the statistical concepts [e.g., constant conjunction or statistical regularity]—provide the foundation of our understanding of causality. But what about poverty as a cause of delinquency? Or maternal education as a causal factor for child survival? I am not questioning the plausibility of those causal statements. Indeed they are plausible, but—I ask—is aleatory causality a viable approach to causality in the social sciences? Salmon thinks it is. However, it seems to me that it is not self-evident to 'see' causal processes and interactions in social science scenarios.

So, does the account apply only to physics? The first possibility is that we accept that the account applies only to physics, and it is useful there in understanding what we are tracking when we are tracking causes *in physics*. The second possibility is that the account says all causality everywhere is transmission of extensive quantities (or conserved quantities, energy-momentum or whichever view you prefer). So whenever there is causality, such-and-such quantities are transmitted, even though most of the time we don't pay any attention to them. When we worry about socio-economic circumstances causing migration behaviour, or a failure in gene expression causing a disease, we pay no attention whatsoever to quantities of physical theory in modelling these systems and trying to change them. However, presumably these systems do not violate physical law, and so in some sense extensive quantities will be transmitted in such systems.

11.3.3 Causal relevance

This brings us to the second, strongly related, problem for the process approach: the issue of causal relevance, raised as a problem for the Salmon-Dowe account. To recall, consider billiard balls colliding. We can see that there is a process where conserved quantities are exchanged, so the process is a real causal process. For Salmon-Dowe the aim is to discriminate causal processes from pseudo-processes, and we have done that. But the account doesn't tell us which properties of the process matter. Yet it seems obvious that the mass and speed of the first ball matters to the movement of the ball it hits, while its colour doesn't.

In more detail, sometimes we want to know which *object* had an effect on which other object. For the billiard balls, that is easy, and knowing this allows us to pick up a rolling ball from the table to stop an unwanted effect. But sometimes we want a more fine-grained description, in terms of the *causally relevant properties*. For the billiard balls, we want to know what the properties of the billiard ball in virtue of which it causes the movement of other billiard balls are. This allows us to do more. We know which object to pick up to stop the effect, but we also know what other objects we can, and cannot, switch in, and still avoid the effect. Switching in differently coloured billiard balls will not work to stop the effect, for example, while switching in cotton-wool balls will work.

Since presumably the relevant properties of the balls are their momentum, the Salmon-Dowe account could lead you to the causally relevant properties in the billiard balls case. But this seems impossible in cases of socio-economic circumstances causing migration behaviour, or a failure in gene expression causing a disease. It's not just that we don't happen to track physical quantities in disciplines like social science, biology and so on. We have a reason for this; that they are not the causally relevant properties in biological and social systems. There are a lot of processes happening in the human body, and in a social system, and a lot happens to them in a day. The problem generally is not to discriminate a genuinely causal process from a non-causal pseudo-process, but to identify a particular causal process from a mess of other interacting causal processes.

Further, in such sciences, we increasingly need to find *quantitative* causal relations. We need to know not just that poverty causes migration, and penicillin cures infection, but how much poverty causes what level of migration, and what dose of penicillin to administer for a given seriousness of infection. None of the accounts here offer any attempt at a quantitative account of processes.

Ultimately this is why the causal process accounts do not apply outside physics. Not only do they fail to apply formally, as the quantities of physical theory are not what we track in biology, social science and so on; more importantly, they do not seem to help with the key reasoning and causal inference challenges of such sciences. In chapters 12 and 13 we see, however, that the most fundamental idea of the causal process theorists, the idea of causal linking, can be of use in such sciences.

11.3.4 Does the approach apply to all of physics?

While the accounts have been designed from a physics perspective, drawing on physical theory, they might not apply to all of physics. This is because physical theory is not unified. And even where the physical quantities used above do apply, it isn't at all clear that they are usefully tracked. The examples above may give the misleading impression that physics deals with straightforward cases. But there is instead an important discussion about causation in controversial areas such as quantum mechanics. Here, however, things get complicated because the focus is on the probabilistic structure of the phenomena at the quantum level, rather than on an explication based on processes. Likewise, work done by the 'Budapest school' (Hofer-Szabó et al., 2013) aims to analyse probabilistic causal structures that satisfy the principle of common cause (see chapter 8) but does not aim to provide an account of causation in physics contexts in terms of processes—that's simply outside the scope of their work. There is also important work on the Einstein-Podolsky-Rosen (EPR) correlations and the notion of modularity and intervention that does not involve the concept of a process as we have been discussing it in this chapter (San Pedro and Suárez, 2009).

11.3.5 Absences

The final well-known problem for the claim that we get causality where and only where we get transmission or propagation of some kind of physical quantity is the problem of absences. We seem to have causal relations where either or both the cause or the effect is an absence. In chapter 9 we examine examples like: 'I am so sorry I missed the meeting, it was because my bus didn't turn up'. The cause was the absence of the bus, and the effect was the person's absence at the meeting. But it seems that a physical process, a world line, or transmission of a physical quantity, cannot link an absence to anything—indeed, nothing could. Absence causality of this kind is extremely important to legal theory, as it is involved in such questions as: Can you be prosecuted for failing to take some reasonable care (an absence), thereby causing harm to your child, or some other person? (For a discussion, see e.g. Pundik (2007) and also chapter 9.)

Causal process theorists don't seem to have any option but to cast cases of causing by absences as somehow secondary causes, pragmatic or linguistic shorthands, whereas what is really going on is positive. So for the bus example, the real cause of the lateness is the actual position of the absent bus—wherever it is. Dowe (2004) takes this approach, saying that real causal process causation is causation proper, and absence causes are a linguistic or pragmatic shortcut, which he calls 'quasi-causation'. (For a very different answer, see chapter 13.)

It is certainly true that pragmatics have a lot to do with absence causation. If we say that absence of oxygen caused death, that is because we expect there to be oxygen present in the bloodstream of a healthy body. Oxygen is normally present, and that is why we can speak of its absence. We don't speak in the same way of absence of arsenic in your bloodstream causing you to live. The difference is simple: we don't expect arsenic in the blood. Biological or social mechanisms are always described against a background of what is normal in the body, or in the appropriate socio-economic context.

The same, interestingly enough, is true of purely physical mechanisms. The mechanisms of supernovae, or of galaxy formation, are stable enough that we can expect similar processes in different cases. We might, therefore, explain unusual behaviour by the absence of some element that is normally present, just as we would mention proceedings being interrupted, say by a collision with another star.

However, there is also reason to think absence causation cannot be dismissed wholesale as somehow pathological. In the philosophical community, this realization came about for two reasons. First, in a couple of papers, Jonathan Schaffer persuaded the philosophical community that absence causation is ubiquitous and important (Schaffer, 2000; Schaffer, 2004). Schaffer argues that shooting someone through the heart technically causes death by absence of oxygen to the brain, and many mechanisms involve absences as causes. He writes:

It is widely believed that causation requires a connection from cause to effect, such as an energy flow. But there are many ways to wire a causal mechanism. One way is to have the cause connect to the effect, but another is to have the cause disconnect what was blocking the effect. (Schaffer, 2000, p285.)

The upshot is that normal, even paradigmatic, cases of causality involve absences. A nice case is discussed by Sandra Mitchell:

There can be suppression of the 'normal' operations of causal mechanisms at a lower level by the control exerted by its being embedded in a hierarchical organization. For example, in a honeybee colony, the normal ovarian development in females is contingent on the presence or absence of a queen. (Mitchell, 2003, p153.)

When the queen is safely present, she secretes hormones that inhibit other females from developing ovaries. But the colony needs to survive if the queen dies, and under these circumstances other females will develop ovaries. The absence cause here is crucial to the healthy functioning of the colony.

The second reason builds on the ubiquity and importance of absence causes. It is reasonable to think that absences cannot be wholesale re-written in terms of presences, such as in the bus case. The reason brings us back to causal relevance: it is the absence that is causally relevant, not the current position of the bus. Lewis writes:

One way or another, we can cook up ersatz absences to serve as relata of the causal relation—though surely they will seem to be the wrong relata, since we don't really think of these ersatz absences as having the same effects (or causes) as the absences they stand in for. (Lewis, 2004, p282.)

This is to say that a lot of the work of science goes into describing phenomena at the right level of abstraction to see the real causes and effects. We might be able to redescribe causal claims convincingly on a case-by-case basis, but not in the wholesale way necessary to save scientific practice. For example, it is true that there are particular different positive mechanisms that operate when there is no queen in the hive. But it's the *absence* of a queen that sets all those mechanisms off. The particular way the queen is lost or dies is wholly unimportant. It is the absence of the queen that situates what happens in both the natural selective explanation, and in the proximal mechanistic explanation.

In summary, causal relevance is the most serious problem for process approaches, as it is causal relevance that makes both the problem of absences, and the applicability problem, so severe. However, the core idea of seeking an account of causal linking, and the methodological turn to looking at scientific practice, are important legacies of the causal process views.

Core ideas

- Process accounts ask what's really in the world, and look to physical theory for the answer. Early process accounts are the first strong turn to the empirical in philosophical work on causality.
- These accounts theorize linking, and worry about what's in between causally related variables. We owe to this tradition important current developments—see chapters 12 and 13.

Distinctions and warnings

- Unfortunately this account doesn't help in all the areas we need such an account. How does it help us with smoking and heart disease? There are lots of ways conserved quantities may be transferred from smoke to the heart (heat for example). These do not all help us figure out the causal link. This is the applicability problem.
- This also illustrates the causal relevance problem. The theory tells us that smoking and heart disease may be related. But it does not tell us what property of smoking may be the vitally important (causally relevant) one.
- Some sciences talk about absences causing, but absences don't lie on world lines, or transfer or receive energy or other conserved or extensive quantities.

Further reading

Dowe (2008); Russell (1948, Part IV, ch5); Salmon (1980a).

CHAPTER 12

Production Accounts: Mechanisms

12.1 Example: how can smoking cause heart disease?

At various points in many sciences we find puzzling correlations—correlations between two variables that we don't understand. For example, scientists found a correlation between smoking and heart disease. That means that people who smoke are more likely to get heart disease than people who don't smoke. But, at the time the correlations were found, it wasn't obvious that smoking could cause heart disease. This was the conclusion drawn by Doll and Peto (1976), when they analysed data from a survey of male doctors between 1 November 1951 and 31 October 1971.[18] They argued that the correlation between smoking and heart disease is different from the correlation between smoking and lung disease. Smoke from cigarettes goes into the lungs, so it is not difficult to think of ways in which smoking could affect the lungs. But how could smoke affect the *heart*? Of course, we know very well that correlation does not imply causation. There may be something else that smokers do more than non-smokers, such as eat fast food, and it is that something else which causes heart disease, rather than smoking. Nevertheless, if correlations are fairly robust, and we don't understand why they exist, we usually want to investigate. We seek an explanation.

In light of the work of Salmon and Dowe (see chapter 11) we might think of these investigations as tracking the processes by which smoking might cause heart disease, but Salmon-Dowe processes seem too simple, and too tied to physics, for this kind of case. Instead, in seeking an explanation, what biomedical scientists wonder is whether there is a plausible *mechanism* by which smoking could cause heart disease. Biomedical scientists talk about mechanisms all the time, as they try to describe all the workings

---

[18] Gillies (2011) examines the history of the research on atherosclerosis and discusses it in relation to evidence—see also chapter 6.

of the body by describing all the different mechanisms responsible for the functions of the body. For example, the mechanism for transporting nutrients and oxygen all over the body is the bloodstream. The mechanism for driving the bloodstream is the heart, which is a muscular pump (but with various special mechanisms to regulate it, so that it doesn't stop). So if we want an explanation of a puzzling correlation but we are not doing physics, or if we want to know whether there could be a causal link between smoking and heart disease, maybe understanding what mechanisms are would help.

Scientists did manage to find out a lot about the mechanism by which smoking affects the heart. For more detailed discussion, see Gillies (2011), because it's an interesting story, with lots of trial and error. Science is never certain, and we do learn from failures, even though negative results often go unpublished. In brief, a lot of heart disease is caused by atherosclerosis, which is the hardening of the arteries due to the deposit of plaques, made at least in part by cholesterol. Plaques can then break off and interrupt the flow of blood to the heart, or increase the formation of blood clots, which do the same thing. This was known, and it is why we are still encouraged not to eat too many fatty foods. It is also known that some elements of the smoke from smoking are absorbed into the bloodstream, through the lung walls. So clearly those elements can travel in the bloodstream to the arteries and the heart itself. But it wasn't known how such elements of the smoke could accelerate the formation of plaques on the artery walls. Smoke, after all, does not contain cholesterol.

What seems to happen is that fat carried in the blood in the form of LDL does no harm. But if it gets oxidized, white blood cells try to destroy it, forming fatty foam cells. If this happens in the artery wall, their movement is inhibited and they stay there, so fatty streaks build up on the artery wall. As yet, this isn't dangerous. There has to be much more build-up to induce heart disease. Smoking increases oxidation, and accelerates the build-up of plaques on the artery wall. This is a very brief summary of the mechanism by which it is thought that smoking causes heart disease.

Here we see the very important place of mechanisms in causal *explanation*. Many sciences explain a phenomenon by describing the mechanism that produces that phenomenon. Sometimes we want to know just from curiosity. But we can also see that mechanisms help us decide what causes what—i.e. they help with causal *inference*. Now that a mechanism is understood, it is much more likely (though *not certain*) that the correlation between smoking and heart disease is causal, whereas if we could not find a linking mechanism, we would still be wondering if there is some common cause of smoking and heart disease. Since mechanisms are concerned with what links or connects putative cause-and-effect variables or events, going beyond mere correlation, they fall within the school of thinking concerned with productive causality, alongside processes in chapter 11, and information in chapter 13: how do causes produce their effects?

12.2 What is a mechanism? The major mechanists

Since mechanisms can help with both causal explanation and causal inference, it seems that understanding what mechanisms are is of interest to science. Mechanisms are becoming a core subject in philosophy of science, and two parallel literatures are being developed. The first strand is the one sparked by the recent works of Bechtel, Craver, Darden, Glennan and Machamer. These authors discuss mechanisms primarily in the context of biology, neuroscience and psychology. The second strand examines mechanisms in the social sciences, for instance in analytical sociology, economics or demography. Important contributions include those of Hedström and Swedberg, Demeulenaere, Elster, Little, Steel and Ylikoski. These two strands are moving in parallel at the moment, but in the course of the chapter we will note where the two strands connect, or could be connected.

All these authors debate many subtle issues about mechanisms, but Illari and Williamson (2012) argue that there is broad agreement that mechanism discovery proceeds in three steps:

(i) Describe the phenomenon or phenomena;
(ii) Find the parts of the mechanism, and describe what the parts do;
(iii) Find out and describe the organization of parts by which they produce, in the sense of bring about, the phenomenon.[19]

Let us consider again the mechanism for smoking causing heart disease. We begin with the phenomenon we are trying to explain. We need the best description of it we can get. When does it happen? In what sub-populations? For example, does it only happen in men? Does it only happen in the old, or does it happen in the young? Do you have to smoke a lot before it happens, or only a little? You might get this kind of information, in this case, from looking at medical records—for instance, aggregated data for a whole country.

Once the phenomenon has been thoroughly investigated, we want to know what parts of the body are involved in smoking causing heart disease. Knowledge of smoking, and background understanding of the mechanisms of the human body, tell us where to start. We know the smoke from cigarettes goes into the lungs, and we know how gases can pass into the bloodstream from the lungs—that is what oxygen does. And of course the blood circulates all around the body, and certainly gets to the arteries and the heart itself. Background knowledge of mechanisms that cause heart disease tells us that the hardening of the artery walls due to the build-up of plaques is a common cause of heart disease. So what happens in the bloodstream in and near the arteries is clearly one area to look. The puzzle was how certain toxic gases, dissolved in the bloodstream, managed to affect the fatty build-up on artery walls. There was still what scientists often call a 'black box' in the crucial place in the mechanism.

In finding the parts, we find entities and activities. Entities are just the parts of the mechanism. We already knew about a lot of these in the case of smoking and heart disease, like the artery walls, and white blood cells. But new entities were also discovered, like fatty foam cells. Activities are the different things entities do, like travel around,

[19] If 'produce' and 'bring about' are synonyms of 'cause', one might raise the question whether what is being offered is a causal theory of mechanisms, rather than a mechanistic theory of causality. If one is primarily concerned with understanding causal explanation in practice in the sciences, this does not matter too much. If one is trying to give a mechanistic account of causality, as Glennan is, it might matter.

or become oxidized, or interact with other entities in many ways, such as by engulfing them, and so on. We also discovered new activities of familiar entities, an example being that white blood cells attempting to engulf oxidized LDL form fatty foam cells.

The final stage is to map out the *organization* of the mechanism. Where things happen, and when are often important in biomedical mechanisms. These are forms of organization. For example, in this mechanism it was important where the formation of fatty foam cells happened. If they formed in the bloodstream they didn't do much damage, at least initially, whereas the formation of fatty foam cells within the artery walls was what set off the real damage. For this mechanism, how fast things happen is important. It takes a long time for the fatty build-up on artery walls to become dangerous, which is why it is worth avoiding eating fatty foods even if your artery walls are fine now, to avoid heart disease in 20 years. We will look more at organization in the next section.

Once the whole picture is built up, and the nature of the entities and their activities and the organization of the whole is established experimentally, we have the mechanism. Broadly the same kind of process can be gone through to find the mechanism for many different kinds of phenomena, including in social sciences.

In chapter 7, we discuss a study run by demographers. Gaumé and Wunsch (2010) studied how people in the Baltic countries, in the period 1994–99, perceived their health, and particularly what the determinants (i.e., causes) of the way they perceive it were. Scientists first needed to isolate and describe the phenomenon of interest, self-rated health, which included fixing the population of reference and relevant facts from background knowledge. For instance, limiting the analysis to the period 1994–99 fixed important facts about the socio-political context, about the policies and societal changes that were in place, and so on. Then they isolated the variables that were likely to play a role in the explanation of the phenomenon. These were isolated both on the basis of background knowledge and theories *and* on the basis of the available data, and included, for instance, alcohol consumption, psychological distress or physical health.

This study used a statistical analysis of data sets to model the mechanism underlying such data. At this level of abstraction, entities are the variables having a causal role—e.g., the above-mentioned alcohol consumption, psychological distress or physical health—and the activities are the (statistical) interactions between these variables—e.g., self-rated health is correlated with these variables. At this level of abstraction, entities and activities are captured by the variables in the model—e.g., the above-mentioned alcohol consumption, psychological distress or physical health—and the (statistical) interactions between these variables.

Once the context and the main factors have been identified, scientists have to 'organize' these factors so that they explain the phenomenon of interest. So self-rated health is brought about by alcohol consumption, psychological distress and physical health, which all have direct paths to self-rated health. However, these factors also interact, for instance physical health also affects self-rated health via alcohol consumption. In quantitative social science statistical analyses, and, more specifically, recursive decompositions (see chapter 7) are used to model the structure, that is the organization, of the mechanisms at hand. Wunsch et al. (2014), commenting on the same study about self-rated health in the Baltic countries, explain that the kind of functional individuation—of the mechanism and of its components—just described for 'natural science' mechanisms is also carried out in social science research.

Discovering mechanisms is also done using qualitative tools (see chapter 2). Steel (2008, p189ff) illustrates with an example from anthropology. Malinowski (1935) studied the population in the Tobriand islands and formulated the mechanistic hypothesis that wealth and influence in the circle of chiefs of the islands was an effect of having many wives. This hypothesis summarizes complex societal mechanisms. In particular, there are two processes that are relevant here. One is the habit brothers have of contributing significant amounts of yams to the households of their married sisters. The other is that chiefs can undertake public projects and political actions precisely because they are financed by yams donated by their brothers-in-law. This is of course a very rough description of the mechanisms. But the idea should be clear: social mechanisms can be identified, described and isolated, insofar as the key entities, such as chiefs, brothers and sisters, and yams and their activities and interactions, such as marrying, and giving yams, are specified. This includes identifying the function of both actors and their actions, other entities and their activities, and ultimately of the complete mechanisms for a given societal structure.

Note that the process of 'mechanism identification' can be described in clean steps, but in actual practice it is messy and iterative. You don't describe the phenomena, and only then find the activities and entities, then move on to finding the organization of activities and entities. You tend to find elements of them all gradually, building up the whole picture. While the phenomenon you are trying to explain sets off the discovery process, it isn't fixed. The process of mechanism discovery helps you understand the phenomenon, and it might lead to significant redescription of the phenomenon itself. Looking for the organization might lead you to an entity you hadn't previously noticed. We see the messy nature of scientific practice in the discussion of exposomics in chapter 24. It is also discussed in Darden (2002).

This core idea of mechanisms and the process of mechanism discovery can help with understanding and articulating what's going on in a lot of science. Illari and Williamson (2012) argue for this in more detail. This will help scientists in different fields work together effectively, being able to describe what they are doing in ways other scientists will understand.

We will now briefly examine the major accounts of mechanisms, to see some of the disagreements.

MDC. 'Mechanisms are entities and activities organized such that they are productive of regular changes from start or set-up to finish or termination conditions.' (Machamer et al., 2000, p3.)

This is in many ways the most influential account. Machamer, Darden and Craver—often called 'MDC'—want to emphasize that mechanisms are *active*. They concentrate on biological and neurophysiological mechanisms, and these are busy things, constantly doing something to create or sustain the phenomenon. For example, the homeostatic mechanism maintaining human body temperature is terribly busy just to maintain a steady state of approximately 37° Celsius. MDC say that activities are just as important as entities. They are just as real as entities in mechanisms, and it is just as important that we find the activities of the entities as that we find the entities themselves. Just finding the parts of the mechanism for smoking causing heart disease is not enough. We also need to know what those parts do.

While there is a raging debate in the mechanisms literature about activities, focusing on what activities really are, Illari and Williamson (2012) argue that a good reason for preferring the language of activities and entities is that they allow unrestricted mapping between activities and entities, which is called the 'arity' of the relation. They write:

> The mapping of entities to activities can be unary, as in a bond breaking, involving no other entity; binary, as in a promoter binding to a strand of DNA; but it can also be 3-ary, 4-ary and so on (see Darden (2008, p964)). The activity of transcription involves DNA, the newly created mRNA, and various regulation and control enzymes, while more highly abstract activities such as equilibrating, or osmosis (Darden, 2008, p277) may involve very many entities, of the same or different kinds, or be such that it is hard to decide on any very clearly defined entity that engages in the activity. (Illari and Williamson, 2012, p12.)

Glennan. 'A mechanism for a behavior is a complex system that produces that behavior by the interaction of a number of parts, where the interactions between parts can be characterized by direct, invariant, change-relating generalizations.' (Glennan, 2002*b*, pS344.)

Glennan emphasizes the 'complex system' aspect of modern work on mechanisms. He is particularly interested in having an account of mechanism that covers all, or at least many, sciences. This is why he initially talks of 'invariant, change-relating generalizations', a concept that he borrows from Woodward (2003), and that we also discuss in chapter 10. In more recent work, Glennan carefully distinguishes his account from manipulationist accounts of mechanisms (Glennan, 2009*a*). Glennan was also the first to note that mechanisms are mechanisms 'for a behavior', which we will return to below. Many scientists find Glennan's characterization intuitively easy to work with, perhaps because it is very general. But note that Glennan is in agreement with the core idea above, adopting a similar characterization in work in progress, although he is keen to emphasize that entities frequently interact with each other in mechanisms. In his published work, he is also happy to talk about entities and activities.

Bechtel. 'A mechanism is a structure performing a function in virtue of its component parts, component operations, and their organization. The orchestrated functioning of the mechanism is responsible for one or more phenomena.' (Bechtel and Abrahamsen, 2005, p423.)

Although Bechtel published an account of mechanisms in 1993 with Richardson, his more recent account above is better known. Bechtel explicitly emphasizes the importance of function, and worries extensively about organization, or 'orchestrated functioning', both of which we will discuss shortly.

After small changes of detail (Bechtel and Richardson, 2010; Glennan, 1996; Machamer, 2004; Craver, 2007; Glennan, 2011), these broad characterizations remain in use by their original advocate(s), and many others. Scientists should feel free to pick the one that seems intuitive to them, or just stick to the core consensus we have described.

While mechanisms and causality are clearly related, the precise nature of the relation is not so clear. All of the major mechanists agree that a mechanism gives a causal *explanation*. They prefer understanding scientific explanation in terms of finding mechanisms to the classical deductive–nomological model. So that is not disputed. (For a discussion in social contexts, see Ruzzene (2012).) It is generally thought that when you have a mechanism, you have at least one causal relation. MDC think that the activities in mechanisms are little specific kinds of local causes. For example, a white blood cell 'engulfs' oxidized LDL, and a fatty foam cell results. 'Engulfing' is the local, specific way in which white blood cells trying to destroy oxidized LDL cause the creation of fatty foam cells.

Only Glennan explicitly gives an account of causality in terms of mechanisms, i.e. C causes E if and only if there is a mechanism linking C and E. This means Glennan is committed to there being no causal relations without mechanisms. For this to be an account of productive causality, it is important that Glennan has distinguished his account from the manipulationist account, which is a difference-making account (see chapter 10). It is also important to clarify in what way the account of mechanism is a causal account. Glennan addresses this in Glennan (2009a) and Glennan (1996). Note that you don't have to believe there are never causal relations without mechanisms to make the relation between causality and mechanism interesting. Even if there is sometimes causation without mechanisms and some mechanisms without causation, if they often come together, that is enough for mechanisms to be useful in causal explanation and inference.

Glennan does think that in fundamental physics there may be causes without mechanisms. Fundamental physics is a special case in science, for its peculiar, mind-boggling results. We have already seen in chapter 8 that quite straightforward descriptions of probabilistic causal structure at the macro-level don't work at the fundamental level. For this reason, philosophers and physicists wonder not only what causal concepts may be appropriate to physics, but whether causal concepts are appropriate *at all*. It is also worth noting that the word 'cause' appears more rarely in physics journals than in biomedical journals. A notable sceptic about causation in physics is Norton (2003), but opposing views, defending the causal stance even at the fundamental level, also exist. See for instance Frisch (2014).

In the philosophy of the social sciences, Reiss (2007) and Steel (2008) hold definitions of mechanisms that share a lot with the core of agreement of the main contenders just described. These also share a lot with Cartwright's 'socio-economic machines' that give rise to 'socio-economic' laws (see for example Cartwright (1995), Cartwright (1999)). In social science there also exist other conceptions of mechanisms. For instance, Mayntz (2002) or Little (2011), conceive of mechanisms as series of events causally connected, while Merton (1964) famously viewed mechanisms as theorizing or reasoning (about social phenomena). The topic of mechanisms and structures in the social realm has been extensively discussed in two important edited volumes: Hedström and Swedberg (1998) and more recently Demeulenaere (2011). Arguably, the social science literature has been preoccupied more with how mechanisms are found or theorized, and with the role they play in explanation and theory, and less with developing a definition that captures the essential elements of mechanisms or that applies to all scientific contexts.

12.3 Important features of mechanisms and mechanistic explanation

We have explained what mechanisms are, and given the best known characterizations. But there has also been a lot of useful work developing thinking about important aspects of mechanisms that can be confusing for anyone trying to think about them, or reason using them.

12.3.1 Function

Mechanisms have functions. Glennan recognizes this by noting that a mechanism is 'for a behavior', while Bechtel and Abrahamsen say that a mechanism is a 'structure performing a function'. This is a problem if we don't know how to understand functions, or if we think such functions require a person observing the mechanism to allocate a function. This problem might be even more troublesome, because the parts of mechanisms also have functions. Parts have the function of contributing to the behaviour of the whole.

The best known analysis of function that doesn't require an observer is the 'selected-effects' function developed by Ruth Millikan (Millikan, 1989; Millikan, 1984). According to this view the function of an entity is the activity it has been selected for in its evolutionary history, which explains why the entity is there. So presumably the function of the heart, on this account, is to pump blood. However, this doesn't help us much with mechanisms. Some mechanisms have nothing to do with natural selection, like the physical mechanisms of supernovae, or mechanisms for the production of gravitational waves. Even in biological mechanisms, it is what the mechanism itself, or the entity within the mechanism, actually *does* that matters. When a newborn baby's head is laid on its mother's chest, it is the familiar thump-thump sound the heart makes that calms the baby, even though the heart was presumably naturally selected to pump blood, with the thump-thump sound a mere side-effect. In social science too it is not obvious that social entities or social mechanisms developed under natural (or perhaps social?) selection in order to perform a given function. Some scholars may argue in favour of such a view, but strong arguments have to be provided. Mechanisms, especially when *explanation* is at stake, need a different account of functions.

The dominant account of function among the major mechanists is Cummins' 'role-functions'. For Cummins, the function of an entity is the role it plays in the overall behaviour of a containing system. He writes:

x functions as a ϕ in s (or: the function of x in s is to ϕ) relative to an analytical account A of s's capacity to ψ just in case x is capable of ϕ-ing in s and A appropriately and adequately accounts for s's capacity to ψ by, in part, appealing to the capacity of x to ϕ in s. (Cummins, 1975, p762.)

This formulation is, admittedly, difficult to follow, but Cummins' definition is equivalent to Craver's 'contextual description' (Craver, 2001, p63), and the idea behind it is quite straightforward. Remember the Baltic study we mentioned earlier. Scientists managed to identify a (complex) mechanism involving different factors in order to explain how people self-rate their health. Each factor in this mechanism has a role, or function. For instance, the function of alcohol consumption is to relieve stress, and this works, in the sense of having the capacity to relieve stress, in the context specified by the scientists. The function attributed to alcohol consumption in this context is supported by the analyses of data, by background knowledge, and also by available sociological and demographic theories.

To return to the heart, we can see the importance of the role in the surrounding mechanism. If the overall capacity of the circulatory system being explained is to move food, oxygen and waste products around the body, then pumping is the role-function of the heart. But if the overall capacity of the system being explained is to calm newborns, then the role-function of the heart is to make a familiar thump-thump sound.

This is exactly right for giving the function of components of mechanisms, where the surroundings set the role of the component. But the overall mechanism may or may not have a role in something else. We can't always specify something. So the 'isolated descriptions' of Craver (2001) are also needed. Isolated descriptions just describe the behaviour of an entity of interest in isolation from its context. The pumping of the heart can be picked out by its role-function as we have seen—its pumping as contribution to the capacity of the circulatory system. But the pumping of the heart can also be picked out in a far more isolated way. The heart contracts. That is something we can observe it doing regardless of context. That the heart beats was known, after all, long before the circulation of the blood was understood. Contracting in this way is an isolated description, of a characteristic activity of the heart. This is exactly what is needed to understand the function of purely physical mechanisms such as mechanisms for the production of gravitational waves.

When massive bodies—stars and galaxies—move in appropriate ways, they will in theory emit gravitational waves, which makes particular kinds of movements mechanisms for the production of gravitational waves. We are doing our best to observe this. There is no need for either an agent or natural selection to confer this kind of function. So what is needed to understand functions of mechanisms more precisely is isolated descriptions in some cases, and role-functions where a containing system is specified.

In the social sciences, 'isolated descriptions' of mechanisms are provided, but, as in the biological case, they are always embedded in some context. Consider again the Baltic study on self-rated health, and focus on the relation between alcohol consumption and self-rated health. This can be characterized in an isolated way: people drink more or less alcohol, that can be observed, and they also report the way they perceive their health. But, in the Baltic study, alcohol consumption is included in the broader mechanisms of socialization. People drink alcohol for many different reasons, such as drinking to conform to social norms at parties, to help deal with stress, and so on. The role-function of alcohol consumption can vary, and that difference is important to different social mechanisms. Some of these will be relevant to studying self-rated health in the Baltic countries, some also relevant to the broader research question of individuals' well-being, etc. In other words, just as for biological mechanisms, the first thing to specify is *what* mechanism is under analysis and in what *context*. An isolated description of activities of entities is possible, but what we need for informative mechanistic explanations is role-functions, which presuppose the identification of a context. In social science, the same factor often performs different roles, or functions in different contexts. Suppose we shift the context to the mechanisms of morbidity and mortality. Here, the context of the mechanisms already suggests that alcohol consumption is included in mechanisms explaining negative effects on physical and mental health, rather than, say, socialization.

12.3.2 Context

Context is important to mechanisms in many ways. We have already seen that we understand the function of mechanism components by their role in the surrounding mechanism. Another important way is that context affects what is considered a part of the mechanism, and what is not. Not everything in the bloodstream is part of the mechanism by which smoking causes heart disease, and not everything in the brain is part of the mechanism of vision, for example. Only some parts are relevant, those Craver and Bechtel think of as working parts. For instance, Craver (2007, p15) writes:

> A component is relevant to the behaviour of a mechanism as a whole when one can wiggle the behaviour of the whole by wiggling the behaviour of the component and one can wiggle the behaviour of the component by wiggling the behaviour as a whole. The two are related as part to whole and they are mutually manipulable.

Craver bases this account on the kinds of experiments we use to find out what things are working parts of a mechanism. Roughly, we intervene to inhibit or excite a part, and see whether the phenomenon is affected. We also intervene to inhibit or excite the phenomenon, and see whether the part is affected. If 'wiggling' the component affects the phenomenon, and 'wiggling' the phenomenon affects the part, they are mutually manipulable, and we have found a working part of the mechanism. But, pragmatically, things are left out that would affect the mechanism. Oxygen being removed from the bloodstream would certainly affect whether you died of heart disease—you wouldn't, because the absence of oxygen would kill you first! But generally we regard presence of oxygen in the bloodstream as a background normal state, and don't include it in the mechanism. Leuridan (2012) criticizes the details of this account, arguing that obligate mutualistic endosymbionts, which are organisms that depend on each other so tightly for their survival that they entirely depend on each other, and indeed come to coevolve, are problems for Craver's account. For example, aphids are tiny insects, and they rely, mutualistically, on the bacteria *buchnera*, as each supplies the other with nutrients the other cannot synthesize. On Craver's mutual manipulability criterion, Leuridan argues, the *buchnera* seem to be components of their hosts. It is important to bear in mind, however, that mechanisms for some phenomena, such as ecological phenomena, might treat endosymbionts as a single entity, while mechanisms for other phenomena, such as for mutualistic interactions, might treat the aphids and *buchnera* as two entities.

12.3.3 Organization

Organization is the least controversial element of mechanisms, being accepted by everyone in the field. But how to understand what organization is, and so inform the job of uncovering organization in building mechanistic explanations, is not easy.

Organization is the final phase of building mechanistic explanations, number three in section 12.2, above. The same entities arranged in different ways might do something different, just as a group of people chatting over coffee may be finalizing a sale, or comparing opinions on the latest film they saw. But there are so many forms of organization that mechanisms exhibit, it is hard to say what organization is. Illari and Williamson argue:

Most generally, organization is whatever relations between the entities and activities discovered produce the phenomenon of interest: when activities and entities each do something and do something together to produce the phenomenon. (Illari and Williamson, 2012, p128.)

This captures simple spatial and temporal relations, and extends all the way to complex forms of organization such as feedback, homeostasis, and limit cycles. This approach leaves it an empirical question what forms of organization are important to what mechanisms, and when. This is right, as there are such a variety of forms of organization being discovered. This is why quantitative simulations of systems are increasingly used to study systems as widely different as, for example, protein synthesis in the cell, the dynamics that lead to supernovae, the mechanisms of urbanization of cities or of relative deprivation rates, etc.

Note that mechanisms can act to change their own organization. For example, protein synthesis mechanisms make their own entities during the operation of the mechanism, most obviously mRNA, and of course it is the protein synthesis mechanisms that themselves create the highly organized environment of the cell which maintains protein synthesis itself. To understand such systems and predict their behaviour, quantitative simulation is increasingly needed. Bechtel calls these quantitative approaches 'dynamical mechanistic explanation' (Bechtel, 2008; Bechtel and Abrahamsen, 2009; Bechtel, 2010).

Some social scientists spell out organization as referring to the practices and norms that govern the relationships between agents. Economists may refer, for instance, to the internal rules and modes of functioning of institutions, say a central bank; sociologists and anthropologists may refer to different practices related to the choice (or imposition) of spouses in different populations or ethnic groups. For some scholars in particular, organization refers to the structure of the mechanism as given by the recursive decomposition (see chapter 7). These two approaches can be seen as complementary, one describing the organization of the mechanism more directly, and often qualitatively, while the other attempts the same kind of description at a higher level of abstraction, frequently aiming to get a quantitative description. They do have in common that structure is supposed to give stability to the mechanisms. This is an important feature because in social contexts mechanisms are very fragile and contextual. So to know how stable a social structure is means precisely to know the boundaries of the applicability and validity of a study.

12.3.4 Applicability

Applicability was the main problem of the account of processes that predated accounts of mechanisms, the major Salmon-Dowe account of chapter 11, as it seemed to apply only to physics. Mechanisms are more widely found than interactions involving conserved quantities. But it is worth wondering how applicable this account of mechanisms is. There are two kinds of applicability worth worrying about.

First, we need applicability across the sciences, so that the account helps us understand scientific practice widely. We noticed at the beginning that theorizing about mechanisms is happening in biology and neuroscience, as well as in the social sciences. So cross-disciplinary applicability is important. Increasingly scientists in different fields collaborate to give an explanation of a phenomenon, building what has come to be known as 'multifield mechanisms' or 'mixed mechanisms'. Craver (2007, p228) writes:

> The unity of neuroscience is achieved as different fields integrate their research by adding constraints on multilevel mechanistic explanations.

This means that we cannot be satisfied with an account of a mechanism in biology, a mechanism in chemistry, a mechanism in molecular biology, and a mechanism in psychology. We have to worry about how different sciences manage to build a single mechanistic explanation of a phenomenon like memory. Russo (2012) also stresses the importance of integrating different fields of research not just to provide better mechanistic explanations, but also to plan better policy actions. She illustrates using psychological and biological mechanisms of obesity that have been exploited to plan actions in the MEND programme. MEND (Mind, Exercise, Nutrition, Do it!) is a public health programme active in the UK that aims to reduce obesity in children by educating them, and their families, to have a healthy life in terms of diet and exercise. One peculiarity of the programme is that officers exploit both biological mechanisms and psychological ones. For instance, they motivate parents by making them aware that obese children lack self-confidence and social skills.

Second, if an account of mechanism is to help us understand causality, its applicability to areas where we find causal relations is also of interest. It is widely agreed that a mechanism gives a causal explanation. So it is natural to ask whether we find mechanisms everywhere there are causes, or whether there are some causes that lack a mechanistic explanation, or have a different kind of explanation.

If mechanisms are used in this way, then the resulting account of causality shares two classic problems of the initial account of productive causality, the process account. These problems are absences, and causal relevance. They are discussed in more detail in chapter 11, and they are not particularly distinctive problems for mechanisms, so we mention them only briefly here. First, we seem to cite absences as causes. For example, the absence of oxygen in Alice's bloodstream causing her death. Since biological mechanisms are contextual, always described in terms of the normal operation of the body, the absences of normally available factors are naturally described as causes. But how can an absence be involved in a mechanism that links cause and effect? Second, mechanisms, as causal explanations, are often described merely qualitatively. This means it can be hard to identify causally relevant properties. Based on a mechanistic explanation, we might be able to identify a number of causes of an effect E. But can we discriminate more important onces, especially when what we need is often a particular level of a particular causally relevant property—i.e. something quantitative? This need is, however, being at least partially met by the increasing turn to quantitative modelling of mechanisms.

These developments make the point of looking for a widely applicable account of mechanism clear. This examination of developments in the literature on mechanisms shows how we have improved our understanding of difficult issues for mechanisms, such as function and organization, to form a core consensus account of mechanism that can help us understand causal explanation in many scientific fields. Understanding mechanisms helps you focus in on how a great deal of scientific work really happens, and how previous scientific work informs new scientific work and allows a framework to help scientists in different fields talk to each other. It is also fit to help us understand how finding mechanisms can help with causal inference too, an issue that is thoroughly discussed in chapter 6.

12.4 What is not a mechanism?

While a widely applicable account of mechanism is useful, it can make it difficult to grasp what is *not* a mechanism. Here we briefly present work by Illari and Williamson (2012), setting out explanations that are not mechanistic explanations. Note that in some of these cases, it is possible that science will develop and eventually yield a mechanistic explanation. The claim is that these kinds of explanations are not yet mechanistic, and may never be. They all miss at least one element of a mechanistic explanation:

(i) Phenomenal description;
(ii) Entities without activities;
(iii) Activities without entities;
(iv) No organization.

Phenomenal description. In cases where there are interesting dynamic behaviours of the system being investigated, a great deal of scientific work can be required to characterize that dynamic behaviour accurately. For example, the Hodgkin and Huxley model of the action potential of neurons (also discussed by Craver (2007)) was a big breakthrough. But what the model gives is a description of the behaviour of neurons; a description that can yield predictions of their behaviour. Important as this is, it is not yet a mechanistic explanation, as Hodgkin and Huxley themselves said. Predictive models are very useful in themselves. But models such as these can also usefully be thought of as a more precise characterization of the phenomenon to be mechanistically explained.

Entities without activities. Darden gives an example of the second case:

The MDC characterization of mechanism points to its operation. Although someone (perhaps Glennan (1996)) might call a stopped clock, for example, a mechanism, I would not. It is a machine, not a mechanism. The MDC characterization views mechanisms as inherently active. In the stopped clock, the entities are in place but not operating, not engaging in time-keeping activities. When appropriate set-up conditions obtain (e.g., winding a spring, installing a battery), then the clock mechanism may operate. (Darden, 2006a, p280–1.)

Remember that nothing is a mechanism as such, only a mechanism for a particular phenomenon. The clock is no longer a mechanism for time-keeping, as the activities required for that are no longer present. However, there will still be some activities happening; heating and cooling, and gravitational forces, for example. So it remains possible that the stopped clock is a mechanism for something else, such as weighing down papers.

Activities without entities. In the third case, sometimes when a scientific field advances, a broad phenomenon is split into more local phenomena. For example, many, many forms of memory have been phenomenally dissociated, such as long-term versus short-term, working memory, episodic and semantic memory, and non-explicit memory such as priming. Discovering these dissociations is important, but this is not yet to give a mechanistic explanation. Some of these forms may—or may not—be activities involved in producing the overall phenomenon of memory. But there are no entities given, and we cannot yet guess at forms of organization. One way of thinking of these kinds of advances is that they attempt to give possible activities that contribute to the production of the overall phenomenon. This can be a useful step towards giving a mechanistic explanation, but it is just one step. For further discussion, see Bechtel (2008).

No organization. There may be no mechanism if there is apparently no organization, the fourth case. The reliable relationship between the weight of a bag of books and the weight of the individual books might be thought of as a mere aggregative relationship because the arrangement of the books does not matter. It is not a mechanism. To take a scientific example, the kinetic theory of gases, which explains both Boyle's Law and Charles' Law, holds that molecules behave, on average, randomly. Does this mean there is no organization? There are two points here. First, note that the assumption that molecules behave on average randomly is an idealization. We put this aside. Second, on the account of organization above, where organization is whatever the entities and activities do and do together to produce the phenomenon, then in this case average random behaviour is the organization by which the entities produce the phenomenon. It may not be the most interesting form of organization, but it is present.

In more usual scientific cases, we can find that we know a good bit about the entities and their activities, but we still don't know what they will do. This happens frequently in, for example, systems studied using omics technologies (genomics, proteomics and so on). We might not know what will happen on the basis of the entities and activities because we don't know yet about the organization. In particular we need quantitative dynamic models of the organization to know what the system overall will do. This is where research is turning. We know a great deal so far, but we cannot give mechanistic explanations of many phenomena.

So the understanding of mechanisms helps us see how some scientific explanations—important as they are—are not mechanistic explanations.

Core ideas

- At heart, mechanistic explanations:
  - describe the phenomenon or phenomena;
  - find the parts of the mechanism (the entities), and describe what the parts do (the activities); and
  - find out and describe the organization of entities and activities by which they produce the phenomenon or phenomena.
- Mechanisms give causal explanations of phenomena, explaining in detail how the phenomenon is produced.
- Some thinkers, such as Glennan, claim that mechanisms exist everywhere there are causes.
- Scientific breakthroughs can be made in better descriptions of the phenomenon, the entities, the activities, and forms of organization of mechanisms. But without all these elements, the explanation given is not (yet) a mechanistic explanation.

Distinctions and warnings

- Mechanisms have functions, as they are mechanisms for particular behaviours. The function of a mechanism without a containing system can be understood as its isolated description. The function of a component in a containing system can be understood as its role-function—the role it plays in producing the phenomenon or phenomena the mechanism produces.
- Context is important to function, and also pragmatically to finding components of mechanisms. We do not usually describe common background factors, such as the presence of oxygen in the bloodstream, as *part* of biological mechanisms, although its absence would affect them.
- Organization can be thought of most generally as whatever relations between the entities and activities discovered produce the phenomenon of interest. Which forms of organization are important to a particular mechanism is an empirical question.
- Applicability is important. We may not find mechanisms in all domains that science is concerned with, and possibly not in all domains where we find causality. But multi-field mechanisms are increasingly important: in some domains we need components of different natures to act in the same mechanism, such as in health economics and social epidemiology. This is where some of the most exciting developments in understanding mechanisms are taking place.

Further reading

Darden (2006b); Illari and Williamson (2012); Reiss (2007); Ruzzene (2012).

CHAPTER 13

Production Accounts: Information

13.1 Examples: tracing transmission of waves and of disease

There is an interesting kind of question that scientists often ask when they are trying to find out whether C causes E. Physicists or electrical engineers might ask: can energy be transmitted between C and E? Can radio waves? Can electrons? Software engineers might want to know whether bits can be transferred quickly enough for a program to run. Doctors might ask: could this virus have been transmitted over such a distance, or so quickly?

These are questions about what can cause what; more precisely, these are questions about *causal linking*. We can rule out possible causes by showing that there's no possible link between cause and effect. As we will show in this chapter, this is reasoning about linking; it is about how cause and effect can—or cannot—be connected, and it seems to be distinct from reasoning about difference-making, which we discuss in chapters 8, 9, and 10. This reasoning is important in daily life, and in science. For example, we might reason: the interference in the sound from the speakers can't be from the TV, it's too far away, or doesn't emit the right frequency of radiation. It must be something else—perhaps the cables to the sound system? Or suppose you just installed your new smart TV, together with the Blu-ray and the home theatre system. You then try out a DVD, and the image appears, but there is no sound. This absence suggests that something has gone wrong with plugging in the cables between the Blu-ray player and the loudspeakers.

Doctors fighting an epidemic can reason in a similar way to decide whether they have two separate outbreaks, or a single virus that has spread to a distinct population. Epidemiologist John Snow famously stopped the cholera epidemic in London in 1854, arguably in this way: by figuring out the 'channels' through which the disease was spreading. To stop such an epidemic it is essential to understand the mode of communication of the disease. This means understanding how a bacterium (or other agent) spreads, and also how the disease is transmitted from person to person. Snow's innovation was to realize that there was water transmission of the disease, at a time when the dominant medical theories suggested only two transmission mechanisms, one by touch (contagion) and one by transmission through the air (miasmas). Snow hypothesized that poverty was at the root of the outbreaks, because of poor hygiene in behaviour and living conditions. The fact that poor people had to eat and defecate in the same place created 'ideal' conditions for the bacterium to spread, because it contaminated water, which was then the mode of communication of disease. Snow thought that this was a plausible explanation, even though the water *looked* very clean, as the agent responsible for the disease was too tiny to be seen with the naked eye. He managed to plot cholera deaths and contaminated water by comparing cholera deaths in different parts of London; it turned out that different water suppliers were active in these neighbourhoods. He managed to convince the authorities to block a suspected water pump and the epidemic gradually stopped (Paneth, 2004). In other words, Snow managed to block exactly what was linking different cholera deaths.

Once we know the route of causal linking, we can also use it to help us predict what might happen next, and also to control problems. For example, public health officials might be able to identify populations at risk, and other populations where a virus or bacterium is unlikely to spread, and so intervene to protect at-risk populations, shutting down means of transmission, such as air travel out of the infected region. This makes knowing about causal linking, a kind of production, useful for causal reasoning, as it impacts on causal inference, prediction and control.

13.2 The path to informational accounts

13.2.1 Information theory and generality

Here we come to a place where philosophy explicitly borrows ideas from another field, in this case, from maths. Philosophical accounts want to use ideas from information theory because information theory is a very general language—and it is a language that was designed to allow us to characterize something like linking. So we shall make a quick detour to explain some key notions from information theory before returning to the philosophical theories.

Information theory is a branch of maths that has influenced many sciences. It is particularly vital to computer science, as it is the language used to describe and measure the information processed by computers and transmitted over the Internet. Without it, we could not say how many bits a computer programe needs, or how fast a connection is needed to download it within a particular time period. The field originated with the work of Claude Shannon (1948). Shannon described an 'information channel' in terms of transmitter, receiver, the signal transmitted and the noise interfering with transmission. This idea is now completely mundane to anyone who uses a mobile phone, but Shannon—and the further work he sparked, allowing us to *quantify* information and describe precisely ideas of information transmission, information loss during transmission, and the idea of noise as something in the information channel that masks the transmitted signal—was essential in getting us to the stage where we could make mobile phones. In particular, Shannon gave us ways of measuring information that are independent of what the information *means*. Thus we can understand information purely formally, as well as understanding it as something that has meaning.

Another crucial move forward in information theory was the development of algorithmic information theory (AIT), based particularly on the work of Andrey Kolmogorov (Kolmogorov, 1965; Kolmogorov, 1983). This allows the formalization of ideas of complexity and compressibility. In its simplest form, the idea is that something, say a car, is more 'complex' than something else, such as a rock, the longer its description *needs* to be. Intuitively, a complete description of a car will be a lot longer than a complete description of a rock—precisely because a car is a more complex kind of thing than a rock. AIT allows us to make these kinds of comparisons because it gives us definitions that are very widely applicable and which can be formulated in comparable ways. For example, anything can be described in a series of yes/no answers. Consider a car: Is it a Honda? Yes. Is it blue? Yes. Is it large? No. This yields a bit string of 110. All computer programmes are basically bit strings. Bit strings are a very, very general way of describing anything, in a comparable format.

This allows us to see how AIT characterizes 'compressibility'. Suppose we add questions to give a complete description of the car. We can then ask whether the resulting string can be compressed—i.e. made shorter. If it can, then a shorter complete description of the car can be given. Only completely random bit strings cannot be compressed; any bit string with a pattern in it can be compressed. Expressed informally, for example, 10101010101010101010 can be compressed as '10 repeated 10 times'. The original bit string can be recreated from the redescription without loss of information. This is the idea that underlies the function of zipping files that ordinary computers can now perform, or behind the compression of long internet addresses into short strips such as <http://www.blq.uh/1q0z3ZU> that you may find in Facebook or Twitter posts. Although there are various ways of choosing the questions to ask to describe something, the resulting measures, especially whether resulting strings can be compressed, are independent of human minds, and so the measures apply easily to the physical world, as well as to the special sciences. When we have the shortest possible bit string giving a complete description of the car—the uncompressible description—we can compare it to the shortest possible binary string description of the rock. If the description of the car is longer, that is what it is for a car to be more complex than a rock.

It is not difficult to imagine that there might be a relationship between Shannon's idea of a channel and causal linking. The relationship between linking and AIT is probably more difficult to imagine. Information theory, including other formal measures of information, is now very important to many sciences, including at least physics, bioinformatics and cognitive science. What makes this possible is that information theory gives us a very general formal framework that can be used to represent and assess *any* kind of process. Anything can be described informationally, from a person to a supernova to a tsunami. The formal framework of information theory ensures that the description, in spite of its unprecendented generality, is not empty or vacuous. Information theory itself is maths, but the maths gives us new ideas, new ways of thinking we didn't have before. With the rise of the computer in its multiple formats, these ideas are no longer alien to everyday life, and they may be useful for thinking about causality.

Other ideas that have their source in informational concepts are also becoming crucial to undestanding developments in the sciences. For example, feedback is a very important idea. A negative feedback loop is a simple control system. One place it is often found is in regulating gene expression by cells. As a gene is transcribed again and again, the level of the resulting protein in the cell rises. If that very protein, or something closely allied to it, then binds to the gene, preventing further transcription, we have a negative feedback loop that acts to keep the level of a particular protein in the cell relatively stable, and allows it to return to its normal level after being diverted from it. This apparently simple idea helps us understand very diverse dynamic systems, as do the various concepts of dynamic systems theory (Katok and Hasselblatt, 1995). We will now examine how informational ideas entered the philosophical literature on causality.

13.2.2 Early views in philosophy: Reichenbach and Salmon

Reichenbach and Salmon were the first to try to express the idea of tracing linking for causality, giving an account of causality as mark-transmission. Their core claim was very simple: a process is causal when, if you mark it at an earlier point, the mark is transmitted to later points in the process (see chapter 11). So, for example, a moving car is a causal process because, if you mark the side of the car at an early point in the process, the mark will be carried along with the moving car, and will be detectable later on. On the other hand, the car's shadow is not itself a causal process, because if you mark the shadow, that mark will not be transmitted, and will not be detectable later.

This view of causal linking is beautifully general, because we can think of so many different kinds of processes as being marked. We can try to alter the signal we think might be interfering with the loudspeakers, and see if the sound they emit changes. We could put floats, or a dye, into a river, and watch to see where the currents take them, to see if the route matches the outbreaks of cholera. The idea of mark transmission applies across many different scientific fields. Indeed, the idea also matches some of the ways we might reason about linking, and try to establish routes of linking.

However, some causal processes, such as those involving fundamental particles, cannot be marked without changing the process, as any change would profoundly alter the process. Some less delicate processes, such as transmission of bacteria, might still be altered by introducing a dye. So causal processes aren't those which actually transmit marks, but those which *would* transmit a mark, if only a mark could be introduced. The counterfactual characterization of mark transmission, presented in detail in Salmon (1984) was criticized by Dowe (1992), which led Salmon to reformulate his theory, now based on processes rather than counterfactuals (see Salmon (1994) and chapter 11). We have seen developments of the original Reichenbach-Salmon idea in chapters 11 and 12, and seen both the advantages and disadvantages of these developed accounts. One disadvantage that is particularly relevant to this chapter is that both accounts lack the very general applicability of the idea of mark transmission. Here, we pursue a different kind of development; the introduction of information theory.

The idea of introducing information theory accords very well with some little-noticed remarks of Salmon. For example, in his 1994 paper, Salmon (1994, p303) comments on his own earlier work:

> It has always been clear that a process is causal if it is capable of transmitting a mark, whether or not it is actually transmitting one. The fact that it has the capacity to transmit a mark is merely a symptom of the fact that it is actually transmitting something else. That other something I described as information, structure, and causal influence (Salmon, 1984, p154–7).

In trying to give an account of causal linking, a major problem is that there are an enormous number of links that we might want to trace, that are of very different types. The examples of causal links that we used above lead us to formulate the question: what do viruses and radio waves have in common? The diversity of worldly causal links is recognized by philosopher Elizabeth Anscombe (1975), who draws our attention to the richness of the causal language we use to describe different kinds of linking, such as pulling, pushing, breaking, binding and so on. It is a real problem to understand what features are shared by cases of causal linking, given how diverse they are.

The views of this chapter all in one way or another hold that the idea of information helps us understand linking. The crude idea is that all these diverse kinds of causal links—energy, radio waves, electrons, viruses and bits—are all forms of information. Put this way, all these scientists are asking a form of the same very general question: Can information be transmitted between C and E? And how? We will also examine how thinking about information *alongside* thinking about mechanisms can help us understand causal linking.

13.2.3 Collier: the first explicitly informational account

John Collier was probably the first philosopher who tried to give an informational account of causality. Since he first published such an account (Collier, 1999), the philosophical community has become much more interested in information, as the philosophy of information is flourishing, including work specifically on causality and information. Collier puts his core idea in very clear terms:

> The basic idea is that causation is the transfer of a particular token of a quantity of information from one state of a system to another. (Collier, 1999, p215.)

Collier fills this out by offering an account of what information is and an account of information transfer. The account of information is given using information theory: the information in any thing is formally, and objectively, defined in terms of algorithmic information theory, as we explain using the example of the car above. Further complexities of Collier's account can be found in his papers, which we set aside here to concentrate on the core idea. This comes in the account of information transfer. It is one thing to describe a static thing informationally, such as a car standing still; another to describe a flow of information, which is something dynamic, something that happens over time, such as a moving car. Collier describes information flow in terms of identity of information:

P is a causal process in system *S* from time t_0 to t_1 iff some particular part of the information of *S* involved in stages of *P* is identical at t_0 and t_1. (Collier, 1999, p222.)

This is refined in more recent work drawing on the idea of information channels (Collier, 2011). In brief, for Collier an information channel is a family of infomorphisms. What Collier means by 'infomorphism' can be understood by supposing you have two systems. Each system consists of a set of objects, and each object has a set of attributes. For example, an object might be a switch, and its possible attributes would be on or off, and another object a bulb, and its attributes also on or off. If knowing the attributes of the first system, the switch, tells you about the attributes of the second system, the bulb, there is an infomorphism. Note that knowing the attributes of the switch might not tell you everything about the state of the bulb, as information might be lost. Nevertheless, for example, in a torch, with the main working components being bulb, battery, switch and case, the information channel is a series of infomorphisms, connecting switch to bulb via case and battery. The state of the switch consistently tells you about the state of the next system, and so on, all the way to the bulb. This allows Collier to give his final view:

P is a causal connection in a system from time t_0 to t_1 if and only if there is an channel between s_0 and s_1 from t_0 to t_1 that preserves some part of the information in the first state. (Collier, 2011, p10–11.)

So for Collier, causal connection is informational connection. In his most recent theory, informational connection is still fundamentally identity, and information flow is given in terms of the identity of information at various stages in the information channel (Collier, 2011, p11–12). This means that, for Collier, the devices we use every day, like mobile phones, are information channels, through which we can track identity of information at the various stages, and it is this that supports the uses we make of them.

This can be spelled out by returning to cases mentioned earlier. Consider Salmon's example of the dented car. The dent in the car is an informational structure, and it will be propagated from the moment of denting to the final resting place of the car. For Collier, though, unlike Salmon, we don't have to think in terms of marks that are introduced, like the dent. For Collier, the car itself is an informational structure, and as it moves, that identical item of information exists at each moment of the process. Information, however, can be lost in an information channel. This is why Collier holds that there is a causal connection so long as some part of the information is preserved in the process over time. Information loss is important to thinking about the transmission of cholera by water. Again, we don't need to introduce a mark, as we can think of the bacteria itself in the sewage system as informational. In this kind of case there will be information loss inherent to the system, as not all of the bacteria will be transmitted from the source to a particular downstream town. Some will die, perhaps be eaten, or be diverted; others will reach different towns. Nevertheless, some part will be transmitted, and so we can construe the sewage system as an information channel. Note that when engaged in causal inference, we will usually think in terms of being able to *detect* the relevant informational structure—the bacterium or the car—only at various points in the route of transmission. However, this is about how we gather evidence of transmission. Collier's idea is that the informational structure exists at every point in the process.[20]

Collier says that a major virtue of his theory is its generality. He has given a view that 'applies to all forms of causation, but requires a specific interpretation of information for each category of substance (assuming there is more than one)' (Collier, 1999, p215–6). So his idea is that he has given a very broad outline, which can also be used to capture differences in different domains, where we can specify kinds of constraints on informational connections, such as constraints we discover empirically about different domains. We will return to this idea shortly. Collier also claims that his view subsumes other theories of causality, most notably the conserved quantities view, simply by interpreting the conserved quantities view as limiting the kind of informational connection we find in its domain of application. But Collier avoids the problem that some physical processes cannot be marked—since they can all be described informationally.

13.2.4 Informational approaches to other issues

While few people have a general informational account of causality, there are various thinkers who are converging on some kind of informational treatment of something closely related to causality. A brief examination of these is interesting, to pull out common motivations. They all seem to be trying to use broadly causal ideas, struggling to express something that requires a generality—a very wide applicability—which seems to be unavailable in other approaches to causality.

The first case of this is two accounts by different structural realists, both of which reach for informational concepts to express their views. Structural realism is a view in the scientific realism debate that says that what is real, what science ultimately tracks through time, is the fundamental structure of the world. It is this structure that is described, for example, in the mathematical expressions that are so important to physical theory. In their theory, Ladyman and Ross (2007) set out an extended attempt to explain how structural realist ideas, originally developed in the philosophy of physics, can actually be extended into the—prima facie very different—special sciences. They are explicit about their reasons for using informational language, and about their influences:

> Special sciences are incorrigibly committed to dynamic propagation of temporally asymmetric influences—or, a stronger version of this idea endorsed by many philosophers, to real causal processes. Reference to transfer of some (in principle) quantitatively measurable information is a highly general way of describing any process. More specifically, it is more general than describing something as a causal process or as an instantiation of a lawlike one: if there are causal processes, then each such process must involve the transfer of information between cause and effect (Reichenbach, 1956; Salmon, 1984; Collier, 1999); and if there are lawlike processes, then each such process must involve the transfer of information between instantiations of the types of processes governed by the law. (Ladyman and Ross, 2007, p210–11.)

---

[20] This has a great deal in common with Salmon's 'at–at' theory of causal influence (Salmon, 1977).

So they are clear that generality is an important reason for using informational language. For Ladyman and Ross, as for Collier, the idea of compressibility is important to their theory, which they call 'information-theoretic structural realism'. Another informational structural realist is Luciano Floridi (2011b), who developed his account quite independently of Ladyman and Ross. Floridi has not developed an account of causality as yet, but his view is that everything can be described informationally (along the lines sketched earlier in the chapter); thus an informational account of causality seems a natural addition to his views. Floridi's motivations are in some ways quite different from those of Ladyman and Ross. He uses informational language in a neo-Kantian effort to describe what we know of the world, with the minimal metaphysical commitments possible. Again, though, it is the generality of informational language, in this case allied to its minimal commitments, that is so attractive.

Ross and Spurrett (2004) are not attempting to give an account of causality itself. They are, however, attempting to understand various problems to do with mental causation, and causation in the special sciences. They therefore need causal ideas that will apply to the special sciences and to the mind. This project forces them to generality, and in the course of that project they explicitly recall both Salmon and Reichenbach:

> How could anyone know, by any amount of observation, which links between processes are causal and which are not? Salmon's answer here is that we can observe something that is precisely diagnostic of causation. That is, we can see that certain processes transmit information about their antecedent stages whereas others do not. Only the former are genuine processes. Following Reichenbach (1957), we can put this in terms of the transmission of marks. [...] We can discover such structures because, as fairly sophisticated information-transducing and processing systems, we can detect, record, and systematically measure mark-transmitting processes. This is a terrifically powerful and, we think, deeply inspiring idea. (Ross and Spurrett, 2004, p618.)

The fact that informational ideas can apply in such diverse sciences is again driving the theoretical turn to information, although here allied to the idea that informational ideas can also make sense of how our brains engage with the world.

Weber and de Vreese (2012) are engaged on a very different project. They are similarly not trying to give an account of causality itself, but they are trying to explore further the arguments of, for example, Menzies (2007), that causal claims are perspective-relative, or contextual. The interesting question they ask is whether this idea of perspective-relativity, usually framed in terms of difference-making approaches to causality, transfers to production approaches, which are usually assumed not to be contextual. To ask this question effectively, they need a production account that has wide applicability. They are forced to hark back to Salmon's older theory to get this kind of generality, and so they use his account in terms of acquired characteristics and mark transmission. Again, broadly informational ideas are used to gain generality, in this case explicitly to ask questions about a production approach to causality.

All these philosophers are trying to address issues that require a better understanding of process or productive causality. They are not trying to address the issue of causal linking, using what is known about possible causal links in causal inference. Nevertheless the nature of the issues they address forces them to turn to a widely applicable understanding of productive causality. It seems clear that they struggle to find an account of causality in the philosophy literature that meets their theoretical needs, which is why they draw on another, not currently much discussed, account. In spite of the fact that the mark transmission account is now quite old, there is clearly still a deep theoretical need for such a concept.

13.3 Integrating the informational and mechanistic approaches

Traditional accounts of production—that we examine in chapters 11, 12, and to some extent 14—do not focus their attention explicitly on linking, and try to address how to conceive of it and how it can help us in causal inference. For example, the primary question of the modern process theory is how to distinguish causal processes from pseudo-processes, while the mechanisms literature is mostly focused on understanding mechanistic explanation. However, in two papers, Illari (2011b) and Illari and Russo (2014) set out to address precisely the problem of reasoning about linking.

Illari (2011b) is interested in how an informational account of causality can be combined with our recent better understanding of mechanisms in order to solve two problems. The first problem is that the informational account has undeniable generality due to its formal properties. Yet, this does not give us understanding of the causal linking in specific domains e.g. biology or psychology. The idea of information, useful enough in characterizing a bare notion of linking, does not obviously relate to the kinds of causal links we actually use in causal inference in the special sciences. Machamer, Darden and Craver provide an example of this kind of objection, although it is phrased as a reaction to Salmon's theory, rather than to an explicitly informational account:

> Although we acknowledge the possibility that Salmon's analysis may be all there is to certain fundamental types of interactions in physics, his analysis is silent as to the character of the productivity in the activities investigated by many other sciences. Mere talk of transmission of a mark or exchange of a conserved quantity does not exhaust what these scientists know about productive activities and about how activities effect regular changes in mechanisms. (Machamer et al., 2000, p7.)

Here, the diversity of activities found in mechanisms in many sciences makes it hard to see what kind of causal connection they could all share. Describing that link informationally allows a very general account, but at the cost of losing rich details of these many different kinds of links—details that are far too useful to discard. What is causality here? How can we answer that question without discarding what we know about the many different kinds of causes? The suggestion of Illari (2011b) is that we can address this problem by relating the informational account to mechanistic explanation in these domains—something we now understand a great deal about, making it useful to relate information to mechanisms. We will return to this point below after examining the second problem.

The second problem is, when scientists look for a causal link, they often speak of looking for a 'mechanism' for the effect. For example, there are many known mechanisms of disease transmission, which spell out how diseases spread. But this raises the question of how we understand mechanisms as causal links. As we saw in chapter 12, mechanisms are activities and entities organized to produce some phenomenon. But this looks like taking a whole, the mechanism, and breaking it up into parts, rather than linking anything. How should we understand such arrangements of parts as linking cause and effect? Harold Kincaid explains the problem:

> In the philosophy of science literature [...] mechanisms are usually thought of as the component processes realizing some higher level capacities [...] I call these 'vertical mechanisms'. However, many requests for mechanisms are about providing intervening or mediating variables between a putative cause and its effect. Call these 'horizontal' mechanisms. (Kincaid, 2011, p73.)

Kincaid uses the terminology of 'vertical' and 'horizontal' mechanisms. These provide, respectively, constitutive explanations (vertical) and aetiological explanations (horizontal). Constitutive explanations consider a system and explain it by invoking the properties that constitute it and their organization. For example, a constitutive or vertical explanation of sunburn would give the mechanisms in the skin that lead to reddening, and pain, then browning, in response to incident sunlight. An aetiological or vertical explanation, instead, considers a system and explains it by invoking the intervening causes (entities and activities) that lead up to some phenomenon. For example, an aetiological explanation of sunburn would give the sun as the source of radiation, and the propagation of the radiation across space to the surface of the skin. So it is not clear how finding a 'vertical' mechanism helps us with causal linking that happens in the 'horizontal' mechanism.

The problem of how to understand information substantively enough for it to become meaningful in the special sciences, and the opposite problem of how to understand causal linking in mechanisms are entangled, and it can be difficult to see any solution to both. Illari (2011b) argues that mechanisms (as characterized by the mechanists discussed in chapter 12) are the channels through which the information flows. On the one hand, this allows us to integrate causality as information flow in the style of Collier with the rich detail of causal relationships we understand from mechanisms. The functional organization of mechanisms structures, or channels, where information can and cannot flow in many sciences. On the other hand, connecting informational causality to mechanisms can allow us to trace the 'horizontal link'—information—across the more familiar 'vertical' or constitutive mechanism. This allows us to ally the resources of our understanding of mechanisms to an information-transmission approach to causality. Note this is in accord with Collier's view. For Collier, causal connection is informational connection, and he notes that information-transmission is always relative to a *channel* of information:

> For example, what you consider noise on your TV might be a signal to a TV repairman. Notice that this does not imply a relativity of information to interests, but that interests can lead to paying attention to different channels. The information in the respective channels is objective in each case, the noise relative to the non-functioning or poorly functioning television channel, and

the noise as a product of a noise producing channel—the problem for the TV repairman is to diagnose the source of the noise via its channel properties. (Collier, 2011, p8.)

Broadly, the mechanism is the thing in the world that we can find, study and describe. But we study it so assiduously because it holds together the conditions for certain kinds of information transmission. In other words, the idea is that when certain kinds of information transmission hold, we will study some structure, the mechanism, that gives us the channel through which the information is transferred. Mobile phones are so amazingly useful because so much of human communication is by voice. Language itself is a means of high-intensity transmission of causal relations, potentially across great distances and long time periods. The information: 'Don't eat that, it's poisonous!' gives you precisely the information you need to avoid death, and is transmitted to you across time and space. Mobile phones can achieve the same thing, right across the globe.

So building up our understanding of mechanisms builds up understanding of information channels—possible, impossible, probable and improbable causal links. This is what we know of the causal structure of the world. We have come to understand many different specific kinds of linking; from radio waves, to hormone signalling in the human brain, to protein receptors on the surface of cancer cells that can be used to signal to the damaged cell to kill itself.

This development is entirely in accord with Collier's idea that there will be multiple ways of constraining the information you find in the empirical domains investigated. The door is now open to categorizing the kinds of information transmission found. In some cases we can even measure them, although much of the time they will be described more informally, as are the many activities in mechanisms. This approach is also in accord with the attempt of Machamer and Bogen (2011) to give an account of productive causality in mechanisms. They argue that pattern replication is not enough, and use the cases of DNA expression and a particular sensory-motor reflex in leeches to argue that we understand the productive causality in these mechanistic explanations only by construing it informationally. The important point here is that the explanations involve seeing how causality is propagated across diverse—distinctly inhomogeneous—systems. Only a very general conception of linking can cope with this kind of causal linking.

Illari and Russo (2014) try to pull together other strands of the causality literature, using exposomics (the science of exposure) as an example. Exposomics will be thoroughly presented and discussed in chapter 24; it will suffice to mention here that this emerging field of research within the health sciences is aiming to push back the frontiers of what we know about disease causation by environmental factors. Exposomics provides useful insights about how reasoning about mechanisms, processes and difference-making complement each other. This has been examined by Russo and Williamson (2012), and Illari and Russo (2014) build on this work. Illari and Russo examine how ideas of causal linking are used in cutting-edge science, particularly when the science is exploring an area with great uncertainty, due to the existence of both known unknowns, and unknown unknowns. Illari and Russo argue that in this case, while known mechanisms are used in study design, too little is known for the possible causal links to be sufficiently illuminated using known mechanisms. Mechanisms can give some coarse-grained connections, but what is sought is considerably more fine-grained linking. Instead of reasoning about mechanisms, the scientists reach for the language of chasing signals in a vast, highly interactive search space. Here, the level of unknowns means that linking mechanisms are generally unavailable. In the discovery phase, and possibly beyond it, scientists also need to conceptualize the linking they are attempting to discover in terms of something that can link profoundly inhomogeneous causal factors. (See more discussion in chapter 24.)

Finally, understanding the relationship between mechanisms and information helps us see why one mechanism supports multiple causes, in both the discovery phase and when much more is known. In the first place, a single mechanism may have more than one function, producing a certain cause effectively. But also if the mechanism malfunctions, it may produce one or a few alternative causes reliably, or cease to produce anything reliably at all.

13.4 Future prospects for an informational account of causality

We have seen how an informational account of causality might be useful to the scientific problem of how we think about, and ultimately trace, causal linking, and so to causal inference and reasoning. An informational account of causality can be useful to help us reconstruct how science builds up understanding of the causal structure of the world. Broadly, we find mechanisms that help us grasp causal linking in a coarse-grained way. Then we can think in terms of causal linking in a more fine-grained way by thinking informationally. An informational account of causality may also give us the prospect of saying what causality *is*, in a way that is not tailored to the description of reality provided by a given discipline. And it carries the advantage over other causal metaphysics that it fares well with the applicability problem for other accounts of production (processes and mechanism). In chapter 24 we will illustrate how the notion of information can help illuminate the scientific practice and conceptualization of causal linking in the emerging field of exposomics research. Here, we relate informational accounts to the objections to other production accounts, discuss possible problems with informational accounts and indicate where future work is needed.

13.4.1 Applicability

Applicability is the prime virtue of the informational account, as might be expected as this is what it has been designed to achieve. Previous accounts that have had a bearing on causal linking are: the Salmon-Dowe theory, focusing on the exchange of conserved quantities; Reichenbach-Salmon mark-transmission; and Glennan's idea that there are causes where there are mechanisms. The informational account is more widely applicable than all three. It does not require the presence of conserved quantities, or the introduction of a mark. It can merge usefully with the mechanistic approach, deepening that account.

The informational account conceives of the causal linking in a way that can be formally defined in terms of computational information theory. But we don't always have to specify the information theoretic structure of a phenomenon. Much of our causal language provides an informal, but meaningful, account for an informational description. This gives the 'bones' of the causal linking, in a way that is applicable to phenomena studied in physics, as well psychology, or economics. So information is a general enough concept to express what diverse kinds of causal links in the sciences have in common.

13.4.2 Absences

The problem of causation by absences has undermined several production accounts. The problem is that everyday language, as well as scientific language, allows absences to be causes or effects. Someone apologizing for missing a meeting might say 'I'm so sorry I wasn't there—my bus didn't turn up.' This intends to claim that the *absence* of the bus caused the person to miss the meeting. Similarly, cerebral hypoxia—i.e., lack of oxygen in the brain—causes brain damage and even death. But how can absences, like missing buses or lack of oxygen, be connected by conserved quantities, or mark transmission, or anything? Absences seem to introduce gaps in any causal connection, that traditional production concepts were unable to account for. Schaffer (2004), for instance, argues that causation by absences shows that causation does not involve a persisting line, or a physical connection.

The solution to this problem that informational accounts offer is entirely novel. Notice, first, that whether or not you think a gap exists depends on what you think the gap is in. There seem to be no gaps in a table, but if you are considering it at an atomic level, well, then there are gaps.[21] Now, information can be transmitted across what, from a brutely physical point of view, might be considered gaps. Suppose the person missing the meeting leaves a message for her boss: 'If I'm not there, it's because my bus didn't turn up.' Then her boss knows about the absence of the bus from her absence at the meeting. Information channels can also involve absences. Recall that a binary string is just a series of 1s and 0s, such as 11010010000001, which can be conveyed as a series of positive signals, and absences of a positive signal. Gaps in information transmission will not be the same as gaps in continuous spacetime. Floridi (2011b, p31) notes that a peculiar aspect of information is that absence may also be *informative*.

However, it is worth noting that this potential is not fulfilled either by the 'at–at' theory of causal transmission of Salmon (1977), nor by the closely allied persistence of the identical item of information through a process view of Collier (1999). According to both these views, brutely physical gaps still interrupt the process, and so seem to break the causal linking, as it is difficult to see how either a mark or an item of information can be continuously transmitted between, say, an absent bus and being late for a meeting. This is in need of future work.

---

[21] If you visit CERN and stop by the shop, you can buy things such as a bracelet with the following printed sentence: 'The silicon in this bracelet contains 99.9% of empty space'. Well... if they say so!

By considering absences, we can see both that information transmission offers a possible novel account of causal connection—causal linking, and also that a novel account is important to solve this rather perplexing problem. An informational account allows greater flexibility, offering the possibility that the kinds of connections that exist in different domains are an empirical discovery, that can be understood as further constraints on kinds of information transmission discovered there.

13.4.3 Vacuity

There is a final problem that arises for the informational account, which is the problem of vacuity. There are so many different ways to describe information. The field of mathematical information theory has flourished since Shannon, so there are even multiple formal measures of information. This is important because it yields the applicability that has eluded previous accounts of causal linking. But it might be a weakness if the account is vacuous; if it doesn't seem to say anything. This might be thought to be the case if there is no one concept of information that is always applied that can be understood as meaning something substantive.

Alternatively, the rich variety of informational concepts available—including quantitative informational measures—can be seen as a huge advantage of the informational approach. There are two points worth noting. First, the formal measures of information available, whatever they apply to, however general, are not vacuous. They are also increasingly connected to information-theoretic methods for causal inference. Second, what is *needed* to make any account of causal linking work is something like a light-touch generality. We need to be able to see causal linking in a way that doesn't obscure the important differences between kinds of causal linking. The informational account offers this; the opportunity to describe—perhaps formally describe—patterns that cannot be described in other ways. Ultimately, the problem of saying something general enough to be widely applicable, while still saying something substantive enough to be meaningful, is going to be a problem for *any* account of production that aims for generality. The challenge is precisely to find a concept that covers the many diverse kinds of causal linking in the world, that nevertheless says something substantive about causality.

In sum, we seem to reason about possible causal linking, and attempt to trace causal links, in many important causal inference tasks in the sciences. Informational approaches to causal production offer a novel approach to conceptualizing causality in a way that meets this task.

Core ideas

- In causal inference, possible causes are ruled in and ruled out using what we know about possible causal links. But what is a causal link?
- An informational account of causality holds that causal linking is informational linking.

- Informational linking can be added to an account of linking by mechanisms, if mechanisms are seen as information channels.
- A very broad notion of information can be used to capture all cases of causal linking, but further constraints could be added to the broad concept to characterize more narrow groups of cases of causal linking, such as in a particular scientific field.

Distinctions and warnings

- If we are to understand reasoning about causal linking in diverse disciplines, we need a very generalized concept of a causal process or causal production.
- Applicability: the major attraction of the informational account is its wide applicability.
- Absences: an informational approach offers an entirely novel solution to the problem of absences, since it offers a genuinely revolutionary conception of linking.
- Vacuity: an informational account may appear vacuous, but it is possible that the variety of informational concepts available will prove to be an advantage, rather than a problem.

Further reading

Salmon (1994); Collier (1999); Illari and Russo (2014); Illari (2011b).

CHAPTER 14

Capacities, Powers, Dispositions

14.1 Examples: systems in physics and biology

In slightly under a century, molecular biology has made stunning progress in revealing the processes and mechanisms happening at the cellular level. Consider for instance the lac (lactose) operon. *Escherichia coli*, commonly known as *E. coli*, has the capacity to metabolize lactose in the absence of glucose. That capacity is triggered by the absence of glucose and presence of lactose. This is what makes the causal claim 'the absence of glucose caused the cell to switch to metabolizing lactose' true. This is a feature of the local system—in this case it is a feature of that cell. Capacities are generally triggered again in the same circumstances, and so we observe regular behaviours of systems with capacities. Generally, cells that have access to lactose but not glucose will switch to metabolizing lactose. This doesn't absolutely always happen. Sometimes a cell will metabolize something else. Alternatively, an *E. coli* cell that has just been blasted at a very high temperature might be too busy dying to do anything else. Nevertheless, we can observe a reasonably general behaviour, and we know a great deal about the mechanism of operation of the lac operon, which helps explain the behaviour, and explain the capacity of the cell to metabolize lactose.

In biology, there are few if any universal exceptionless laws. Instead, there are descriptions and models of particular systems, like a kind of cell, or a particular species of squid. And what applies to one cell, or one species of squid, might not apply to another cell or another species. The language of powers, capacities or dispositions of these local particular systems is very common. What makes causal claims true here? Perhaps it is some property of the local system, like a capacity.

In physics, on the other hand, laws are much more important. If you read physics textbooks or papers, you'll soon notice that there are lots of mathematically expressed laws. These give you functional relationships between variables of interest, such as between force, mass and acceleration. So you may be induced to think that causality has to do with very widely applicable causal laws that can be expressed mathematically like $f = ma$ (force equals mass times acceleration) or $PV = nRT$ (expressing the functional relations between pressure, volume and temperature of an ideal gas).[22] Then maybe it's the existence of these very general laws of nature that make even homely causal claims like 'pushing the trolley made it start to move' true. Perhaps when we find such laws, we find out something about the deep causal structure of the universe. Yet even in physics there have been attempts to move the focus away from laws and on to the properties of systems. For instance, this is how Anjan Chakravartty (2008, p108, emphases added) describes the very simple physical phenomenon expressed in the ideal gas law:

> A volume of gas that comes into contact with a source of heat may expand in virtue of the *disposition* afforded such volumes by *properties* such as temperature and pressure, and by doing so will come into contact with other regions of space. The property instance present in these new regions together with those of the gas will determine how both are further affected and so on.

Chakravartty is right that properties of systems are important to physics, as well as laws. In fact, mechanistic explanation is often used in physics, such as astrophysical mechanisms of star formation and supernovae, discussed in Illari and Williamson (2012), and finding mechanisms certainly involves finding capacities of local systems. While there is no doubt that laws are important to physics, so are capacities of systems.

So now we have at least two scientific disciplines where the language of capacities, powers or dispositions is used. Yet such things are a bit puzzling. We can't directly see a capacity of an object. Further, the capacity of an object might never manifest: an *E. coli* cell might never metabolize lactose. How do we know about such things? What are they? Let us see how this might be understood. We will see that thinking in terms of capacities can structure our causal reasoning, and also affect our practices of causal inference, if capacities are what we are looking for evidence of.

14.2 The core idea of capacities, powers and dispositions

Traditional philosophy of science was very concerned with laws of nature. Many philosophers traditionally held that laws of nature are what make our causal claims true. They also concentrated on 'categorical' properties, that is properties like shape that are always present in objects, and don't need particular circumstances in order to show up. The recent philosophical literature has explored a different path. We have realized that most of the sciences do not use universal, exceptionless laws at all. We have also realized the importance of dispositional properties or capacities in science, such as the

---

[22] Some will object that these are functional laws, having nothing to do with causality. Russell (1913) famously held such view (see also chapter 17).

capacity of *E. coli* cells to metabolize lactose, whether or not they are actually metabolizing lactose. So recent philosophy of science has become much more interested in the idea of capacities, powers or dispositions, and in exploring their relation to causality.

Philosophers exploring these issues have slightly different versions of the view, and the words 'disposition', 'power' and 'capacity' are all used in both the philosophical literature and in science. In most of the discussion of this chapter we will focus on the common core of the view. This is the idea that a capacity–power–disposition (CPD) is a property of a particular thing or system; a potentiality that has effects under certain conditions. First, we introduce the main views very briefly.

Dispositions. Alexander Bird (2007) calls his view 'dispositional essentialism'. He holds that all of the fundamental properties in the world are dispositional. Properties like the *E. coli* cell's disposition to metabolize lactose is dispositional in the sense that cells, in the presence of the stimulus conditions of absence of glucose and presence of lactose, have the disposition to begin metabolizing lactose. Bird holds that this kind of property could be fundamental—fundamental properties are not all categorical, such as the cell's possession of the lac operon. For Bird, dispositional properties account for all laws of nature, including causal laws. Indeed, laws of nature inherit their force, or 'necessity' from the manifesting of dispositions.

Powers. Mumford and Anjum (2011) develop a theory of causation that hinges upon the notion of power. In their view, it is properties of objects or systems that have the power to do the 'causal work'. For instance, a causal model relates events—encoded in variables—such as smoking and cancer development, but if smoking causes cancer it is because of some of its powerful properties, for instance the power of tar and other chemicals in the cigarette to damage the cells of the lungs. Powers like this are what make our causal claims true. Indeed, many powers theorists point out that as powers are the causally interactive properties of things, they are the *only* properties we can observe and detect with scientific instruments. Traditional categorical properties of objects such as shape are observable—apart from anything else they have the power to produce perceptual images on our eyeballs. But this is not so for molecules. For example, figuring out the shape of proteins is a fiendishly complex process, involving advanced technology and extensive data processing—i.e. we *infer* the shape of proteins from their causally interactive properties. Chakravartty (2008), defends the view that what we find out about in science, what we can be realist about, are ultimately powers. He argues that causal processes are basically agglomerations of powers, extended through time, and it is causal processes that 'realize' causal relations.

Capacities. Nancy Cartwright, since the publication of her influential book, has been defending the view that causal claims, in particular causal laws, are really about capacities (Cartwright 1989). Cartwright argues that this is even true of causal laws in physics, such as Newton's second law of motion. The law states that the acceleration of a body is directly proportional to the net force acting on the body (in the same direction as the acceleration), and inversely proportional to the mass of the body. For Cartwright, this law expresses the *capacity* of a body to move another body. Note that for Cartwright, it is *things* that have capacities, whereas Bird, Mumford, Anjum and Chakravartty focus on properties themselves. We will see that Cartwright is still applying this work today, arguing that often what we seek evidence of in science is a capacity claim, which she sometimes also calls a 'tendency' claim. For example, she says that randomized trials (see chapter 2) test drug efficacy, that is the *capacity* or tendency of drugs to relieve or cure symptoms, in a specified population. An important task in science is precisely to find out about these capacities, to express them in a quantifiable way, for instance through laws, and to find out where else they hold.

The literature on powers and dispositions has a long tradition, that we cannot cover here, descending from Aristotle (see for example Marmodoro (2013) or Armstrong et al. (1996) for a discussion). We have prioritized brief explanations of the core concerns of some of the major philosophers still working on these views right now. You can follow up the details of these views in the references we give. For the rest of this chapter we will focus on the core idea of capacities–powers–dispositions, to show how it is of interest to science. We will talk of 'capacities' as that is a friendlier term than 'CPD', but much of what we say here applies to all of the CPD views.

Capacities have intuitive appeal because we make claims about capacities that range from the completely mundane to important discoveries of science. Alice has the *capacity* to pick up a glass of water, even when she doesn't actually do it. *E. coli* cells have the *capacity* to metabolize lactose in the absence of glucose, which is crucial to their survival if their supply of glucose is interrupted. They retain this capacity even when it is not triggered—when there is plenty of glucose available.

What these authors all share is that the idea that capacities are what make capacity claims, and some general claims in science, true. For authors in the CPD tradition, capacities are real and local. We will look at these two claims in turn.

Reality. First, consider the claim that capacities are real. Capacity claims can seem obscure, when considered in a particular way. Capacities or powers have been thought to be mysterious occult Aristotelian things clearly antithetical to modern science. After all, how could a capacity be *real*? Alice can pick up a glass of water, and when she does, this can be clearly observed. But where do we observe her *capacity* to pick up the glass? When she is just sitting doing something else, her capacity is not obvious. When an *E. coli* cell is busily metabolizing glucose, where is its capacity to metabolize lactose? Until it does metabolize lactose, what is there to observe? Similarly, we make counterfactual claims. If Alice had tried to pick up the glass of water, she would have managed it. If we had removed the glucose from the cell's surroundings, and given it lactose, it would have begun to metabolize lactose. Why do we think we know something that doesn't systematically appear?

One good reason for worrying about what capacities are, and understanding reasoning about them in science, is to answer this possible problem. The claims above seem perfectly sensible, and it would be a shame to eschew them if there is no need. Cartwright in particular is interested in how we get evidence of capacities—evidence for claims about capacities, if you like.

Locality. The second feature is the locality of capacities. The question of locality, and how biological causes and explanations work locally, is pretty important. This is not just for biology. We build mechanistic explanations in some places in physics, and those can be local and fragile (i.e. they may cease to work in slightly different contexts) just as in biology. A relatively slight change in the initial conditions for a dying star will yield different types of supernova, for example, one of which involves blowing apart completely, while the other involves collapsing to a neutron star. For example, an important difference is whether the core mass of the star reaches the Chandrasekhar mass, which is approximately 1.2–1.4 solar masses. This is very large, but it is a property of that star.

It is important to realise that in many such domains, scientists are not seeking laws. They are certainly seeking generality—repeatable patterns—wherever they can be found. There are different types of supernovae, capturing certain kinds of regularities, and that is useful to find. But often extensive generality cannot be found. In these cases finding more local and fragile patterns is pretty useful.

For example, the lac operon, which only operates in the less-than-ideal circumstances where glucose is absent, is vital to the survival of the *E. coli* cell. Its operation is fragile. It cannot exist without the cell machinery, which maintains conditions within the cell, creating all the things the cell needs and expelling waste products. Significantly disrupt these processes, and cells cease to function. Nevertheless, we understand how the lac operon functions, and most of the time it functions remarkably reliably, in its normal context. And it is just as well for *E. coli* that it does! It is responsible for the survival of large numbers of cells when their supply of glucose is temporarily interrupted. So the local fragile systems we study are pretty important.

So the idea that local dispositional properties are real and important to science and to making regularities in science true is an interesting one. Note that this idea of 'making regularities true' is a metaphysical kind of making (see also chapter 21). The idea is that it is the thing having the capacity, that is the thing in the world, the existence of which means that the claims we make are true. That is, it is many *E. coli* cells really having the capacity to metabolize lactose that makes the general regularities—describing their general behaviours—true.

This kind of metaphysical making may not immediately be of interest to science. It has little to do with prediction, explanation, control, inference or reasoning—at least, it has little to do *directly* with prediction, explanation, control, inference or reasoning. However, we will see that the idea of capacities and their relation to causality can help us sharpen up our reasoning, especially when discussed in connection with evidence. In this way, it might also help us indirectly with explanation, and possibly inference too.

14.3 Capacities in science: explanation and evidence

The theories of Cartwright, Bird, Chakravarrty, Mumford and Anjum all have multiple aims, and involve complex philosophical arguments. Here, we discuss the most critical points, relevant to causality in the sciences.

14.3.1 Explanation

A common way of presenting the importance of capacities is in their relevance for explanation. We see regular behaviour in nature, but is there anything beyond the mere regularity itself, something that explains the regularity? For those holding CPD views, it is important that capacities give us something in the world beyond mere regularity. They give something as the basis of our ability to build scientific knowledge—to build explanations, including explanations of the surface regularities we observe more directly.

This is what we try to do when we try to build mechanistic explanations. The parts of mechanisms have capacities, and part of building mechanistic explanations is in seeing how putting parts with certain capacities together builds explanations of further regularities and capacities, as in the case of the lac operon. This is why evidence of capacities is important and then feeds forward into further evidence of capacities. If a cell has the lac operon mechanism, then that is reason to believe it has the capacity to metabolize lactose even if it *never does* metabolize lactose.

An objection that has been raised to the role of capacities in explanation is known as the 'dormitive virtue' objection. The idea is: why do sleeping pills cause sleepiness? Well, because they have a capacity to induce sleepiness, a 'dormitive virtue'—which is just to say they cause sleepiness because they have a sleep-making property! If this is all that using capacities in explanation amounts to, then that's no good. But of course there is much more than that to the claim about the lac operon, and to Alice's capacity to pick up a glass of water. In making these capacity claims, other mechanisms and processes are engendered. For instance, if we say that various substances in tobacco smoke have the capacity to cause cancer, there are scientifically interesting questions about the *other* mechanisms and processes that are triggered by contact with these substances in the body. These might include cleaning mechanisms in the lungs, first alteration in cell structure, apoptosis when the cell cannot repair itself any more, etc.

In sum, capacities can appear occult and mysterious, but they seem to have some place in our scientific explanations.

14.3.2 Evidence of capacities

With some exceptions (Cartwright, 1989), much less attention is paid to how thinking about capacities can help us think about evidence in science. The ubiquity of capacity language in science, and the idea that capacities might have something to do with explanation of surface regularities, should help show that capacities are of some interest to science. But probably the best way of seeing that we should take capacities seriously, even if they appear occult and unobservable at first sight, is to consider what could count as *evidence* for capacity claims—both mundane and scientific.

What is the evidence for Alice's capacity to lift a glass of water? Well, surely, the many occasions when she actually does it! We might also count as evidence occasions when she does something relevantly similar, such as lift a glass of orange juice—people who can lift a glass of orange juice can also lift glasses of other similar fluids.

Scientific capacity claims such as the capacity to metabolize lactose are also very interesting. Naturally, we can study both cells that are actually metabolizing glucose, and also those that are actually metabolizing lactose, and this is evidence that these kinds of cells have the capacity to do either. But the cells' capacity to metabolize lactose is particularly interesting, as this is a case of a capacity that has a *mechanistic explanation*. The *E. coli* cell has mechanisms for metabolizing glucose, and this is what cells with available glucose usually do. But those mechanisms also change rapidly if glucose is absent. So evidence for all these lower-level mechanisms counts as evidence that the cell, even while cheerfully metabolizing glucose, retains the capacity to metabolize lactose. If a cell is currently metabolizing glucose, but it possesses all these mechanisms, and nothing stops them from operating, then we can be pretty confident that it has the capacity to metabolize lactose.

Even in physics, we do not always look for universal exceptionless laws. In many areas, we also build mechanistic explanations. For example, we give mechanistic explanations of supernovae—the explosions of exhausted stars that create all the heaviest elements in the universe. We can thus have evidence that stars have the capacity to explode long before they actually do.

Notice that we also make perfectly sensible negative capacity claims. Alice doesn't have the capacity to fly to the moon, for example, at least, not without the aid of a pretty sophisticated spaceship, and many other people. This is a claim we can be pretty sure about! This is so, even though saying what counts as evidence for such a claim is tricky. Nevertheless, it is clearly an empirical fact—merely one with a complex evidence base. Evidence for it includes things like our evidence for how much energy it takes to get a physical body from the earth to the moon, knowledge that a normal human being has no such source of energy, and evidence that Alice is a normal human being. It also includes our evidence for our knowledge of the physiological needs of a human body, such as air at a reasonable pressure, and the damage a body would suffer under the kinds of stress required to get to the moon. So this is a complex, but still scientific, claim.

This means that evidence for capacity claims can perfectly sensibly be given, even if the evidence is varied. In chapter 6 we discuss the multifaceted nature of evidence for causal relations. We can observe actual things in the actual world that give us excellent reasons to believe that things, like human beings and cells, have capacities even when those capacities are not currently being expressed. Negative claims have an interesting function, here. If an actual object is relevantly similar to another actual object actually exhibiting some capacity, and nothing exists to prevent the object exhibiting that capacity, then we take this as evidence that the object has the capacity that it is not exhibiting.

We can return to the question of what makes claims like this true. If there are universal exceptionless laws of nature then they are an obvious candidate. But there aren't any relevant to Alice's behaviour, or the lac operon. Instead, it seems to be properties of the local system that are relevant. So as an account of the truthmakers of causal claims, capacities seem very sensible. They are also good for supporting counterfactual claims. We know the cell would have shifted to metabolizing lactose if we had taken away all the surrounding glucose, because it has the capacity to do so. We can know this by looking to see if it has the mechanism for the lac operon.

So an important way in which capacities are of interest to the sciences is precisely to reconsider how much of the scientific enterprise involves searching for evidence of capacities. For example, capacities are a large part of what we know about chemistry.

14.3.3 Masking

We now have the idea that reasoning about capacities is interesting to reasoning about causes, particularly to our reasoning about causal explanation and evidence for causes, and thereby also to causal inference. There is a great deal more to say about evidence for capacities—and so for local, fragile (i.e. easy to disrupt with small changes of context) causal claims. In this way answering the truthmaker problem is really of interest in helping our causal reasoning—in this case our reasoning about evidence. We have argued already that we can give evidence for capacities, and this shows capacities are not mysterious or occult. With the exception of Cartwright, other advocates of capacities, powers or dispositions views say comparatively little about how we find out about capacities in science, but we do reason about them, look for them and find evidence of them in science.

One of the major problems of evidence is how the locality and fragility of capacities make it difficult to get evidence of them. Two interesting problems that affect this were recognized a long time ago in the philosophical literature on dispositions: the problems of masks and finks (see in particular Martin (1994) and Bird (1998)).

A capacity is masked if another, opposite, capacity hides it. Take a very simple example: suppose drinking Coke makes you put on weight, due to its high sugar content. However, Coke also has a high caffeine content, and we know that caffeine makes you move around more, which of course burns energy—which tends to make you lose weight. Coke seems to have both capacities, and we don't know which one will 'win'—which one will mask the other—until we investigate further. In this case, the weight-gain capacity probably masks the weight-loss capacity, as there is much more sugar than a little extra movement will burn. So overall, Coke has the capacity to make you gain weight. But this is due to the masking of one capacity by another.

In the case of masking, both capacities actually act, but the weaker one is drowned out, hiding it. But it is possible to discern both. For example, the sugar in Coke will be digested, and eventually get to the body's cells, and so will the caffeine. Finking is a variant of masking where the masking capacity actually 'prevents'—finks—the operation of the masked capacity. A simple example is a residual-current circuit breaker (RCCB). This device is designed to disconnect a domestic power circuit if the current in the two conductors (live and neutral) becomes unequal. This will happen if someone touches a live wire and some current is conducted through their body to earth. The circuit breaker is designed to react fast enough to prevent injury. If it works correctly, the wire never does carry enough current to harm a person. The RCCB is supposed to prevent—fink—the wire from ever exhibiting the potentially harmful capacity.

Although these examples are simple, many naturally occurring systems have similar features. Biological systems have extensive backup systems, such as the lac operon, which kicks in when the usual metabolizing of glucose is impossible. There are multiple mechanisms to repair your DNA if it is damaged—or if it is copied incorrectly. These features are essential for you biologically, but they make it hard to find evidence of capacities.

This is most famously understood in gene knock-out experiments, which often fail! The idea of a gene knock-out experiment is simple: we think we have found the gene responsible for a certain capacity of the cell, or the organism, such as the lac operon. We try to confirm this by knocking out the gene, and observing that the organism or the cell no longer has the capacity—perhaps it can no longer metabolize lactose. Unfortunately for the evidence, a lot of the time the organism and the cell retain the capacity! What has happened is that some other mechanism, previously unactivated, has activated to yield the same or a very closely similar capacity. These are biological masks and finks. These ideas, coming from the philosophical literature, can help us sharpen up our language when talking about searching for evidence in science.

14.3.4 Interactions

Obviously masks and finks make finding evidence a pain. Instead of simple experimental confirmation that we have found the gene normally responsible for a particular capacity, we must try to build up an understanding of the operation of multiple systems so that we also understand the way they interact—how one is a backup that activates when the first one fails. Third and fourth backups are not uncommon, particularly in the most vital biological mechanisms, such as DNA copying and repair mechanisms. This means that the contextual operation of capacities is vital to causal inference in science.

The core reason for this problem is that capacities interact in their production of effects. Human cells have the capacity to metabolize glucose, and make DNA—and to do a thousand other things! They also have many emergency situations. For example, if the cell's DNA is damaged, the normal cell cycles are halted to allow DNA repair mechanisms to operate. There are many reasons why we might not observe the effects we expect—because all these capacities interact in the overall functioning of the cell. The human myocyte or muscle cell has the capacity to contract, allowing people to move, but it needs the physiological machinery of the body to allow it to do so. Muscle cells generally need nerve cells telling them when to contract, and the body to keep them supplied with calcium ions, potassium ions and glucose. We will not usually observe contraction without all this support.

We focus now on this interaction aspect, which is very important. What we are looking for is evidence of capacities, and what we are reasoning about in building at least some explanations, is local and fragile capacities that interact. The local and fragile aspect of their interactions make it very difficult to find evidence for them. Explanations are contextual, as the phenomenon will only be produced in the right context, when the interactions of different capacities come out the right way.

It is not always clear that a capacity is a capacity of a particular object, exactly, rather than of a particular object in a particular system. Take one of Cartwright's favourite examples: the capacity of aspirin to relieve headache. You have to do something rather specific with the aspirin for it to relieve headache. You have to swallow it for it to work, and you have to swallow enough of it—but not too much of it. So it's worth wondering whether it is the aspirin, all by itself, that has the capacity to relieve headache, or it is a capacity of the aspirin, when swallowed by a person with a headache in a sensible dosage.

This makes the holism of Chakravartty (2008) appealing. For Chakravartty, powers are a feature of a *system*, not of a particular object. The muscle cell, then, has the capacity to contract when in the body, where it is supposed to be, directed by nerve cells, and when that body is functioning well and so doing all the things it usually does to support its muscle cells. The cell on its own doesn't generally have the capacity to contract. Indeed, a solitary cell will usually die. Solitary cells have very few capacities.

If it is the function of reasoning about capacities and looking for evidence of them to recover something relatively stable from the local, fragile and patchwork nature of biological systems, then perhaps thinking of them as properties of objects *in a system*, or even of the system itself, rather than as isolated objects, is a promising way to integrate capacity reasoning into scientific practice. Claims about capacities are often said to hold 'ceteris paribus', which means that background context is taken into account (see for instance Steel (2008, ch6) or Reutlinger et al. (2011) for a discussion). This form of holism will also help decide where to look for evidence of such capacities. It is also interesting in terms of what we are explaining—i.e. a phenomenon maintained only in a particular context, where it is important to understand features of that context, and how they can impede the phenomenon if they fail to hold.

Ultimately, capacities do lots of interesting work for us. In a philosophical analysis, they are what make causal claims true. In scientific practice, reasoning about capacities helps us ask and answer many other interesting questions, such as what evidence supports a capacity claim or what context allows or inhibits a given capacity, and so on. Some of their advocates do seem to want more from them. For example, it seems that Bird (2007) wants his account of dispositions to yield an account of causality, as Mumford and Anjum (2011) demand of their account of powers.

We have shown how CPDs can help us understand things about causal reasoning, causal explanation and evidence for causality, and that they complement other existing accounts of production, e.g. mechanisms or information. We discuss this complementarity further in chapter 24, where we develop the idea of a causal *mosaic*.

Core ideas

- Capacities–Powers–Dispositions (CPDs) are local truthmakers of causal claims.
- The idea of truthmakers answers the philosophical question of what—metaphysically—makes causal claims true.
- The idea of CPDs as truthmakers might help us sharpen up our causal reasoning; to be more precise about what we're reasoning about and how we get evidence of it.
- CPDs can also help us think about causal explanation, as the things which underlie and explain the surface regularities.

Distinctions and warnings

- CPDs can be masked, when an opposite CPD acts and hides it, as in the Coke case. They can also be finked, when another CPD prevents the first from operating at all, as when RCCBs disconnect to break the circuit and prevent injury.
- Masking and finking create problems for finding evidence of capacities, as does the highly interactive nature of many capacities.
- CPDs, note, aren't accounts of causality in the same way as most of the other accounts in this book, but are accounts of what grounds causal claims—what makes causal claims true; what the truthmakers are.
- The 'dormitive virtue' objection to CPDs as explainers is the following: it is no explanation to offer for why a drug makes people sleepy that it has a 'dormitive virtue' i.e. a sleepy-making property! Scientific practice seeks to offer more than this, such as the mechanistic explanation for the capacity of cells to metabolize lactose.

Further reading

Choi and Fara (2012); Kistler and Gnassounou (2007); Chakravartty (2008, chs2, 3).

CHAPTER 15
Regularity

15.1 Examples: natural and social regularities

Why do we think there is causality in the world at all? Well, the world isn't completely random, with unpredictable chaotic events just endlessly occurring, and nothing repeated, nothing stable. Things frequently surprise us, but we also come to expect a lot of things. We have examples of regular behaviour in the natural world. The earth keeps turning and the sun rises every morning, fairly reliably. Astronomy can describe the movement of planets and celestial bodies with great precision, how these occur regularly in time, and therefore predict future events, such as the appearance of comets or astronomical alignments.

Regular behaviours also characterize many social phenomena. Some people are grumpy every morning, while others manage an inexplicable cheerfulness; and mothers like flowers on Mothers' Day. These are simple observations that any of us can make in everyday life. But social scientists assume—and in a sense rely on—social regularities in order to study society (Goldthorpe, 2001). Consider studies on immigration (see e.g. Parsons and Smeeding (2006)). Immigration has similar characteristics across different populations and times. In Italy this is a relatively new phenomenon as immigration from the Maghreb and sub-Saharan Africa started only in the nineties, whereas in countries like France or Belgium it started much earlier. Many aspects are different, but many are the same, and this helps social scientists in drawing their conclusions.

The observation of regularities in our daily lives as well as in the natural and social world may make us believe that this is what causality is: repetition of regular patterns. We might think causality is the source of the patterns in the world that we observe, which creates the patterns we expect to persist, and which we rely on in planning our lives. There is some truth in this idea, and some form of regularity is important to our practices of causal inference, prediction and causal reasoning. But the story is also much more complex.

15.2 Causality as regular patterns

David Hume is the most famous proponent of the 'regularity theory' of causation, which means that the regularity theory was one of our earliest modern theories of causation. A whole book would not be enough to present Hume's thought properly, including the multiple interpretations given by Hume scholars. In modern times, Hume is one of the most cited historical philosophers working on causality, as he took pains to discuss the problem of causality thoroughly.

Hume's most important concern was: How do we acquire knowledge? As one of the great empiricists, he held that we gain knowledge of matters of fact from the world. Knowledge gained from experience, from interactions with the world, is called 'empirical' knowledge. Hume was interested in sense data, our perceptions, the ideas formed in response to our perceptions, and how we make connections between ideas. Causality is a problem because it seems to go beyond what we simply observe. We think that some of the things we observe are connected to other things that we observe. How? The principle of causality allows us to order ideas, believing that certain external things are causally connected, and that allows us to have expectations about the future. How do we get this idea from the world?

Hume's answer comes in two parts: a story about what we observe, and a story about what goes on 'in our heads' in response to these observations. Today, we would frame this in terms of the *evidence* we gather about a putative causal relation, and the way we *reason* about that evidence. Note, however, that Hume never discussed evidence and considered passive observations of the surrounding world. Hume says that we observe a regular succession of things that are constantly conjoined, and these are the things we come to think of as causally connected. In the *Enquiry Into Human Understanding* he writes:

[...] similar objects are always conjoined with similar. [...] [A cause is] an object followed by another, and whose appearance always conveys the thought to that other (Hume, 1777, Section VII, Part II, Paragraph 60).

A stock example in the literature is the following: we observe instances of smoke regularly following instances of fire, therefore we tend to infer that fire causes smoke. On a billiard table, instances of collision between balls precede changes in direction, so we come to think that collisions cause changes in direction.

Strictly speaking, this regularity, this constant conjunction, is all Hume thinks we actually observe. But it cannot be the full story. The principle of causality allows us to infer that fire causes smoke and that we should *expect* smoke next time there is fire, and collisions cause changes in direction, and this is what we should expect the next time we see a collision. But this principle, this expectation, is not something we *observe* in the external world. So the question becomes: what grounds such expectations? Causality, argues Hume, cannot be grounded in logical necessity—this comes from his criticism of rationalist philosophies. He believes that the only access to causal relations is empirical, through observed regular successions of events. How can Hume account for causality then?

Hume says that custom or habit leads the mind to form expectations, on the basis of observing regularity. We then observe our own new expectations. This is what leads us to the idea that regularly associated objects, or events, are causally connected. Note that this makes an aspect of causality *mind-dependent*. The principle of causality that helps us order what we observe arises by custom or habit in our own minds. It is in this sense that Laudisa (1999) reconstructs the history of causality as a 'model to shape knowledge', or, borrowing a medieval formula, *scientia est cognitio per causa* (knowledge is understanding via the causes). But a consequence of this is that an important part of causation is 'in our heads' rather than 'in the world', and in this sense causality is mind-dependent: whatever we deem causal is *causal* because of some cognitive processes of the epistemic agent on the material observed. We discuss in chapters 17 and 18 how this feature has also motivated the development of other accounts.

We should immediately defuse a worry, though. To say that causality is mind-dependent does not imply denial of the existence of the world or the empirical basis of our knowledge. Don't forget, Hume was an empiricist, and would never have denied this!

This gives the core of Hume's account, and we will not attempt to give a historical exegesis of his ideas. Historical exegesis of Hume is far beyond the scope and goals of this book. (If you are interested in Hume scholarship good places to begin are Stroud (1981) and Blackburn (2008).) But we will try to show that the core idea of regularity in our observations is of interest to science. It shows how utterly fundamental our search for causes is in our deepening understanding of the world; both scientific and everyday.

Hume's thought has been very influential, and a number of other accounts are based in one way or another on his view. One is J. L. Mackie's that we discuss in chapter 4. In fact, Mackie's starting point is Hume, as he builds on a Humean view of causation, developing a more refined account. The counterfactual account developed by David Lewis, and also discussed in chapter 9, is another account based on Hume. Lewis is explicit: another famous quote from Hume's *Enquiry* states that a cause is

[. . .] an object followed by another, and where all the objects similar to the first are followed by objects similar to the second. Or in other words where, if the first object had not been, the second never had existed. The appearance of a cause always conveys the mind, by a customary transition, to the idea of the effect. (Hume, 1777, Section VII, Part II, paragraph 60.)

Lewis took the second definition as quite unlike the first, and this incited him to develop his counterfactual account. Finally, Helen Beebee's work in chapter 18 springs from Hume as projectivist, namely from his considerations about what to expect based on the observations of regularities.

Regularism has been defended again in the recent philosophy of causality. For instance, Stathis Psillos is interested not just in reconstructing Hume's view, but also in drawing some conclusions about the metaphysics of causality (Psillos, 2002; Psillos, 2009). Psillos is also interested in the idea that we only observe regular succession in the world. We do not observe any relation that we are entitled to call causation. In other words, causation is not a production relation (which we discuss in chapters 11, 12 and 13). Nevertheless, Psillos still says something about the metaphysics of causation. He critically engages with the metaphysical question of causality, namely what kind of thing causality is. Baumgartner (2008), to give another example, offers a reassessment of a 'regularity theoretic' account and argues that much criticism of regularism is ill-posed because it deals just with oversimplified cases and does not use the proper tool—that, Baumgartner suggests, is extensional standard logic.

15.3 Updating regularity for current science

According to Hume's view of causality, as presented here, what actually happens in the world is regular succession. We mentioned simple examples above, such as smoke regularly following fire, or change of direction following collision of billiard balls. However, there are objections to the view which we now discuss.

15.3.1 What Humean regularity can't achieve

In the light of our current scientific knowledge, and our understanding of scientific method, the Humean view of regular succession does not provide a faithful understanding of causality in science. This is not a criticism of Hume per se, but rather an assessment of what his philosophical view can or cannot achieve in the light of the challenges of science now. For instance, we widely agree now that there can be causality without regular succession, although in Hume's time, cases of indeterministic or probabilistic causality had not been discussed yet. Consider a stock example in the recent philosophical literature: smoking causes lung cancer. The noxious effects of smoking are important and widely known, which leads many people to alter their behaviour. But many smokers do not get lung cancer! While the causal relation between smoking and cancer is not contested, its 'regularity' is not univocally established. There can also be regular succession without causality. Night invariably follows day, but day does not cause night. So if what is meant in Humean regularity is 'same cause; same effect', we now understand that causes are just more complicated than the account allows.

Because of these complications, the idea of 'same cause; same effect' does not suit many medical contexts. In fact, epidemiologists have been working on multifactorial models of disease since the seventies (see for instance causal 'pie charts' discussed in chapter 4). Multiple causes of the same effect are not uncommon in everyday contexts either. We see in chapter 4 that Mackie provides an account of *complex* regularities, where there are several causes acting at the same time, a number of background conditions and possibly negative causal factors as well. That regularity needs to be modified to cope with increased complexity is also clear from probabilistic accounts discussed in chapter 8, as many causes we are interested in do not invariably follow their effects, such as the smoking and lung cancer case.

Further, regularity seems to tie causality to the *generic* case (see chapter 5). We are of course interested in 'regular' causal relations, so that we can establish that 'smoking causes lung cancer' in some population of reference. But we also worry about whether Alice smoking caused Alice's lung cancer. This is a single-case instantiation of a generic causal relation. If causality is regularity, then Alice smoking causing Alice's lung cancer, as a unique case, cannot be causal. Or, at best, it is parasitic on the general case. Conversely, if we give primacy to single-case causal relations, then regularities at the generic level may just be generalizations, whose *causal* character is then undermined. These types of problems are further discussed in chapter 5.

This means that it is hard to establish convincingly that regular succession exhausts what causality *is*, i.e. that it gives an account of the metaphysics of causality. But regularity is still important to how we *find out about* causes; the epistemology and methodology of causality. We make progress on this question by shifting the focus of the discussion to the evidence that supports claims about causality (see also chapter 6) and by asking what role regularity plays in this business of evidence.

15.3.2 Where to go?

Let's finish by discussing regularity in the terminology of current scientific contexts, for instance those of probabilistic causality (see chapter 8) or quantitative causal methods (see chapter 7). When we say that causes raise the chances of their effects ($P(E|C) > P(E)$), or that effects are functions of their causes plus errors ($Y = f(X) + \epsilon$), we claim that there is a (statistical) dependence between the cause and the effect. What such a statistical dependence tells us is that the cause-variable is correlated with the effect-variables, and the stronger the correlation, the higher our confidence usually is in its causal character (if other conditions obtain, of course—see chapters 2 and 7).

Rephrased in these terms, regularity is about frequency of instantiation in time and/or space of the same causal relation. In other words, in the data set analysed, cause- and effect-variables will show a *robust dependence*. This is the idea that regularity conveys. And this provides an important constraint on the (statistical) relationships between variables. There are also important presuppositions underlying this notion of regularity. One is that our sample is big enough to be able to detect *significant* dependencies—we know that statistics works better with large, rather than small, numbers. Another is that data have been collected with sufficient precision and accuracy, such that 'nonsense' correlations can be excluded.

So it seems that we need to adjust the regularity view, adapting it to the language of current science. Quantitative social scientists or epidemiologists will be familiar with the idea that robust dependencies capture the (frequent) occurrence of a causal relation in a data set.

Why do we have the intuition that regularity is important? Well, it is no accident that an 'adjusted' sense of regularity is at the core of multiple theories of causality: see for instance necessary and sufficient components in chapter 4, probabilistic theories in chapter 8 and variational reasoning in chapter 16. The idea of some kind of pattern, even if it does not amount to the old 'same cause, same effect' idea, is still very important to our idea of causality. Regular behaviour of some kind is also important to establish *generic* causal relations, or to help causal assessment in the single case using regular behaviour as one of the relevant pieces of evidence (see chapter 5). While the regularity theory does not exhaust the task of a causal theory, its core idea, updated, is an important component in many theories.

Core ideas

- Hume is concerned with how we get causal knowledge. His answer is that we connect causes and effects in our mind because we observe, by experience, that instances of E regularly follow instances of C.
- Humean regularity has been and still is a very influential theory of causation, as it sparked several other important accounts, for instance counterfactuals, INUS, etc.
- Current causal methods urge us to rethink what regularity may mean in the quantitative analysis of large data sets.

Distinctions and warnings

- Regularity can be interpreted as a metaphysical thesis about causation or as a component of a causal epistemology and methodology.
- The notion of regularity has great intuitive appeal, but it is difficult to nail down what regularity really means in specific scientific contexts.

Further reading

Hitchcock (2011, sec1); Psillos (2002, chs 1, 2); Psillos (2009).

CHAPTER 16
Variation

16.1 Example: mother's education and child survival

Demographer John C. Caldwell was a pioneer in studying the relations between education of women and child survival in developing countries. He analysed rural and urban areas in Nigeria. Previous studies emphasized the influence of sanitary, medical, and social and political factors on mortality. In analysing the impact of public health services, Caldwell noted that many socio-economic factors provide little explanation of differing mortality rates, whereas mother's education turned out to be of surprising importance. Similar evidence had also been gathered from surveys run in Greater Bombay (1966), Ghana (1965–6), Upper Volta (1969) and Niger (1970). Caldwell's analysis was essentially based on two surveys. The first (May–June 1973) collected data on about 6,606 women aged 15–59 in the city of Ibadan (Nigeria). The second (June–July 1973) collected data on about 1,499 Yoruba women aged 17 or over in Nigeria's Western and Lagos States. Caldwell analysed child mortality by levels of maternal education, in order to see whether or not the causal hypothesis—that mother's education is a significant determinant of child mortality and must be examined as an important force on its own—is correct.

Data was then organized into tables. One table showed, for instance, the proportion of children who died by levels of maternal age (15–19, 20–24, ..., 45–49) and by different levels of maternal education (no schooling, primary schooling only, at least some secondary schooling). The proportion of children who died was always higher in categories of less educated women. This doesn't change significantly when other factors, e.g. socio-economic status, are taken into account, reinforcing the hypothesis that maternal education has significant causal effectiveness on its own. Table 4 in his paper, in fact, shows only those characteristics that might explain the relationship between child mortality and maternal education—specifically mother's place of residence, husband's occupation and education, type of marriage (monogamy or polygamy)—and how child mortality varies for each of them. Caldwell comments on Table 4 (Caldwell, 1979, p405, emphasis added):

The figures in Table 4 demonstrate that, at least in terms of child mortality, a woman's education is a good deal more important than even her most immediate environment. If any one of these environmental influences had wholly explained child mortality, and if female education had merely been a proxy for them, the CM [child mortality] index would not have *varied* with maternal education in that line of the table. This is clearly far from being the case.

Education, argues Caldwell (1979, p409), mainly serves two roles: it increases skills and knowledge, as well as the ability to deal with new ideas, and it provides a vehicle for the import of a different culture. Educated women are less fatalistic about illness, and adopt many of the alternatives in childcare and therapeutics that become available. Educated women are more likely to be listened to by doctors and nurses. Also, in West African households, education of women greatly changes the traditional balance of familial relationships with significant effect on childcare.

The way to detect causal links between maternal education and child survival lies, in Caldwell's approach, in an analysis of variations between variables of interest. In this chapter, we will see that reasoning about variation is a common feature of causal reasoning across the sciences. It impacts most obviously on our practices of causal inference, but also in our predictions, and efforts to control, particularly to control social science causes.

16.2 The idea of variation

In this chapter we investigate the question of finding a scheme of reasoning—a *rationale*—for causality. It will be drawn from our causal *epistemology*, which asks different questions than do causal metaphysics or conceptual analysis (see chapter 22). A scheme of causal reasoning can be extracted from different causal methods (see chapter 7). Finding a core rationale will help deepen our understanding of these more diverse methods, by capturing how we can gain causal knowledge, if at all. Once this is achieved, this core rationale can inform our reasoning about causality in other ways, helping us reason from causal knowledge we already have, to prediction and control. We can make the starting-question of this chapter more precise: when we reason about cause–effect relationships, what notion or notions guide this reasoning? Is this notion 'regularity'? 'Invariance'? 'Manipulation'? Or 'variation'? Or is it some combination of all of these? The question can be asked for experimental and non-experimental methods alike. Non-experimental methods are things like observational studies where we can't experimentally alter the possible cause of interest (see chapter 2). For example, we are not legally allowed to force people to smoke to see whether they get lung cancer, and we cannot turn a woman into a man to see if her pay improves. But we can observe populations of men and women, smokers and non-smokers—and mothers with infants, as above—and see what *actually* happens in those populations. We try to see whether some things are different (the effects) when other things are in place (the causes). Philosophers and methodologists examined in this chapter hold the view that the logic behind diverse methods in observational contexts, such as the design of experiments or model building and model testing, relies on the same rationale.

16.2.1 Variation for pioneering methodologists

There is a long tradition of methodologists thinking about how to gain causal knowledge. We could go back as far as Francis Bacon, Galileo Galilei or Isaac Newton. In modern times, John Stuart Mill systematized scientific methodology, giving an account of how to make causal inferences. Mill's methods are largely but not solely about experimental contexts. They are also about what we need to *observe* in order to establish what causes what. We will now briefly present Mill's four methods. While you read, ask yourself: what is it that Mill wants us to track? Regularities? Invariant relationships? Variations? What?

Mill's notion of cause is presented from within an experimental approach, but we shall see that Mill's methods are not limited to experimental settings. In the *System Of Logic*, experimental inquiry is seen as the solution to the problem of finding the best process for ascertaining what phenomena are related to each other as causes and effects. Causal analysis, says Mill, requires following the Baconian rule of *varying the circumstances*, and for this purpose we may have recourse to observation and experiment.

Millian experimental inquiry is composed of four methods. Suppose you have a simple recipe to bake a cake. You have used it successfully for years, but all of a sudden the cake does not rise properly. You may want to try Mill's methods to find out what's wrong with it. This is how one of the authors (Federica) used Mill's methods.

(i) Method of Agreement: comparing different instances in which the phenomenon occurs.
   The last three times I baked the cake, it didn't rise. What are the common factors in these cases? Same yeast? Same oven temperature? Same ingredients?

(ii) Method of Difference: comparing instances in which the phenomenon does occur with similar instances in which it does not.
   I used to bake the same cake, and it rose nicely. In Canterbury, I had a ventilated oven. The yeast used in Canterbury was different from the one in Brussels. I can't imagine standard flour or sugar would make a difference.

(iii) Method of Residues: subducting from any given phenomenon all the portions which can be assigned to known causes, the remainder will be the effect of the antecedents which had been overlooked or of which the effect was as yet an unknown quantity.
   I searched the internet. Using a static or a ventilated oven can make a difference to how cakes rise. So, using a static oven I should bake the cake longer. No other differences seem to matter, since I made sure I didn't use self-raising flour.

(iv) Method of Concomitant Variation: this method is particularly useful when none of the previous methods (Agreement, Difference, Residues) is able to detect a variation of circumstances. For instance, in the presence of permanent causes or indestructible natural agents that are impossible either to exclude or to isolate, we can neither hinder them from being present nor contrive that they shall be present alone. But in such a case a comparison between concomitant variations will enable us to detect the causes.

I could experiment on most of the putative causal factors for properly baking a cake. However, I also investigated a larger set of instances by asking my sister, who also bakes with the same recipe, about her baking experiences. I could not intervene on the factors she mentioned, but I could record concomitant variations of the rising of the cake with other factors: Do variations in the type of yeast lead to different results in rising? Does the type of oven? Did she notice other factors that may have a causal role?

You might want to know whether Mill's methods led to any useful result. They did. It seems the problem was using a certain food processor which has blades that can cut when they turn clockwise, or just mix when they turn anti-clockwise. A chemist friend of mine hypothesized that blades may break yeast molecules, thus affecting cake rising. (Notice that this is a mechanism-based explanation, see chapter 12; notice also that such an explanation uses both evidence of difference-making and of mechanisms, see chapter 6.) I also tried the stability of this explanation by using a traditional mixer, which does not cut, and also using the original food processor, but setting the blades to turn anti-clockwise. (Notice, this amounts to testing invariance under intervention, see chapter 10.) And the cake rose properly again!

The fourth method is the one that most prominently makes use of the notion of variation, as Mill uses the actual word. But the other methods, considered carefully, are also about variations. In comparing instances in the Method of Agreement and of Difference, we look precisely at things that vary, and at things that stay the same. Causal reasoning in Mill's methods recommends that we do more than simply observe the effect (failed rising) following the cause (to be determined!). The four methods tell us to look for how things vary or stay the same, when other factors change or stay the same. Whether these observed variations are regular—and what that means—is a further question, and we address it later in this chapter and also in chapter 15.

Mill says that the four methods are particularly well suited to the natural sciences and that the experimental method is inapplicable in the social sciences (Mill, 1843, Book VI, chs VI–VII). Yet he also specifies that between observation and experiment there is no logical distinction but only a practical one (Mill, 1843, Book III, ch VII). We could read this as a statement about a *unified* logic underlying causal methods. The rationale of variation (see also later) goes precisely in this direction. However, the problem of causality in observational studies is not overtly addressed in Mill's work, as he does not tell us what alternative method, in line with his four methods, would be suitable for observational contexts. So Mill just *states* that there is logical unity and does not address the difference in practice. We explain how theorizers of variations fill in the details in the rest of the chapter.

It is at this stage that the work of another pioneering methodologist comes to the rescue. Emile Durkheim argues against the view that the four methods cannot be applied in the social sciences (Durkheim, 1912, ch VI). According to Durkheim the comparative method is the only way to make sociology scientific. However, not all of the Millian four methods will do. Only the Millian Method of Concomitant Variations will, for the obvious reason that in the observational contexts typical of social science we cannot really change the putative factors. Durkheim claims that the method of concomitant variations is fruitfully used in sociology. Witness Durkheim (1912, 160–1) (Our translation and emphasis.²³):

There is only one way of proving that a phenomenon is a cause of another, that is comparing cases where they are simultaneously present or absent, and looking whether the *variations* they exhibit in these different combinations of circumstances prove that one depends on the other.

Durkheim has a point. Recall the type of studies briefly presented in chapter 2. They are about comparing situations, trying to establish what varies with what, and what makes things vary.

16.2.2 Current variation accounts

Federica Russo (2009) offers an account of causality as variation. Note that Russo's account is epistemological. Russo offers the account not as an account of the nature of causality, or as an account of what our causal language means—although the account may have implications for those questions. The account is meant to make precise the rationale that captures our epistemology of causality, which of course impacts directly on our design of methods to find out about causes. Russo intends the domain of application to be in the first instance the social sciences, but it is an interesting question whether the account extends beyond the social sciences.

Russo starts by noticing that we design, and should design, causal methods in the social sciences to uncover *variations*. Simply put, we do this by examining, in a data set, correlations between our target variable (the effect, such as child mortality) and possible causes (such as maternal education). Clearly, only some variations are of real interest to us, as correlation does not imply causation. Background knowledge and preliminary analyses of data will suggest what variations are worth further examination. For instance, we will typically consider those (joint) variations that are regular (see chapter 15) and invariant (see chapter 10). These are the variations that are strong enough and that stick around when some things change. We find these roughly by searching for correlations which remain across chosen partitions of the population being analysed. Also, these are the variations that are likely to be of some use to us in setting up policy. We find out which variations are regular and invariant by building and validating models for the population of reference. But there are many other important tests that help establish which variations are causal, for instance goodness of fit or exogeneity, as discussed in chapter 7.

The rationale of variation implies a certain holism in causal inference. You often find a bunch of causes together by *validating* your model, which almost always will include more than one possible effect variable, and you need to think about the different stages of model building and model testing in this process.

---

[23] In the original: "Nous n'avons qu'un moyen de démontrer qu'un phénomène est cause d'un autre, c'est de comparer les cas où ils sont simultanément présents ou absents et de chercher si les variations qu'ils présentent dans ces différentes combinaisons de circonstances témoignent que l'un dépend de l'autre".

16.3 Variation in observational and experimental methods

The rationale of finding variation certainly fits some work in the social sciences, as theorized in Russo (2009). The idea of variation is present in other authors' work, such as Gerring (2005), although they are not as explicit as Russo. We will explain further by looking more closely at current observational and experimental methods and an interesting question will be whether the rationale of variation also applies to experimental contexts.

We focus initially on statistical models used for causal analysis in contexts where experimental alterations of the putative cause-variables are not performed, for ethical or practical reasons. This is typical in the social sciences such as in the maternal education studies, but it also happens in astronomy where, for example, we typically cannot alter the orbit of planets. We discuss the features of these statistical models in greater detail in chapter 7. Here, it suffices to recall that many statistical models can be expressed in what scientists call the 'reduced form'. It is 'reduced' because it condenses a whole complicated statistical model into a seemingly simple formula: $Y = \beta X + \epsilon$. This formula says that the effect Y is a function of the causes X plus some errors ϵ due to imperfect measurement or to other chancy elements in the causal relation.[24] (There is of course a vast literature on the meaning of the errors, which we do not need to enter here. For a discussion, see e.g. Fennell (2011).) Once we read the equation causally, we also state that it is not symmetric, which means it cannot be re-expressed algebraically in terms of X instead of Y. That is, $X = \frac{Y-\epsilon}{\beta}$ does not have a causal interpretation. Another important point to note at this stage is that while we can use the same 'form' for different models, what we plug in for Y and X is an empirical matter and the whole model-building phase is devoted to establishing precisely which variables have to be included, to estimating the parameter β, to checking whether the assumption of non-correlation between the ϵ and the Xs holds, etc.

There are three possible readings of the reduced form: variational, manipulationist, and counterfactual.

Variational reading. Variations in the putative cause X are accompanied by variations in the putative effect Y. How much Y varies in response to the variation in X is quantified by the parameter β. Standard tests (goodness of fit, exogeneity, invariance, etc.) will help decide whether and to what extent these joint variations are chancy or causal. Notice that this reading makes sense whether or not X has been experimentally altered. In non-experimental social science, joint variations between X and Y are often 'passively' observed in the given data set, with no experimental alterations at all.

Manipulationist reading. From this basic variational reading we can derive the manipulationist one. In experimental contexts, experimental alterations or *manipulations* on

---

[24] Actually, this formula says that the function is linear. Many causal relations are not linear, so an even more general reduced form would be $Y = f(X) + \epsilon$, which says that Y is *some* function of the causes X plus the errors. But for simplicity, let's stick to the linear form.

X make X vary, such that Y varies accordingly. (On manipulation, see chapter 10.) Thus, in a controlled experiment, joint variations in X and Y are due to 'provoked' manipulations on the putative cause-variable X, unlike in observational studies.

Counterfactual reading. The counterfactual reading is also derived from the basic variational one. Here, the equation is read as saying that *were* we to change X, Y *would* accordingly change. On this reading, joint variations between X and Y are not simply observed in the data set (non-experimental contexts), nor are they observed in response to performed manipulations (experimental contexts). They are instead hypothesized, assuming that the equation faithfully represents the causal relation between X and Y.[25]

Variation, here, is the most basic reading of the reduced-form equation, showing the common core that remains in more sophisticated readings. Model building and model testing are guided by the idea of observing and performing tests on joint *variation* between variables of interest. Preliminary analyses of data and background knowledge will provide some information about what variables jointly vary, as in Caldwell's maternal education case discussed at the beginning of the chapter. Model testing then helps us establish whether these joint variations between variables of interest are causal rather than spurious or accidental. For instance, it is important to check for exogeneity and regularity of occurrence, the most important test being, admittedly, invariance (see chapters 7 and 10).

We will now consider experimental methods more explicitly. Even in experimental contexts, our *reasoning* about causal epistemology often turns around the notions of 'change' or 'variation'. Take gene knock-out experiments. Suppose we have sequenced a gene, a coding sequence of DNA, so we have some idea what polypeptide chain it allows the cell to produce. But we may remain unsure about the function of the resulting protein. By manipulating the gene, knocking it out (often using something which binds to it to prevent it from being transcribed), we provoke, induce, produce *changes* and then we observe whatever consequences ensue. In the case of genes we want to know whether the cell or the organism changes in its functioning. Does it die because

---

[25] These counterfactuals share the same intuition behind Lewis's counterfactuals (Lewis, 1983; Lewis, 2000), namely 'had the cause not occurred, the effect would not have occurred either', and the basic property of not being back-tracking (see chapter 9). But they are different in some important respects. One is that Lewis's counterfactuals are sentences in the conditional form, the antecedent of which is known to be false. These counterfactuals are usually single-case: 'Had I taken an aspirin half an hour ago, my headache would have gone by now'; 'Had Bob watered my plant, it would have survived', etc. On the contrary, 'statistical' counterfactuals are still *models* and aim to establish some generic facts about causes and effects of, for instance, a treatment and recovery. Testing invariance under the counterfactual reading of the structural equation is far from being straightforward. Sometimes we miss the relevant reference classes or even the plausible ones, which generates the indictment of not being based on empirical data (see for instance Dawid (2001)). Some have even come to the conclusion that the results of counterfactual interventions are parasitic on the existence of laws of Nature, which justifies why it is the case that, for instance, Ohm's law turns out to be invariant under counterfactual interventions (see for instance Psillos (2004)).

it is unable to produce some kind of vital protein? Why? Is it merely abnormal in some way? What way? Or does nothing very much happen at all? Note that we look for joint variation: in cells without the gene, does something similar happen that doesn't happen to cells with the gene? The difference for experimental studies is that we provoke the changes, rather than simply observing them; but the core reasoning about variation is the same.

Thus, this account spells out an *epistemological* difference-making account. Variation does not tell you that all there *is* to the nature of causation is difference-making. It says that the way you *find out* about causal relations, the rationale by which you design methods, is tracking difference-making, and then substantiating the correlations you find with other sorts of arguments, such as mechanisms, exogeneity and invariance, in ways that depend on the scientific context—see also chapter 22.

This idea accords well with other views of causality that are concerned with our causal epistemology and methodology. For instance, in probabilistic approaches, we compare the unconditional (or marginal) probabilities of the effect and its probability conditionally on the cause. This means that the cause makes the probability of the effect *vary*. Alternatively, manipulationist theorists require that *changes* in the cause-variable yield *changes* in the effect-variable; so manipulationists must agree that what we track, at the most basic level, are variations. Counterfactuals establish causal links by reasoning about what would have been *different*, had the cause not been present; so, again, counterfactual approaches recommend that we reason about variations.

It is worth comparing the variation account in a bit more detail with some closely related accounts.

Regularity. Variation is probably most closely allied to the regularity account (see chapter 15). To see the difference, the variation account pays attention to what we have to check for regular properties, namely joint variations between variables of interest. Such joint variations must show some regularity, in the sense that the joint variations show up 'often enough' in the studied data set. In statistical modelling, this is often expressed with the concept of 'strength of association'. Regularity can be in time, for instance when we analyse time series, but regularity can also concern factors other than time. For instance, the joint variation between cigarette smoking and lung cancer is remarkably regular, in the sense that the association has been found to be strong in most populations.

Manipulation. The division of labour between the idea of variation and the idea of experimental alteration or manipulation is important. In the context of this chapter, manipulation is a *tool* to establish what causes what, that can be used in experimental contexts. In performing manipulations, we make the putative cause-variable vary and observe what variations occur in the putative effect-variables. We do not do this in observational contexts, where we cannot use this tool. Nevertheless, we still look for variation—a variation in the effect due to a variation provoked in the cause (holding fixed anything else). So, a rationale of variation underpins manipulations, just as it does the methodologies we employ in observational contexts (see also chapter 10).

Invariance. Invariance is important, as not just any joint variation between variables counts as causal. Russo's account aims to be very broad. For example, it is broader in scope than 'manipulationist' invariance. In observational contexts, where we cannot intervene, invariance is not tested against ideal or actual manipulations, but across chosen partitions of the population being analysed. The way in which the partitions are chosen very much depends on our background knowledge and on the data actually available. For instance, in the demography example discussed above, it is important to partition the population according to relevant socio-economic factors, including appropriate stratification of the variable 'maternal education'. The major advantage of this view is that it helps you see how invariance, a very useful idea when it comes to causality, applies to domains like the social sciences, and lots of issues in the biomedical sciences, where intervening in order to establish causal knowledge is not possible. Non-manipulationist invariance is therefore particularly useful when we have to establish causal knowledge that underpins and justifies policy programmes.[26] In order to set up a policy programme, in fact, we should aim to act on a causal relation that is as invariant as possible. Invariance thus helps us establish causal relations that are valid for the population as a whole; that is we are interested in establishing generic causal claims (see also chapter 5 and Russo, 2014).

To understand the import of the rationale of variation, consider the following objection. What if there *aren't* variations? This is an important point, which leads to a refinement of the rationale of variation. Before checking for meaningful joint variations between variables, we also need to check for variation *within* variables. To explain this idea, let's consider Giere's account of probabilistic causation (Giere, 1979).

Giere thinks that when we consider a causal claim of the type 'smoking causes lung cancer' in a real population, we have to make a comparison with two hypothetical populations. One hypothetical population is one in which *nobody* smokes and consequently lung cancer rates are (hypothetically) lower; the other hypothetical population is one in which *everybody* smokes and consequently lung cancer rates are (hypothetically) much higher. Giere thinks we can grasp the meaning of a (probabilistic) causal claim by comparing what would happen to the two hypothetical populations above. In these two hypothetical populations, notice, there is no 'internal variation'. But now think: suppose you are an epidemiologist and receive data about a population where everyone (or nobody) smokes. What can you infer about the relation between smoking and lung cancer? If the population was really so homogeneous with respect to the smoking factor (in either direction), how would you detect an effect on lung cancer? In *practice*, that would be impossible, as we need to compare different instances of the putative cause with different instances of the putative effect (remember Mill's method). It is true, however, that counterfactual reasoning about the hypothetical population can help isolate hypotheses in a heuristic way (see chapter 9).

This is not just speculation, but a real practical problem. Here is an example. Norris and Inglehart (2003) examine an extensive body of empirical evidence relating to the debate on the 'clash of civilization thesis' (Huntington, 1996). According to

---

[26] Needless to say, if we can gather evidence from experimental studies, this will make, ceteris paribus, causal conclusions more reliable, and policy programmes too.

this thesis, the 'line' between Western Christian countries and Muslim and Orthodox countries lies in democracy and in the role people give to religious leadership. Norris and Inglehart, however, put forward the idea that the clash lies somewhere *else*, namely in gender equality and sexual liberalization. How can they prove that? Would it be enough to examine *just* Muslim countries? The answer is no. To establish whether and to what extent gender equality and sexual liberalization make a difference to culture and democracy, they have to compare Muslim and non-Muslim countries. In fact, they analyse data from 75 countries, of which nine are predominantly Muslim. If there is no variation within the cause-variable, within the effect-variable, and then jointly between the two, there isn't much we can infer from data. Of course, this is not to say that variation suffices to establish causation! Chapter 7 discusses other characteristics that joint variations must have, such as exogeneity or invariance, and chapter 6 discusses the different pieces of evidence needed to establish a causal relation.

The main purpose of the account of variation, as the core rationale unifying our causal epistemology, is that it is worth holding on to the common concept at the core of all of these. Manipulationism, regularity and counterfactual interpretations of the reduced form equations are all layers of sophistication that are laid on over this common core. The concept of invariant variations is a relatively simple, but important, one.

Note that the variation account is quite specific in scope (see also chapter 22 on narrowing down the scope of causal questions). It aims to say something about causal epistemology, and it aims initially to apply to observational contexts in the social sciences. We have shown that it also offers an interesting core rationale that extends to experimental contexts in social sciences, and even into the other life sciences. Although it may not directly impact on other questions about the metaphysics of causality (unlike many of the other accounts we discuss), it sheds light on aspects related to causal methodology (see chapter 7) and the role of evidence of difference-making in establishing causal claims (see chapter 6).

Note that it can be a virtue of an account not to overreach and attempt to solve any and every problem! Adopting the rationale of variation does not require you to throw away other causal notions. Once you realize that causal reasoning tracks variations, it does not follow that you reject manipulations as a valid tool for *testing* causal relations. Manipulations remain an important testing tool, in addition to a way of *acting* on causal relations. Indeed, you should intervene when you can. Interventionist experiments are frequently very useful. But they are also often impossible. And in many social science or medical cases, what you mean by saying mother's education is a cause—say, of child survival—is clearly not that if only we could go back in time and intervene to change a mother's own childhood education, then her child would be more likely to survive. One meaning is that, if mother's education influences child survival, we could base a policy programme on women's education and change child survival. Whether this is a plausible guess, however, depends on what actually happens, and for this we need to study variations between education of women and child survival.

It is our view that the different causal notions discussed in this part of the book all have their place and role within the mosaic of causality. This idea is thoroughly developed in chapter 24. Here, the important idea to retain is that we need to give the appropriate answer to the question being asked. The importance of the rationale of variation is to remind us that, amongst the complex alternative methods discussed in chapters 2 and 7, there is a common structure of reasoning about causal epistemology.

This should help us with the scientific problems, particularly causal inference and structuring our reasoning concerning what evidence can be gathered in particular domains, and what that evidence can justify in terms of prediction and control.

Core ideas

- To establish causal relations we track what varies with what, either by observing changes, or by provoking them in experiments.
- The notion of variation underpins causal reasoning in experimental and observational methods alike.
- Pioneer methodologists such as Mill and Durkheim described their methods in a way that is permeated with 'variational reasoning'.

Distinctions and warnings

- 'Variation' spells out one aspect of difference-making discussed earlier in the chapters about difference-making accounts, namely that to establish what causes what we track variations (whether they are observed probabilistic relationships, due to manipulations, etc.).
- The rationale of variation is not in conflict with manipulationism, when manipulations are construed as methods to establish what causes what and the rationale of variation underpins them.
- The question of what epistemology underpins causal reasoning is compatible with pluralistic positions about, for example, evidence. Mechanisms are also useful to causal inference (see chapter 6).
- It is important to distinguish the scope of different questions (on which see also chapter 22):
  - Defining causality / causation or establishing truth conditions for causal claims are metaphysical / conceptual questions.
  - Establishing what evidence supports causal claims lies between epistemology and methodology.
  - The rationale of variation answers the epistemological question of how we reason causally.
  - Manipulations and tests such as invariance are tools to establish causal relations and fall in the remit of methodology.

Further reading

Durkheim (1912, ch6); Russo (2009, ch4); Wilson (2014, sec5).

CHAPTER 17

Causality and Action

17.1 Example: symmetry in physics; asymmetry in agency

Before we do any scientific work at all, we experience causality as agents. Our actions, as well as our intentions, reasons or volitions, do bring about effects—we all know this. Is there causality without human agents? Can we understand causality without human agency? This might sound mad. But some philosophers have denied something like this, and many people have thought that human agency is vital to scientific knowledge. We will explore this idea in this chapter.

It seems that the fundamental laws of the universe—or at least some of them currently described in physics—can be written as dynamic laws, which are just laws that relate what happens to variables in the laws over time. A peculiarity of these laws is that they are 'time symmetric'. This means that they still describe the evolution of a system correctly, even if we 'displace' the phenomenon in time. One important displacement is time reversal. For example, the relation between force on a body such as a car, the mass of the body and the resultant acceleration of the body is commonly written as $f = ma$. But this equation just gives a functional relationship between force, mass and acceleration, in an equation that can be written with time flowing either forward or backwards. This may seem counterintuitive, but physics provides the justification that this type of time reversal makes sense for the *local* description of the movement of a body, not for the global description of a system (for instance, we cannot make entropy decrease by reversing time in the relevant equations).

There are also other ways in which we can rewrite $f = ma$: $m = \frac{f}{a}$, and $a = \frac{f}{m}$. Let us abstract from the description of motion in time. This equation describes the functional relation between three quantities: f, m and a. Is this functional relation *also* telling us something causal? What is the cause? Is it the force, the mass or the acceleration? Intuitively, the cause of a car accelerating is the force exerted by the engine, when the accelerator pedal is pressed. But from the equation, we can work out what will happen if each of force, mass and acceleration is altered. This is because we can start with values for each, and work forward—or backward—in time. If the body is decelerating, for example, we might conclude that a force *was* applied against the direction of travel.

Yet human beings seem to experience causality as going in one direction only—forward in time. We can affect the future, but the past is fixed. Once Alice pushes the pedal, the car accelerates. There is no way Alice can change the velocity of the car, affecting whether she already pushed the pedal in the past. This is standardly expressed by saying that causality is asymmetric. This is debated, but it seems uncontroversial that causality is at least asymmetric a lot of the time. The question is: is this asymmetry a feature of the world itself, or only of human experience of the world? Indeed, why do human beings have a concept of causality *at all* if the fundamental laws are symmetrical? Perhaps we have evolved it to allow us to navigate around the world to let us have effective agency. So the idea is that the laws are symmetric, but causal asymmetry is recovered once agents act.

Physicists really worry about this problem about where the arrows of time and of causality come from. Indeed, the symmetry of laws and how to evade it—'symmetry breaking'—has been very important in recent physics (see Brading and Castellani (2013) for discussion). Philosophers, and some physicists, have also worried about the place of causality in physics. Here we examine a tradition in philosophy of holding that the arrow of causality comes from human agency, and this is what makes causality and time go—or perhaps seem to us to go—in the same direction.

In addition to this debate, agency theories are also important because they shifted our thinking about how we find out about causal relations—profoundly shifted it. A tradition originating with Hume, discussed in chapter 15, held that the problem of finding causal relations arose from our observation of the world. It took the world as distinct from us, and us as trying to work out causal relationships from *observed* regularities. We still do this; many observational studies are very important, but this is not our *only* means of finding out about causal relations. We also mess with the world! From birth, to maturing into advanced scientists, people poke at the world and see what happens. In this way, agency theories are important forerunners of theories discussed elsewhere in this book, particularly in chapters 7 and 10, although these theories do not describe themselves as 'agency theories' in order to distance themselves from an anthropocentric notion of causation.

17.2 Early agency theorists

The idea of an agency theory of causality has a lengthy pedigree. We will briefly present the views of Collingwood, von Wright and Gasking. While reading, you will easily see similarities with the manipulationist view discussed in chapter 10. This is true, and manipulationist theorists acknowledge a conceptual debt to, or at least an inspiration from, these agency theorists.

Collingwood (1940) first held that some types of causation involve agency as an essential element, but he didn't produce a full agency theory. Collingwood (1940, p285ff) distinguishes three senses of cause: (i) historical, (ii) practical and (iii) theoretical. The *historical* sense refers to cases in which the cause and the effect are human activities.

The two other senses have less to do with agency, though. The *practical* sense refers to cases in which natural events are considered from the human point of view, namely when the cause C is under human control and it is used to control the effect E, as for instance in medicine or engineering. Finally, the *theoretical* sense refers to cases in which events are considered theoretically, i.e. as things that happen independently of the human will, but not independently of each other, and causation designates exactly this physical dependence of events.

For Collingwood, sense one is the earliest of the three, sense two is a development from the first and sense three is a development of the second. Collingwood's reply to an objection also suggests that these different notions of causation are not easily reducible to each other:

'What you have distinguished are not three senses of the word 'cause', but three types of case to any one of which that word is appropriate, the sense of which it is used being constant.' But, I shall try to show, if you ask what exactly you mean by the word on each type of occasion, you will get three different answers. (Collingwood, 1940, p288.)

This opens the interesting question of whether causality has one meaning, or several different ones; whether there are different things in the world that correspond to 'causation', etc. We address this issue in chapters 18 and 23.

Much later, von Wright (1971) was also interested in the connection between agency and causality, but he was more concerned with how we find out about causal relations than with an attempt to say what causality itself is. In *Explanation and Understanding* von Wright develops at length the idea that by manipulating causes we bring about effects (von Wright, 1971, secII). However, in the preface, von Wright also explains that his book has grown out of his interest in the philosophy of action. Von Wright agrees that the notion of cause he is interested in is tied to the idea of action and, translated into a scientific context, this ties causation to experimentation too. Yet, he is not sure that action or agency is 'what is involved in the interpretation offered by many theoretical physicists of their mathematical formalism' (von Wright, 1971, p36).

It is instead in *Causality and Determinism* that von Wright (1975) argues more powerfully for a link between causation and manipulation, although he points out that such an analysis does not fit the case of human and social sciences. He says:

Independently of our stand on the metaphysical questions, it will be readily admitted, I think, that the idea of experimentalist or manipulative causation has important applications in the natural sciences—and also that its applicability becomes debatable when we move to the human (including the social) sciences. If we wish to identify causation as such with manipulative causation, we could then say that the category of causation is primarily at home in the (experimental) natural sciences and basically alien to the human sciences. If again we do not wish to make this identification we may instead distinguish types of causation and say that causation in the natural sciences is primarily of the manipulative type, whereas in the human sciences another type (or other types) of causation are prominent. (von Wright, 1975, p58.)

Von Wright is suggesting that we cannot give a reductive account of causation in terms of manipulation. Granted, his account of causation in experimental and natural science somehow relies on his account of causality for human actions, but this does not imply that manipulation will do in every circumstance. Again, this raises an interesting question concerning pluralism about causation (see also chapter 23).

In the fifties, Gasking (1955) produced the first full-blown agency theory of causation. His view was that we have a causal relation between C and E when agents have a *recipe* for bringing about E by bringing about C. This is just to say that we make E happen by doing C. Gasking held that this is true even though many events which we take to be causally connected do not come about by an agent actually producing C. Note that Gasking's theory introduces asymmetry into the relation between C and E, because, according to him, we can produce E by producing C, *but not vice versa*. So we can produce the acceleration of the car by pressing the accelerator, but not produce the pressing of the accelerator by accelerating the car. This is his way of solving the original problem of working out which one of force, mass and acceleration is the cause.

What about situations where manipulation is not possible or feasible? Here Gasking says that in these cases we most often want to establish a theoretical point, rather than give a 'recipe' for bringing about effects. However, the plausibility of such claims comes from *some* analogy to other situations where we, agents, are able to produce effects. He says:

> One often makes a remark of the form 'A causes B' with the practical aim of telling someone how to produce or prevent the event B, but not always. Sometimes one wishes to make a theoretical point. And one can sometimes properly say of a particular happening A, that it caused some other particular event, B, even when no-one could have produced A, by manipulation, as a means to produce B. For example, one may say that the rise in mean sea-level at a certain geological epoch was due to the melting of the Polar ice-cap. But when one can properly say this sort of thing, it is always the case that people can produce events of the first sort as a means to producing events of the second sort. For example, one can melt ice in order to raise the level of water in a certain area. We could come rather close to the meaning of 'A causes B' if we said: 'Events of the B sort can be produced by means of producing events of the A sort'. (Gasking, 1955, p483).

To put it otherwise, the notion of cause is closely tied to the notion of manipulation because this is what we experience as human agents, and we try to extend this notion of causation beyond agency to justify other sorts of claim.

17.3 Agency and the symmetry problem

The most prominent current agency theorist is Huw Price, although he has developed his views by working with Peter Menzies (who no longer seems to be an agency theorist) and, more recently, with Richard Corry and Brad Weslake. In various places, Price argues that the symmetry problem described above is serious, and argues against various attempts to solve it (see particularly Price (1996)). One prominent solution that Price discusses is to impose the arrow of time on the symmetric functional laws, and stipulate that events earlier in time cause events later in time, and not vice versa. We press the accelerator first, and then the car accelerates, so the pressing of the accelerator is the cause. The details of such arguments are not important to understand

Price's views—except that Price thinks that they cannot work at the micro-level. Indeed, applying causal notions at the micro-level may well be a mistake. For Price, an agency theory of causality is the only prospect for solving the symmetry problem, particularly without merely stipulating that the causal arrow follows the arrow of time. For this reason, Price thinks we should investigate the action of agents on the world. From this, causal asymmetry at the macro-level emerges.

In 1993 Menzies and Price gave the first agency theory since Gasking. Their formulation clearly owes a debt to Gasking's theory. They write:

> An event A is a cause of a distinct event B just in case bringing about the occurrence of A would be an effective means by which a free agent could bring about the occurrence of B. (Menzies and Price, 1993, p187–9.)

They give more detail, and two points are particularly important, because of objections first made in response to Gasking. Agency theories are supposed to be full-blown reductive theories of causality. But if the notion of 'bringing about' E by bringing about C is just another way of saying 'causing' E by causing C, then the theory is not reductive. It is just saying that two causal notions are related! This is the circularity problem for agency theories.

Menzies and Price seem to have two agency notions: 'effective means' and 'bringing about'. The idea of directly bringing about an event is, for them, primitive; something which every agent is familiar with, such as pressing the accelerator or sipping from a cup of coffee. They suggest that the other, 'effective means', can be spelled out in detail in decision-theoretic terms. Decision theory is used effectively in many sciences, particularly in psychology and economics. It is a way of characterizing the most rational thing for an agent to do, among a range of options, to get what they want. In the broadest terms, decision theorists say you should choose the course of action *most likely* to get you what you want. Decision theorists spell this out in terms of probabilistic equations, which don't include any—any overt—reference to causality. This can be done for any action, such as your decision to accelerate the car by pressing the accelerator, rather than, say, hitting the brake, or getting out and pushing.

The second important development by Menzies and Price is to recognize that decision theory alone cannot give a non-causal notion of 'effective means', as many have argued that decision theory requires covert reference to causality to recommend the right action (Cartwright, 1979; Joyce, 2002). But according to Menzies and Price a non-causal, purely probabilistic, account of an effective means can be given by introducing a new notion of an 'agent probability' (Menzies and Price, 1993, p191). Agent probabilities are 'probabilities, assessed from an agent's perspective under the supposition that the antecedent condition is realized *ab initio*, as a free act of the agent concerned' (Menzies and Price, 1993, p190, emphasis in original).

Using decision theory and agent probabilities adds a vital probabilistic element which is a new and interesting development for agency theories. This development also makes the idea of *agency* really central to agency theories of causality, because only agents assign agent probabilities.

Price continues to develop these ideas, right up to the present day. He thinks that the centrality of the agent's viewpoint can explain the relation many people think exists between causing and making probable. It is because agents ought to act to render the results they desire most probable that causing E involves making E more probable. So Price's early theory is very like Menzies and Price's joint theory, but more radical in important ways. Price believes that when agents try to decide what to do, when they deliberate they must see themselves as free to act. This means agents have to treat actions they are considering as possible. They do this by 'breaking them off' causally or probabilistically from their causes. This means agents ignore causes of their own actions beyond their deliberation, and consider only what will follow once the actions are done. This is what Price means when he talks of allocating agent probabilities. Price thinks this is *required* of agents, and introduces a forward-looking asymmetry into agents' experiences that wouldn't be present otherwise, so offering an agency theory of causality that solves the original symmetry problem. Price says his view is, while updated, in line with a broadly Humean projectivism about causality (discussed in chapters 15 and 18).

17.4 Agency and action

In 2005 Gillies offered the only other agency theory, that he calls an 'action-oriented theory of causality'. Gillies takes seriously Price's concern that causality may be absent from fundamental physics, and thinks we do need to explain the preponderance of causality in other sciences. Perhaps because of this, Gillies presents his view as a variant of Price's, but Gillies makes much more moderate claims than in Price's latest work. Gillies writes:

> The basic idea behind the action-related theory of causality is quite simple. It is that causal laws are useful and appropriate in situations where there is a close link between the law and action based on the law. The concept of cause has evolved in such a way that the transition between a causal law and carrying out actions based on that law is particularly easy and simple. (Gillies, 2005a, p827.)

Gillies then seeks to use this idea to explain the source of asymmetry in many causal laws—those closely related to agency. For example, it is natural that we are interested in penicillin causing recovery from bacterial infection, as we can use penicillin to kill bacteria, but cannot use bacteria to affect penicillin. This is why we have a causal law in medicine in a simple asymmetric form. Gillies suggests that fundamental physical laws are much further removed from human action than those of, say, medicine, and this is why they appear so different in form.

Interestingly, Gillies suggests the action-related theory can also explain why it is useful to give functional laws such as $f = ma$. Such laws are economical, as we can work out lots of effective actions from them. In this case, obvious actions include pressing the accelerator pedal to get the car to go faster. But we can also work out other actions, such as making cars lighter, so that they accelerate faster per unit force, thus saving both petrol and the environment. Gillies notes that actions we are interested in as agents include both actions to produce something, as Menzies and Price focus on, but also actions to avoid something. We can also avoid using the accelerator where possible, to avoid burning too much petrol.

Gillies' theory is much more modest, partly because he doesn't share Price's views on decision theory, which are the source of Price's most controversial claims. For Gillies, the action-related theory of causality is more an explanation of the source of asymmetry in causal laws from the point of view of the use we make of scientific discoveries to guide our actions, and it is not a reductive theory of causality. Gillies accepts that there is overlap between the idea of a causal law, and the idea of an agent bringing something about. This means that there seems no prospect of Gillies' theory explaining why the arrows of time and of causality agree, as Price claims his theory does.

Paying serious attention to the relation between our discovery of causes and how we want to use them in action can be a useful way of thinking even if you are not interested in a full-blown agency theory. Gillies could be interpreted as offering this as a kind of 'agency-lite' theory.

17.5 Problems for agency theories

There are two classic problems for full-blown agency theorists. They were both raised as problems for Gasking's account, and they are an ongoing problem for current agency theorists such as Price. These are the circularity problem and the scope problem.

We have met the circularity problem already. It is that unless we have an account of agency that is not dependent in any way on an account of causality, the resultant account of causality in terms of agency is circular. Price evades this problem by using a novel theory of agency using agent-specific probabilities to characterize an effective means in a non-causal way.

The scope problem is about the range or scope of causal relations that an agency theory can allow. Recall that an agency theory is one that requires using a real notion of agency to explicate causality. Can such a theory, then, allow for causal relations that are quite independent of agents? For example, can it allow that a distant star going supernova could cause the destruction of its satellite planets? Also, we think the history of the universe included no agents, for a very long time. Was there no causality in the world then? Can we say that the Big Bang created the universe?

It is not clear that this problem will worry Price, as he seems to be deeply deflationist concerning causation. However, it has worried many commentators. Responses to the scope problem have been broadly in line with Gasking's original response: we recognize these causes by analogy with more familiar examples, such as local explosions that we can control destroying nearby objects. It seems to be true that this helps us recognize such events, and this kind of response might be fine for theories such as Gasking's. Gillies' strategy to deal with the scope problem is to follow Russell (1913) and eliminate causality from physics. Outwith physics, though, we can still derive 'avoidance' actions from causal laws that fall beyond the scope of agency. For instance, movement of tectonic plates causes earthquakes. While we cannot manipulate tectonic plates, we can still avoid the consequences of earthquakes by keeping away from the most sensitive areas, or constructing buildings designed to resist them.

But for a fully reductive theory, like Price's, there is a problem with making any similar move. The problem is that for Price, the solution to the circularity problem is to move further and further into something distinctive to agents—assigning agent probabilities—that nothing else but agents ever does. Making features of real agents crucial in this way separates our understanding of causality from the worldly features agents respond to, at least more so than the theories of Gasking or Gillies. The upshot is to damage efforts to solve the scope problem by claiming that we recognize agent-independent causal relations merely by analogy with more familiar ones. In such a case we seem to be pointing to something about the worldly relation that we recognize—something that exists in the relation whether or not agents can or could use the cause as an effective means. This might not concern Price himself, but for many others there is tension here, and it is not clear how the agency theory could recover.

Moving on, what might be said about the problem that is the core problem for modern agency theorists; the asymmetry of our experience of causality in a world where the fundamental laws seem to be symmetric? There are two different explanations we might seek.

First, there are other places to look for explanations of the emergence of asymmetry from symmetry. This can be seen by moving away from considering this as a problem specifically about causality. The last century of developments in physics might be referred to as the 'century of symmetry'. Our understanding of many things has been advanced by understanding how non-symmetry arises from symmetry. For example, it may be that at the level of fundamental particles, the division between bosons and fermions arises out of symmetry breaking. This has nothing to do with agency, or observation. If this is so, one might think agency is likely to be the wrong place to look for an explanation of the emergence of asymmetric causal laws from symmetric fundamental laws. The solution is more likely to arrive along with a general solution to the problem of symmetry breaking.

This does not seem to be what Price is claiming. He can agree with the points above, as he wishes to explain why our *concept* of agency—the concept we all have as agents—is asymmetric. Why do agents experience the world as causally asymmetric? A successful explanation of the apparent asymmetry of our experience of a fundamentally symmetric world does seem to require a substantial explanation of what, exactly, is the feature of us as agents that explains this asymmetric experience. In particular, it looks as though Price must think this explanation adds something to the purely physical kinds of explanations mentioned briefly above. At its simplest, Price thinks that agents are required to assign asymmetric agent probabilities. But why are we required to do that? Without a really convincing explanation, it is tempting to plump for the answer: well, so that we accurately track a real asymmetry in the world. This may return us to the problem of the worldly structure explaining our experience as agents, rather than our experience as agents explaining the apparent—but fundamentally misleading—structure of the world.

This debate remains unsettled. Finally, it is worth noting that the place of causality in fundamental physics is also still hotly debated, with philosophers such as Frisch arguing that causality, and causal reasoning, is important even in physics (Frisch, 2014).

17.6 Merits of agency theories

For clarity, it is worth reiterating that agency theories are *not* the same as another group of theories they are easily confused with: manipulationism such as that espoused by Woodward and collaborators (discussed in chapter 10), and various causal methods, such as those espoused by Pearl and collaborators (discussed in chapter 7). Each in their way is a descendant of the agency view, but they are not agency theories. In particular, the early views of Hausman and Woodward (1999) and Woodward (2003) aim to overcome the objection of anthropomorphism raised against agency theory. If, the argument goes, the notion of causation is intimately connected to agency, then causation will lose any objective, in the sense of agent-independent, character. In chapter 10 we examine an attempt to develop a notion of intervention that is not agent-based. Notice, however, that developing a non-anthropomorphic account of causation is somewhat orthogonal to developing an action-oriented (or 'agency-lite') account such as Gillies'. In fact, one might still argue that our actions will succeed to the extent that they are grounded in established or valid causal relations, which may be unrelated to agency, such as the causal relations involving earthquakes.

Yet, manipulationism and causal methods such as Pearl's are inheritors of agency theories in the sense that they take very seriously the importance of intervention into the world, alongside undisturbed observation of the world, to causal discovery in science. But for these theories, it ultimately doesn't matter what intervenes in a causal system. It could be an agent, or a meteor arriving from space, *so long as it bears the defined relation to the causal system being investigated*. What crucially distinguishes agency theories is that they use a notion of real agency to explicate causality.

Independently of the problems discussed earlier, considering the relation between agency and causality is worth doing at some point as it helps sharpen up causal reasoning, primarily about *how* we find causes, and what we use them for—and the connection between the two (for discussion see Cartwright (2007)).

In terms of causal inference, it was the agency theorists who gave philosophical thinking a hard shove away from inferring causal relations purely from patterns observed in a world separate from the observer. Hume, discussed in chapter 15, thought we could only observe such patterns, which he called 'constant conjunctions', and that we *assume* a connection between them that we don't directly observe—an assumed connection we call 'causality'.

Agency theories and successor theories pointed out that we have another source of information about causal relations. We do not merely observe accelerator-pedal-pressings and car-speedings-up; we can ourselves *press* the accelerator and see the car speed up. We can also see that the harder you press the pedal, the faster the car accelerates—up to a limit, of course. This resulted in a better understanding of the difference between different experimental methods, particularly between observational studies and interventional studies. Note, however, that observational studies are still a vital source of our causal knowledge; the difference between observational and interventional studies is important, and needs to be treated carefully—see the discussion in chapters 2 and 7.

Related to this, it is useful to ask: does the evidence I have support the action I want to take? For example, does the fact that a drug, or a social policy, worked in a trial on the other side of the planet, with a population significantly younger than, or with a different social structure from, the one you have in mind, mean that it will work for you? Does the evidence I have about which pedal to press to get a car to go faster, based on my experience in one country, transfer to another, or would I be best advised to test it out gently in the car park before driving off in my new rental car abroad? These are all valuable questions. We also discuss this in chapter 6, particularly Cartwright's worries about evidence-based policy, which are about how to *act* in response to causal knowledge, and also apply more widely to the problem of extrapolation.

Core ideas

- Modern agency theories attempt to explain why our concept of causality is asymmetric—we experience causal relationships that only go from past to future—even though fundamental physical laws are symmetric in time.
- A reductive agency theory of causality is one that claims that causal relationships are relationships that agents can use to get what they want. Price is the most prominent reductive agency theorist, holding broadly that C causes E when an agent thinks bringing about C is an effective means to bring about E.
- Theories such as Gillies' offer an account of the relationship between the actions we want to take and the laws we express in science, rather than a reductive theory of causality.
- Thinking of the relationship between finding causal relations and our need to take action can be useful, particularly for suggesting nuanced treatment of evidence of causality.
- Manipulationist and causal method approaches descend from agency theories, as they use the idea of intervening into the world, but they are not agency theories as they no longer require actual agency.

Distinctions and warnings

- The first classic problem for reductive agency theories is circularity. If we give an account of causality in terms of an account of agency that is itself dependent on causal notions, this account of causality is circular.
- The second classic problem for reductive agency theories is scope. In some versions, agency theories might be committed to, for example, there being no causality in the world before there were agents. If so, could we say that the Big Bang created (i.e. caused) the universe?
- Reductive agency theories need a serious explanation of what, exactly, the feature of us as agents is, which explains the asymmetry of our concept of causality, that isn't really a fundamental feature of the world.

Further reading

Gillies (2005a); Menzies and Price (1993); Norton (2003).

CHAPTER 18

Causality and Inference

18.1 Example: combatting the spread of AIDS

Modern medicine is increasingly stretching the boundaries of the traditional biomedical domain, with some diseases such as the obesity epidemic having social causes, and requiring policy intervention on an international or even global scale for effective management. A particularly good example of this is management of HIV and AIDS, discussed by Hutchinson (2001). Hutchinson lays out the reasons for HIV being such a challenge even to our most advanced medicine. First, the HIV virus itself is extremely simple, as it piggybacks on the body's cellular mechanisms. This means that it can mutate very easily without losing functionality, creating many, many strains of the virus even within a single body, making it harder to find ways to kill all of them. Second, within the body, the virus directly attacks the body's usually effective defence system, the immune system, eventually disabling it to create full-blown AIDS. Finally, almost all research done is on the strains affecting richer countries, with very little done on those strains causing far more damage in poorer countries, and both effective measures to prevent transmission and existing effective treatments require a certain amount of wealth and education, such as accurate clocks to stick to treatment regimes.

This kind of case shows the sheer complexity and interdisciplinarity of modern science and science policy. To manage HIV effectively, we need to know a lot of different things. We need to know about the nature of the virus and its mechanisms for replication, and the reasons it is highly mutagenic and so resistant to direct attack. We need to know about the mechanisms of the human immune system, how HIV attacks them and how to slow it. We also have to know about the social causes that make transmission prevention such a problem, and how we can combat those—how to target education as well as distribute free condoms.

We also need to do a lot of different things. It is clear that there is no single, simple cure. The reason we can now greatly extend the lives of those infected with HIV is because we have treatment methods that target the virus everywhere we can, stopping it reverse transcribing itself, stopping it leaving infected host cells, slowing its attack on the immune system, and so on. We have also greatly slowed transmission in the richer, more educated world, and need to work out better how to get these methods to work in very different social contexts.

Causal reasoning is ubiquitous in the fight against HIV. A great deal of causal inference is involved, finding multiple causes—biochemical, medical and social—as the upshot of thousands of studies. Many mechanisms have been discovered, yielding causal explanations. Prediction is important, and HIV requires quantitative prediction. We do not seek to kill the virus, as we cannot, yet; we only slow it. But giving someone 20 more years of life is a very significant gain. Population models are important to understand transmission and target efforts. Control is our ultimate goal: slowing HIV in every way we can, and, ultimately, eradicating it.

The case of HIV really demonstrates how complicated it is to find and use causal relations, and how causal reasoning needs to be integrated across multiple scientific and policy domains. And happily we have a new hope, with very recent cases reported in the media of babies born with HIV now testing as HIV negative. If this can be replicated, this holds out the glimmer of a promise of a complete cure.

18.2 Different sorts of inferences

Traditional philosophical approaches to causality attempt to reduce causality to some other concept, for instance to probabilistic or counterfactual relationships (chapters 8 and 9), or to physical processes or mechanisms (chapters 11 and 12). The rationale is to explain causality in terms of another concept that is non-causal in character and that already belongs to our ontology—something we already think is in the world. But the approach of this chapter is altogether different in that it attempts to explain the concept of causality not in terms of what there is in the world, but in terms of what we can infer and explain about the world.

In this approach, causality is treated not as an element of our ontology, but as a category that we use to chart the world, to make successful *inferences and explanations*. This is like some variants of agency theories (see chapter 17), but causality here is about our inferences and reasoning more broadly rather than just about our actions and deliberations.

This approach highlights the kinds of evidence (see chapter 6) and the methods (see chapter 7) used to infer causal relations. It does not try to give an account of causal production. Indeed, it is not meant to. There are a variety of particular theories falling under this approach, and we will look at them briefly. What is interesting is whether this approach can give us everything we want in a theory of causality. For a start, does it imply that causality is in the human mind, not in the world; a form of anti-realism about causality? And what might follow from that? An interesting question to keep asking of these theories is: how do we get agreement about causal structures, across different people, or for the same person over time? While we want our view of the causal structure of the world to evolve as we come to know more, we don't want wild disagreements between different people—at the very least we want a reasonable possibility of convergence. We will examine what the views in this chapter imply.

18.2.1 Beebee's Humean projectivism

The approach of this chapter is very new, in the sense the proponents of it discussed here are all still research active and refining their views. But there is a sense in which the approach descends from Hume and therefore is in this sense quite traditional. We can see this by examining the views of Helen Beebee, which she takes to be a modern interpretation of Hume, along what are called 'projectivist' lines. Beebee explains projectivism:

> According to the projectivist interpretation, Hume holds that our causal thought and talk is an expression of our habits of inference. On observing a, we infer that b will follow, and we 'project' that inference onto the world. (Beebee, 2007, p225.)

In brief, we see a cause, C, and we infer that the effect, E will come along soon, and making that inference is treating C as a cause of E. We 'project' this inference out into the world, to think of C and E as themselves causally connected.

Beebee's view falls in this chapter because she is highlighting the importance of inference to our ascriptions of causality. Note that according to this view the inference is quite restricted. For Beebee, following Hume, the inference is just from one C to a following E, and it is a simple matter of prediction. We predict E on the basis of C, and prediction is all there is to the inference, although this prediction leads us to think that E must happen on the basis of C. In brief, it is the 'must' we project into the world. Beebee suggests an interesting role for that projection. She suggests that we make more successful inferences by projecting the must, more successful than agents who merely predict, because it is easier to track that sense of must. She writes:

> Is there an important difference between being able to think or say that the black ball must move, and being able to think or say only that the black ball will move? I think so. For, in saying or thinking that the black must move, we conceive of ourselves as having good reasons for thinking that the black will move. (Beebee, 2007, p245-6.)

Note that not just any inference will work; the 'good reasons' are important. Alice can't just buy a lottery ticket, and expect as hard as possible that there will shortly be lots of money in her bank account! However hard Alice tries to expect money in her bank account, she cannot deliberately create the sense of 'must'. Beebee says that the inferences have to be *rational*. Presumably this will explain why our projections so often converge. Alice's friends and family can also try to expect Alice will shortly be extremely rich, but they cannot fake the necessary sense of 'must' either. So long as everyone is rational, nobody really expects buying a lottery ticket will cause Alice to become rich. This helps, but there doesn't seem to be much more to say about why our projections so often coincide. Since an important part of Beebee's project is to interpret Hume, she is restricted in what she can add to the theory. But the epistemic and inferential views are not so restricted, and add some interesting features to a similar kind of approach.

18.2.2 Williamson's epistemic theory

Like all the approaches of this chapter, the epistemic theory seeks to answer the question of what causality is by looking to our inferential practices. The epistemic theory, which emerges in the work of Jon Williamson, focuses attention on the practices by which we find out what causality is—but we will see there is also another side to the theory.

The motivation for the epistemic theory is interesting. Williamson identifies two possible positions in the current literature, neither of which he favours. The first is to choose one of the many theories currently available (see the rest of the book) and try to make it work in the face of the well-known problems which they all have (see the 'Distinctions and warnings' at the end of each chapter in Part II for a guide to these). The second is that one can opt for causal pluralism (see chapter 23) and give up the idea that there is a single concept of causality. Williamson suggests that the epistemic theory offers us a unified concept of causality, but one that can still accommodate the diversity in causal methods (see chapter 7) and in sources of evidence (see chapter 6) that we should not ignore.

Williamson suggests that we will find what unity exists in the concept of causality in our inferential practices. We have a plurality of evidence for causality, but we still have one, admittedly eclectic, concept of causality. He writes:

> Under this view, certain causal beliefs are appropriate or rational on the basis of observed evidence; our notion of cause can be understood purely in terms of these rational beliefs. Causality, then, is a feature of our epistemic representation of the world, rather than of the world itself. This yields one, multifaceted notion of cause. (Williamson, 2006, p69.)

Calling the theory the 'epistemic' theory, writing about 'causal beliefs' and pointing out that many theories of causality pay inadequate attention to the epistemic aspect of causality suggest that it is only how we *find out* about causes that matters to the epistemic theory. This impression is enhanced by quotes such as, 'Causal claims have two uses: they are used for inference on the one hand and explanation on the other' (Williamson, 2006, p70). However, *action* is also important to Williamson, so 'used for inference' should be interpreted as including making decisions about actions, for instance policy actions. This is becoming even stronger in current work (Williamson, 2013).

Williamson expands the kinds of inferences used by the approach beyond those discussed by Beebee. However, yet again, not just any inferences people happen to make will do. Williamson also requires that the inferences—both to causal beliefs and to decisions—be rational. Williamson's account is:

> According to the epistemic theory, then, the causal relation is characterised by the causal beliefs that an omniscient rational agent should adopt. (Williamson, 2006, p78.)

So in this account the causes of HIV are what the omniscient rational agent comes to believe are the causes, based on the many varieties of evidence for those beliefs, and the actions that they suggest are useful in attempting to control HIV.

This is quite a strict limit on the causal structure of the world. Williamson suggests that the beliefs of the ideal agent—and this is the other element in the story—can be formally represented using, for instance, Bayesian networks (see chapter 7). Williamson argues that formal methods will not be fully constrained by evidence, but they will be largely constrained. So rational agents with access to the same evidence will have a good deal of agreement about the causal relations in the world. Williamson doesn't say a lot about the agency side in the earlier work, but in current work he makes it clear that successful and unsuccessful actions will frequently help to decide between competing causal beliefs:

> The inferential target provides a success criterion: ... causal beliefs are rational to the extent that they lead to successful PEC[Prediction, Explanation and Control]-inferences. (Williamson, 2013, p268.)

So we also learn by implementing policies to prevent, for example, HIV transmission, and studying why they do not work. Williamson thinks that unsuccessful results in scientific practice do teach us something about causal relations, and this is why, in the hypothetico-deductive methodology that he sketches, failures enter the next round of modelling (Williamson, 2005, ch8).

18.2.3 Reiss' inferentialism

A closely similar approach is the theory Julian Reiss calls 'inferentialism'. Reiss is similarly motivated by sustained counterexamples to existing standard theories of causality, and also seeks to resist certain varieties of pluralism.

Reiss takes the idea of inferentialism from a general theory which says that what sentences mean is constituted by their inferential connections. Due to this, Reiss concentrates on sentences, where Williamson concentrates on beliefs, but this is not a difference that will concern us. Reiss makes clear that he is not defending a general inferentialism, just inferentialism with respect to causality. Also, he writes:

> I understand inferentialism about causation less as a mature theory of causation than as a theoretical framework within which one can address various issues concerning causation in a philosophically fruitful way. (Reiss, 2012a, p770.)

Reiss also endorses the view that the inferences we are interested in enter various scientific practices:

> I will argue that causal claims are (typically) inferentially related to certain evidential claims as well as claims about explanation, prediction, intervention and responsibility. (Reiss, 2012a, p769.)

Reiss explicitly recognizes two main categories of sentences relevant to the meaning of causal beliefs: causal claims follow from statements of evidence for causal relations (the 'inferential *base*') and statements that follow from them, such as beliefs and actions (the 'inferential *target*'). Again, causal claims about HIV follow from statements about our evidence regarding HIV, and also get some of their meaning from the beliefs and actions they suggest.

As in the previous accounts, not just any inference agents actually make will do. Reiss writes:

I propose the following formulation: a sentence is inferentially connected with a causal claim (and therefore part of the claim's inferential system) if and only if the scientific community that asserts the claim is entitled to infer it from the causal claim or to infer the causal claim from it (in conjunction with other sentences the community holds). Three aspects of this characterisation require commenting: the nature of the scientific community who is entitled to make the inferences; the nature of the entitlement; and the nature of the inferences that are being made. (Reiss, 2012a, p771.)

For Reiss, there are no external global standards for the relevant inferences. A possible implication of Williamson's ideally rational omniscient agent is that there is one and only one causal structure, if there is only one ideal set of beliefs about the world. However, Reiss thinks that the standards of entitlement are always and inescapably local and contextual, not just because they are specific to given scientific fields where the inferences are drawn, but also because they are specific to the purposes of inquiry (Reiss, 2012a, p772).

In this view, disagreement across communities may be irresolvable and useless. But useful disagreement is possible as long as it is contextually motivated. Reiss then concludes:

Causation is fundamentally subjective in that what warrants the appearance of the term 'cause' or a cognate in a claim is the claim's being inferentially related in certain ways to certain other kinds of claim, and to be inferentially related is to play a role in—human—reasoning practices. But, and this is what makes causation partly objective, these reasoning practices may not arbitrary [sic] but rather shaped by objective facts.' (Reiss, 2012a, p775.)

Reiss's view allows for community convergence on causal structures, and successful interventions, but there is no longer a single causal structure of the world. If there are scientific disagreements on aspects of causes and possible preventative measures for some features of HIV, in the inferentialist theory, there are conflicting causes.

18.2.4 Spohn's ranking functions

This philosophical tradition establishing close links between causation and inference also includes, according to Reiss (2012a), the work of Wolfgang Spohn. Spohn's theory of ranking functions is ultimately about a kind of *epistemic* probability attached to causal relations, and in this sense the account can be placed next to Reiss's, Williamson's and Beebee's.

Spohn has been developing his theory of ranking functions for some time (Spohn, 1983; Spohn, 1986; Spohn, 1994; Spohn, 2002; Spohn, 2006; Spohn, 2012). His theory is meant to be an alternative to existing accounts of causation, especially the counterfactual account (see chapter 9), exploiting an idea that ranking functions theory (i) is very akin to probability raising (see chapter 8), (ii) makes causation somewhat epistemic and (iii) starts from necessary and/or sufficient conditions (see chapter 4) and is able to subsume the probabilistic case. Clearly, Spohn is similarly disinclined to choose a single available theory of causality, instead trying to bring together successful elements from these.

Spohn starts from the following definition:

C is a cause of E iff C and E both occur, C precedes E, and C raises the metaphysical or epistemic status of E given the obtaining circumstances.

Spohn requires that the cause is always prior in time to the effect. What he means by 'metaphysical or epistemic status' intentionally equivocates between some objective (viz. metaphysical) features of causality such as an objective probability, and some epistemic (viz. subjective or agent-dependent) features of causality such as a subjective probability or Spohn's degrees of belief in the ranking function theory. Spohn thinks that we should embrace the subjective turn, but the work of the theory is done by ranking functions, rather than a particular interpretation of probabilities. Ranking functions offer a theory of the dynamics of belief. Ranking functions behave in a way that is very similar to probability measures and so, Spohn says, we can also read ranking functions as an attempt to analyse deterministic causality as analogous to probabilistic causality with probabilistic theories. Spohn explains this formally, but we will not examine those formalities here.

Spohn's subjective relativization of causality to the epistemic subject is in a sense a Humean move, as the nature of causal necessity is not to be found in the world, but in the way the agent associates ideas about sense data. This is similar to Beebee's sense of 'must'. That causation is a matter of what the epistemic agent comes to believe is also the core of the epistemic theory, and related to inferentialsm. Like these theories, Spohn also needs to impose some restrictions on the beliefs agents can come to have about causality. He does this by attaching beliefs in the ranking functions to propositions, which are assumed to have some objective truth conditions, which are grounded in the *realization* of the object of the function. In these ways Spohn has a great deal in common with the other approaches of this chapter. He has done a remarkable job in giving structure to intuitive puzzles in the stock philosophical literature, such as the Billy and Suzy saga (see chapter 20), highlighting that these examples are actually more complex than might be thought and require proper formal tools of analysis.

18.3 Does inferentialism lead to anti-realism?

The idea of causality as in some sense constituted by what we, people, need it for, makes the inferential views of this chapter strongly related to agency theories (see chapter 17). Although this view does not attempt a reduction of causation to experiences of agents, it is a form of anti-realism about causality. In this view, causality as a reasonably coherent *concept* is strictly speaking a construction of the human mind. Williamson is explicit in using the expression 'mind-dependent' in earlier work.

This has become a very sophisticated anti-realism, however. It does not strictly imply anti-realism about the *metaphysics* of causality. Williamson (2013, p272) also notes:

This brings us to the question of realism. The epistemic account of causality in terms of causal beliefs is in principle compatible with either anti-realism or realism about causality [...], since the claim that causality is best understood epistemically leaves open whether causal relationships are really 'out there', in the sense of being characterisable without invoking epistemological considerations.

So Williamson leaves open the possibility of an alternative possible conceptual reduction of causality. See chapter 22 for exploration of the relation between concepts and metaphysics.

Even conceptually, there is no suggestion that individual people can think whatever they want about causal relations. No theory here allows that kind of arbitrariness. The element these theories add over the agency element helps enormously with this. This is the story about how we come to have causal beliefs, or make causal claims or causal inferences: the evidence we seek. We run studies, and observe populations, and do experiments in labs; and try to figure out what causes what, try to predict what will happen without our interference, and try to figure out what to do to get what we want.

Beebee says that we project the sense of 'must' into the world, based on the regularities we observe. And for Reiss and Williamson, the story can be more complex. There is something in the world that we are tracking; the correlations in the HIV positive and negative populations we study, the biochemical mechanisms of replication of the HIV virus, the chemical reactions that inhibit the action of reverse transcriptase. It is just that these are diverse. We don't have much reason to think they are all similar. It is we who recognize that they are similar—because they lead us to have causal beliefs about HIV and guide our predictions and interventions in combatting it.

To get all this to work, it is crucial that there is some convergence, some means by which the worldly things we track constrain what our causal beliefs and claims are, and so ways people find causal beliefs that really do help combat HIV—and come to agree on a lot of these beliefs about the causes of HIV.

18.4 The heart of inference

This leads us to examine what is really the heart of the theories of this chapter. Inference is their explanatory core, and it is also what generates convergence, as identifying the right kinds of inferences is what bars an implausible wild variation in causal beliefs.

To recall, Beebee constrains allowed inferences by admitting only *rational* inferences, but doesn't say much about how to interpret 'rational'. Williamson admits the causal beliefs of the *ideal agent with exhaustive evidence*, with the aim of allowing a strongly constrained causal structure. Reiss focuses on *entitlement* to make the inference, which he says is a social standard, admitting critical disagreement, but only from within the right context. Reiss seems to allow a great deal more divergence in causal beliefs than Williamson does. So one upshot of earlier work seems to be: if you like this theory but want constrained causal structure, you will prefer the epistemic theory; if you like divergence, you will prefer inferentialism.

However, both allow a certain amount of disagreement. Williamson notes: 'This characterisation leaves room for a certain amount of subjectivity as to what causes what, where two ideal causal epistemologies disagree' (Williamson, 2013, p270, footnote 5). It is clear that the story about how we come to have causal beliefs, and how we use them, is absolutely crucial. And it is a very difficult story to tell—partly because a great deal of the evidence we seek for causality is indeed highly diverse. But this is the core of the explanatory story regarding causality that is offered by these approaches.

One thing worth noting is that these theories could make more of the agency element than appears on the surface. Reiss and Williamson both more-or-less explicitly allow the building of causal beliefs to be a dynamic, rather than a static affair, modified gradually over time as extra evidence is constantly gathered. As unsuccessful actions impact everyone regardless of what they believe, they can be an important source of convergence. Successful action can create a feedback loop, encouraging convergence. Reiss and Williamson do seem to intend this, at least including scientific and possibly policy interventions. Indeed, in recent work Williamson argues that the way the epistemic theory treats the inferential target is an improvement over inferentialism. Broadly, his claim is that, for inferentialism, causal claims have to vary with different evidence bases—the inferential base—while the epistemic theory takes evidence into account, but ties how we understand the claims themselves more closely to the inferential target. He writes:

> While for the inferentialist the inferential base and target act together to determine the meaning of a causal claim, under the epistemic account of causality the roles of the inferential base and target are strictly separated. As we shall see, the inferential target is used to provide an independent normative success criterion for the way in which causal claims are posited from the inferential base. (Williamson, 2013, p270.)

This means we can use the inferential target as an independent standard to judge how well the inferential base is used. Certainly, as our ability to create entirely new worlds, such as the internet, continuously expands, such success will be an important source of confidence in our understanding of causes.

Finally, allowing disagreement on causes might not be so bad for these approaches. Sensible, rational people actually do disagree a lot about the causal structure of systems even in the face of the same evidence. This is one important focus of scientific debate. No philosophical theory of causality can eradicate scientific disagreement of this kind, or should seek to do so. Perhaps our theory of causality in the sciences should respect that, and should be developed in tandem with existing accounts in social epistemology. In the end, these approaches do focus attention on the right problem, even if it is a very difficult one. A theory of causality—even many theories of causality—can help us sharpen up our thinking about causality, and hopefully make it easier to select and seek evidence. But no theory could possibly make the search for causality easy. Perhaps requiring the theories of this chapter to settle a single causal structure of the world is a wholly unreasonable requirement in light of their own starting points.

Core ideas

- Inferentialism seeks a path between pluralism about causality—based on the diversity of causal evidence, methods and causal practices—and unsuccessful monistic theories of causality.

- The idea of inferentialism is that the unity in our concept of causality comes from our inferential practices with respect to causality, which have two varieties:

    Our evidence base for causal claims, such as knowledge of causal mechanisms and correlations.

    The actions that are suggested by the evidence we have.
- It is good to engage seriously with the extent of the diversity in our evidence for causality, the different things we want to know, and the many different actions we may take on the basis of available evidence.

Distinctions and warnings

- This is a sophisticated form of anti-realism about causality. Strictly speaking, on these views our concept of causality is dependent on human practices, including our minds.
- The success of these theories strongly depends on a substantive account of inference, ruling out irrational beliefs and actions.
- If this account is successful, these views do not imply that what causes what is arbitrary; instead they are able to identify causes in a sensible way.

Further reading

Beebee (2007); Brandom (2000, ch1); Williamson (2005, ch9); Reiss (2012a); Spohn (2012, ch14).

PART III

Approaches to Examining Causality

In this part of the book we reconstruct the intellectual journey that led to the 'Causality in the Sciences' approach and its methodological underpinnings: the use of examples and counterexamples, and the role of truth and validity. We maintain that causal questions fall into the categories of epistemology, metaphysics, methodology and semantics, which all feed questions about the use of causal knowledge.

CHAPTER 19

How We Got to the Causality in the Sciences Approach (CitS)

19.1 A methodological struggle

In the twentieth century, there was a great deal of philosophical work on causality. Simultaneously, many of the philosophical literatures have been struggling to find their place in a broadly naturalistic, empiricist world view. This world view holds that we get all our knowledge—or at least our really important knowledge, which is scientific knowledge—from engagement with the world. This engagement with the world is called empirical engagement, which is why the view is called 'empiricism'. As philosophy is not usually empirical, conducting few observations and experiments[27] and making no new technology, its place as part of academic inquiry has often come under attack. Many are happy to accept that philosophy should have a place alongside music and art, as a distinctively human activity we should support as important to human life. But can philosophy contribute to inquiry? In this chapter, we will offer a quick trip through the struggle within the philosophical causality literature, which may help you if you are puzzled about the different approaches philosophers take towards the accounts or views of the chapters in Part II.

In this chapter, we'll look at three episodes in the history of philosophical work on causality. To some extent we describe these three episodes in a way that artificially separates them, as philosophical practice regarding causality has usually involved aspects of all of the styles we look at. But we offer here our understanding of how the current focus on causality in the sciences came about, and why we came to form the causality in the sciences (CitS) community, and to write this book. As such, this chapter will

---

[27] See the emerging field of 'experimental philosophy', as discussed by Alexander (2012).

help scientists and philosophy students to understand how CitS philosophers see their aims and method, and how they see these relating to science. We will also see that the CitS community shares a great deal with other philosophical communities such as the Philosophy of Science in Practice and the Philosophy of Information.

19.2 Causality and language

We begin with the attack of the logical positivists on metaphysics, in the first half of the twentieth century. Logical positivism was a movement that developed around the 1920s and 1930s in Europe and later around the 1940s and 1950s in the US, as many intellectuals had left Europe during World War II. A common interest of logical positivists was science and scientific methodology, together with scepticism about the value and usefulness of 'traditional' metaphysics. They thought that an important test for the *meaning* of ordinary as well as scientific claims was whether they are grammatically well-formed and structured and whether they pass strict 'verificationist' criteria. This last means that claims will only be allowed as legitimate if there is a method for testing their truth. This was called 'verificationism', and tended to be limited primarily to empirical methods of verification. Traditional metaphysical questions such as 'what is causality?' were rejected as uninteresting by the logical positivists, as there was no way of verifying answers to such questions.

However, aspects of traditional metaphysical questions survived in the philosophy of language. Language was still recognized by philosophers in this period as an interesting empirical phenomenon. Since in everyday language competent speakers, philosophers and scientists all make what are apparently causal claims, explicitly, like, 'smoking causes lung cancer' and implicitly, like, 'the noisy kids gave me a headache', these remained a legitimate target of investigation. Philosophy of language gradually paid attention to other aspects of meaning and truth that are particularly anchored in our experience as speakers (see for instance Austin's or Grice's work on speech acts and on speaker meaning). Philosophy of language thus began developing a broadly Wittgensteinian approach according to which to understand the meaning of a word is to understand its use in different contexts. This view became quite widespread.

Important work on causality was carried out in this tradition. Anscombe (1975), for instance, noticed that there are many words and concepts expressing causation, e.g. pulling, pushing, blocking, sticking, breaking, binding, lifting, falling, etc. We customarily use these in ordinary language, which means that our ordinary language is thoroughly laden with implicitly causal terms. But these words also mean different things, which raises the question of whether there is anything these terms have in common, something that explains why these are all 'causal' terms. We analyse Anscombe's position in more detail in chapter 23.

Mackie (1974), who is examined in chapter 4, is explicit about his belief that we need to study language to understand the nature of causality. He explains that his philosophical analysis is about ontology, i.e. about what causal relations are in general, while the sciences find out particular causal relations and causal laws. Conceptual analysis and analysis of language, in his view, give us a rough guide to what is out there.

That's why he spends a lot of time analysing propositions in the form of causal conditionals. Another example of this approach, also considered in chapter 4, is the classic work of Hart and Honoré (1959). They say that their starting point is the analysis of language, as this was the dominant approach at the time they originally wrote the book. They claimed this approach as valid because law uses causal concepts borrowed from ordinary talk, albeit refined and modified. Both Mackie and Hart and Honoré use this approach successfully to discriminate different kinds of causes.

This approach kept work on causation going, even when metaphysics was out of fashion in philosophy, but it is no longer the dominant approach to work on causality. This is primarily because language has many highly contingent features. Causality is an important concept for how we structure our understanding of the world. But how we speak about causality is at least partly an accident of where we are born, as there are many languages, and any language is constantly changing according to social conditions that have little or nothing to do with causality itself.

Nevertheless, the tradition of linguistic analysis remains influential. Causal claims made in everyday language are still of interest to philosophers of causality, as are the causal claims made in the sciences. Part of the turn to the sciences is because the language of ordinary speakers is likely to miss at least some of what is interesting about causality. This will be the case if it is science that is driving conceptual innovation in the concept of cause, and ordinary language lags behind scientific innovation and changing scientific language, or entirely misses scientific innovations of which non-scientific folk remain forever unaware. And in so far as examining scientific language helps us understand scientific *practice*, we should certainly not ignore it.

For example, one current approach clearly influenced by the philosophy of language tradition is Reiss' inferentialism, discussed in chapter 18. Reiss is concerned about the meaning of 'cause' in various occurrences in scientific talk. Inferentialism aims to provide a semantics for causal claims in terms of the inferences that are valid in a given context. Their validity, in turn, depends on their evidential basis and on the chosen inference target. For instance, we claim that drug C causes recovery from disease D because, in the randomized clinical trial conducted, the drug proved effective in the population examined and the inference is therefore justified. Certainly, inferentialism is allied to the linguistic tradition, but is extending it to examine other aspects of scientific practice, such as evidence, needed for causal claims, and the actions we wish to take on the basis of causal claims.

19.3 Causality, intuitions and concepts

Philosophy generally wished to investigate more than the contingencies of causal words or claims in a particular language. The core thought behind the approach in philosophy which sees itself as involved in conceptual analysis is that there is something beyond language itself that is more stable—the intuitions or concepts expressed in language.

The most important account of conceptual analysis in philosophy—which also informs approaches to causality in particular—is the 'Canberra Plan', articulated by

Frank Jackson (1998). The Canberra Plan was also an attempt to characterize the place of philosophy in a naturalistic world view. Canberra Planners try to break down barriers between the methodology of philosophy and the methodology of science, to see the two methodologies as lying at different places on a spectrum from the empirical to the theoretical, but as engaged in a broadly similar enterprise.

The Canberra Plan envisages philosophical investigation as having two broad stages. The first is conceptual analysis, and it is *a priori*, or non-empirical. Philosophers are thought of as taking the big philosophical concepts, which structure our understanding of the world, such as value, consciousness and causality, and clarifying our folk intuitions about those concepts. The second stage is where philosophers look to reality—sometimes to what science is saying about reality—to find out what in reality, if anything, fills the role that has been clarified in stage one. This approach likens philosophy to science, as it considers philosophical concepts as like the theoretical posits of science, such as electrons and mass. Since philosophical concepts like causality structure our thinking about the world, they are treated as posited and then used in our theories, in our attempts to explain the world. Thus, like the theoretical posits of science, these concepts are likely to be interdefined in our theorizing but, if our theorizing is successful, filled by something real.

So what happens in approaching causality this way is analysis of our intuitions regarding causality—the folk psychological concept—with reference to cases of causal relations. Science is used, but it is not particularly special. Any competent speaker will have a concept of causality, and lots of examples of cases of causality to draw on. This project is different from the ordinary-language approach in that it tries to examine concepts behind language, and also in that it tries to figure out what the worldly realizers of the clarified conceptual roles might be. Huw Price is an example of a philosopher working on causality who was influenced by this approach in offering arguments concerning causality and agency (see chapter 17). But Price also raises some interesting concerns for applying the Canberra Plan specifically to causality. We often treat scientific posits, like electrons, as fillers of a causal role in a theory that accounts for empirical phenomena. But we cannot so easily take the theoretical role of *causality* to be a causal role! Instead Price argues we should think of causality pragmatically, as essential to our practices as agents. He holds 'the view that a philosophical account of a problematic notion—that of causation itself, for example—needs to begin by paying close attention to the role of the concept concerned in the *practice* of the creatures who use it' (Price, 2001, p2–3, emphasis in original).

In philosophical method according to this story, then, suggested accounts of causality are put forward, and assessed to see whether they apply to cases of causality. Different philosophers put forward different accounts, and argue for them, and the ultimate tribunal is the philosophical community, which accepts or rejects different views. A great deal of debate proceeds by intuition based arguments, with opposing philosophers offering examples as 'intuition-pumps' to try to alter each other's intuitions, or, perhaps more successfully, the intuitions of the audience.

A constant challenge for work in this tradition is the question: what intuitions or concepts should we base our philosophical theorizing on? Just as natural languages are subject to the contingencies of their histories, so the intuitions of people are subject to the contingencies of their evolutionary history, their personal developmental history and of course even of fashion. After a while, even the practice of doing philosophy can warp your intuitions. Philosophers have proven adept at not giving up their intuitions, so much so that there is even an informal name for the strategic move of accepting the counterintuitive implications of your view: it is called 'biting the bullet'!

There is a need, then, for some kind of constraint on what intuitions are admitted to help us build theories of concepts such as causality. An increasing number of philosophers are concerned about this, and agree that it is in engagement with the world that we test our intuitions. Notice, though, that in actual practice philosophers following the Canberra Plan do not strictly separate the two phases of conceptual analysis and seeking the role-fillers. Generally, there isn't a purely a priori stage solely about pre-existing folk concepts, because examples of role-fillers are characteristically constantly used to motivate suggested accounts, adjustments to accounts or rejections of accounts (see also chapter 20). So in a naturalistic approach to philosophy, the a priori or conceptual enterprise is not entirely separate from turning to the empirical world.

For philosophers in the CitS tradition, a natural segue from this insight is to look at science as the place where we test our intuitions in their most rigorous engagement with the empirical world—especially our intuitions concerning core scientific concepts such as causality. It is also the place where concepts under pressure change most quickly, and so we think it is the most fruitful place to look for examples. In the introduction to a book on meta-metaphysics, Manley uses Theodore Sider's work to characterize 'mainstream metaphysics':

> Competing positions are treated as tentative hypotheses about the world, and are assessed by a loose battery of criteria for theory choice. Match with ordinary usage and belief sometimes plays a role in this assessment, but typically not a dominant one. Theoretical insight, considerations of simplicity, integration with other domains (for instance science, logic, and philosophy of language), and so on, play important roles. (Quoted in Chalmers et al. (2009, p3–4, closely similar to p385.))

In some ways, the CitS tradition is not in sharp disagreement with this view, but there is a significant difference in emphasis and in domain of application. For CitS philosophers, science is not one constraint among others but deserves a much more important place in philosophical theorizing—particularly theorizing about such important scientific concepts as causality. If we get the important influence of worldly engagement on concept formation straight, then clearly how we use examples in philosophical work is extremely important.

The tradition begun by the Canberra Plan remains extremely influential. In some way or other, all philosophers are still concerned about concepts. Another important example of recent work in this style of conceptual clarification is Ned Hall (2004), which we use at many points in this book. Hall gives us two concepts of cause, 'production' and 'dependence' (also often called 'difference-making'), and argues that these two concepts are distinct by asking what we naturally think of an everyday case of causality. When Billy and Suzy throw rocks at a bottle, Billy throws first, shattering the bottle, but Suzy's back-up rock whistles through the flying shards of glass an instant later. Hall suggests Billy's throw *produces* the shattering of the bottle. But the bottle shattering does not *depend* on Billy's throw, as Suzy's rock would have shattered it anyway. Using this case Hall argues that there are cases of production without dependence. He uses another case to argue that we sometimes get dependence without production. Billy and Suzy are now pilots in World War III. Suzy sucessfully takes off, flies abroad, and bombs an enemy town. Suzy clearly produces the bombing. Billy is her escort fighter pilot. He flies around, and shoots down enemy fighters trying to shoot down Suzy. Hall argues that the successful bombing depends on Billy: if Billy had not shot down the enemy fighters, they would have shot down Suzy. However, Billy does not produce the bombing, as he is not connected to the process that did produce the bombing. Hall argues that, while for most cases of causation the effect is both produced by the cause and depends on it, this is not always true. Sometimes we have production in the absence of dependence; and sometimes we have dependence in the absence of production. Since these are genuine cases of causality, Hall says, our concept of causality has to capture them. Hall argues that this shows we have not one but two concepts of causality.

The intuitions, and the examples that Hall used to motivate them, stood the test of time. The Billy and Suzy cases were successful intuition pumps. They gathered consensus in the philosophical community, and the distinction between production and dependence has also been discussed in specific scientific contexts (see e.g. chapter 6). There is no consensus about what to do about the problem, of course, and some philosophers wish to collapse the concepts, or dismiss one of them—but there is agreement that Hall has presented a genuine, and important, problem!

19.4 Causality in the sciences

The approach of the Canberra Plan was an important move towards naturalism for philosophy, breaking the apparently immutable barrier between empirical and non-empirical, or a priori, work. In many ways, Causality in the Sciences (CitS) is a development of this movement, moving further and further into science. As a result, CitS has more blurred frontiers, as it becomes more diverse in response to the messy and diverse nature of scientific practice itself.

CitS philosophers can probably all agree wholeheartedly with a remark by Price (2001, p104):[28]

> But progress is surely possible by a kind of reflective equilibrium, and informed interplay between the two kinds of constraints, philosophical and scientific. The difficult task, the task which falls especially to the philosopher of the special disciplines, is to open and keep open the dialogue.

This is the heart of the difference. Many philosophers coming across the Canberra Plan think it implies strict temporal stages; that we should do conceptual analysis first, and only then turn to cases of causality—possibly but not necessarily including scientific cases.

But, as we have said, these episodes were seldom really sharply separate, especially in work on causality. Nevertheless, it is a common interpretation, and a natural reaction to it is to reverse the supposed temporal order, and say: now we look to *science first*, looking to science to generate ideas about causality, and only then proceeding to

---

[28] Although we should note we can speak with complete confidence only for ourselves!

conceptual analysis. But this is an overreaction. We suggest instead successful philosophy of causality is *iterative*, moving freely between science and philosophy, neither first, often simultaneously studying both literatures. It proceeds, ideally, in a dialogue between philosophy and science—which is sometimes, but not always, a dialogue between scientists and philosophers!

Philosophers of science of our generation and before will also notice that another route to CitS can be traced. Hans Reichenbach was a most influential philosopher who initiated a tradition in philosophy of science, and notably in philosophy of causality, that was taken up by Wesley Salmon, who studied with him. Salmon was a key figure in the philosophy of causality, and his thought is still influential—see chapter 11. He is mainly known for his 'process' theory, but in fact he also discussed causation with respect to explanation, statistical modelling and probability, physics, and experiments. Other branches of philosophy of science, such as the philosophy of mechanisms, originate from his work, because people were pushed to assess his account in other fields, such as biology or the social sciences (see chapter 12). In sum, the CitS approach we present and explain here is certainly an heir of the Reichenbach–Salmon tradition by building effectively on their deep interest in scientific problems.

There are three ways in which science is crucial to the philosophical enterprise: generating new ideas, problem selection and testing philosophical accounts of causality.

Generating new ideas. The enterprise of science and technology changes how we see the world, and because of this these practices put stress on the concepts we use to understand the world. This is true of specifically scientific concepts, like 'mass' and 'charge' which change in meaning as scientific theories change, but scientists also put stress on causal concepts by trying to apply them every day when they search for causes, or use causal knowledge to predict. They can do this either explicitly (actually using the word 'cause') or implicitly (using concealed causal concepts such as 'binding', 'colliding', or possibly even 'risk factor', which is currently in a limbo between the realm of probability and that of causality).

In finding new kinds of causes, and struggling to explain and model new kinds of systems, science and technology have been driving innovation in our concept of causality for years. For example, it was the advent of apparent genuine indeterminism in quantum mechanics that convinced philosophers to relax the element of causality that held that every event had a sufficient cause. This element had previously been treated as an unshakeable a priori truth. It was Newtonian mechanics that so profoundly affected the concept of mechanism—a concept that is still changing in current science. But the distinction between the innovation driven by science and technology and the change in our conceptualization of causality or the world is a relatively recent one. Surely Newton was at the same time a scientist and a philosopher. But today's specialization, both in the sciences and in philosophy, has the consequence that what we now need is a synergy between a 'philosophically minded' science and a 'science-oriented' philosophy. This is why philosophy of science often works from case studies—interesting episodes in science that illuminate causality and related concepts (see chapter 20).

While scientists do sometimes notice tension in their use of concepts, including causal concepts, most current scientists are under enormous pressure to get on with their studies. This limits their time for engagement with broader methodology and epistemology; even the grounds for a causal interpretation of their models, and far less examination of underlying metaphysical assumptions. Philosophy, on the other hand, has a history of examining exactly these kinds of problems, and can make itself extremely useful. Philosophers might have to infer assumptions scientists make about causes from how they phrase their conclusions, or how they set up their experiments. But philosophers can usefully ask questions like: facing a problem with inferring causation with known unobserved causes, and unknown unobserved causes, does the counterfactual theory lead us right? Does the probabilistic theory? Can manipulationism or Bayesian networks? If this is right, then a great deal of useful philosophical work is also scientific work, and vice versa. There is overlap between philosophical and scientific concerns. It would greatly enhance the scientific project for every scientist to reflect, at some point in their career or training, on these issues. On assumptions they themselves make about causality, about what they are searching for and how, and on what they will do when they find it—and do those match up? It would greatly enhance the philosophical project of understanding causality to see how causal concepts are used—and warped—in practice.

Problem selection. In the CitS approach, looking to the sciences is crucial to the choice of philosophical problems to address. For CitS, the most useful problems to look at are often problems generated by the sciences—an idea already found in Popper (1963). This can happen due to change in the sciences creating conceptual change, or a lacuna where a new concept is needed. As science does this, it can constantly generate new problems in a way in which philosophy does not do all by itself. As a source of original problems, science is extremely interesting. While it frequently takes philosophical work, whether done by philosophers or by scientists, to uncover these problems, meaning that of course the philosophical literature is sometimes the first place such a problem is articulated, ultimately it is the sciences which are the ongoing source of this kind of novelty.

This is a further use of case studies in philosophy of science; to suggest what problems exist, what might be worth solving. So philosophers of science search for practical applications that illuminate theoretical tension in causality. This is the way of *finding* the problems that are both philosophical and scientific. For example, how do we make sense of causality in an indeterministic world? Since we seem to see indeterministic effects only at the sub-atomic level, can we safely assume determinism in, say, medicine? Should we? But note this kind of uncovering of the real nature of the problems often involves moving between case studies from science, and the philosophical literature trying to dig out conceptual problems.

Testing accounts. The final use of science is to test philosophical accounts of causality. Ultimately, the test of the accounts is in whether they do solve the problems they were designed to solve, which are generally problems of reasoning about causality when the concept is under stress in the sciences. These are the five problems of causality we introduce in chapter 1 and use throughout the book.

Right now, philosophy works in such a way that the final assessment about the 'success' of a philosophical account is generally done by other philosophers of science, in that tribunal that we call peer review! But ultimately the concern is to see whether we can solve problems for science. It is also increasingly common for philosophers of science to work in close collaboration with scientists, even within scientific institutes. In this approach, we favour causal concepts, ideas and methods according to how applicable they are to scientific problems, and whether they show some hope of solving them. But developing the account is conceptual work, even if it is tested by going back to the case studies—either those that suggested the novel moves, or illuminated the problem, or others, in an attempt to see how widely the solution might apply.

In this way, CitS is not science-first or philosophy-first, but iterative, constantly looking for conceptual problems in the sciences, designing accounts to help solve them, and looking back to the sciences to test them. This enterprise is also of interest to the public understanding of science. Indeed, all the humanities disciplines studying science are now important because science is so powerful. The CitS approach will, for example, complement investigation of the social dimensions of scientific knowledge. Sociologists of science are interested in studying the effects of science on society, and society on science, for instance by looking at how techno-science changes our lives, or how environmental concerns arise in response to the use of science-based technologies, or at the role and place of women in science. A sociological perspective on science also looks closely at the development of scientific theories in their historical contexts or as results of intersubjective practices (see Longino (2013)).

The CitS approach also works towards offering a more sophisticated understanding of the relation between science, technology and expertise—through causal knowledge. Consider for instance debates about 'evidence'. The relation is currently treated as 'just' a matter of the 'kind' or 'amount' of information that justifies or does not justify some decision or action. CitS aims to work with various evidence-based movements and make evidence a question for causal analysis that deeply involves modelling, but also reasoning and conceptualization of mechanisms, and all this taking into account the *practical* constraints under which bodies such as NICE or IARC have to come up with texts that summarize the available evidence and provide a recommendation as a consequence. This means working with current evidence-based theorizers to enlarge their methods, broaden their concepts and integrate different approaches to causal questions.

CitS falls in the spirit of 'philosophy of science in practice', which seeks to examine philosophical issues in the sciences in engagement with the practice of science, and with 'history and philosophy of science', which seeks to understand philosophical issues with respect to science, provide sensitivity to the historical context of those scientific episodes, and understand history of science with respect to philosophical theorizing about scientific 'progress'. As science produces knowledge, CitS aims to contribute to this knowledge production. Scientific method and results are also made of a conceptual apparatus, of argumentative structures. This is where CitS contributes to producing more solid science. Yet, this is not just a posteriori reconstruction of scientific reasoning, but a synchronic construction of scientific reasoning, *with* scientists. In this way CitS also shares a great deal with the 'philosophy of information', which considers philosophy at its most vibrant when it is engaged with many current debates, including scientific but also ethical, political and policy debates.

In this spirit, CitS often does not aim to generate concepts that are eternal, instead aiming for a 'timely' philosophy, in the sense of Floridi (2011b). It offers philosophical views and accounts that may be of interest for addressing current problems. They may be discarded as ideas move on.

The introduction to CitS offered in this chapter can be followed up in the other chapters in Part III: on examples in 20, truthmakers versus modelling approaches in 21 and various questions in philosophy of causality in 22. The whole book exemplifies what scientifically informed philosophy and philosophically informed science can achieve. We have done our best to systematize the literature, given the *current* challenges of the sciences. The chances—and hopes—are, in 20 years, or perhaps even in ten years, this discussion will have altered significantly, thanks to progress in the sciences and in philosophy.

CHAPTER 20

Examples and Counterexamples

20.1 Examples of examples!

Billy and Suzy: the next episode. Once upon a time we met Billy and Suzy (chapter 19), throwing stones at a bottle. Billy's stone smashed the bottle, while Suzy's stone followed behind it. In that case it seemed that Billy smashed the bottle, even though the smashing didn't depend on Billy's throw. We meet Billy and Suzy again, now grown up international assassins. As it happens, they are independently contracted by the Syndicate to assassinate the president during a visit to a remote, hot location.[29] While the president is visiting the facilities, Billy poisons the water bottle intended for the president to use while walking in the desert. A little later, while the president's attendant is distracted by examining the mechanism of a toilet flush, Suzy makes a hole in the water bottle, hoping the president will expire of thirst. The water drains away, so the president does not drink the poison. However, the secret service accompany the president at the last moment. So the president, despite being hot and annoyed with an empty water bottle, makes it to the conference centre to give the speech.

Independently, Billy and Suzy decide they will have to assemble their high-powered rifles. They happen to shoot simultaneously at the president. A freak kick to the fire alarm from a passing squirrel turns on the fire sprinklers, and the president is shocked by the water into jumping aside at the last minute, so that both bullets miss. While the secret service, shouting for backup, all pile on top of the president to make a shield, a passing angel, worried that the humans will catch cold now they are all soaking wet, turns off the air conditioning. But the air conditioning has malfunctioned recently, as it has not been maintained properly, and it starts sucking out the oxygen in the main conference hall. Merlin, a magician who had agreed to

---

[29] We do not specify which president. Spoiler alert: we advise you to fill in your least favourite, because the president is not going to make it.

keep an eye on the dodgy air conditioning while the president was in the building, is unexpectedly recalled to Camelot and ordered by the captain of the knights, and by King Arthur, to protect Camelot from Morgana's attack. Merlin is busy at the critical moment and the air conditioning continues to malfunction. The hosts realize what is happening, and rush to open the doors and cut the power to the air conditioning. The president alone, however, is under a pile of burly men still determined to make a bullet shield, with rather compressed lungs. Everyone else survives, but the president dies from oxygen deprivation.[30] *What caused the president's death?*[31]

If this kind of 'toy' example leaves you rather bemused, you might try to turn to real scientific examples instead to illuminate our ideas about causality. Real cases where we try to figure out causes in the sciences will solve all the problems, right? Well, looking to science might be a good idea, but it certainly doesn't solve all the problems. Our intuitions about what to say about the president's death might waver, but real science is complicated, and also contains significant disputes. Further, it requires a certain level of expertise even to assess scientific examples.

The bewildering first philosophy conference. Alice, although she is now working in the causality literature, remembers finding herself in the following scenario as a fresh young graduate student. Here she relates the tale: The speaker, Bob, is talking about physics, about a follow-up of string theory. He is trying to combine string theory with quantum mechanics, presenting the equations that supposedly unify these theories. However earnestly he tries to explain, I don't understand it so I can't really disagree with him. He's saying that there aren't really any entities, only mathematical structure, which seems really odd to me, but I'm unable to formulate a question that makes sense. Also, I know that quantum mechanics and string theory are very complicated anyway, and there is a lot of dispute among physicists, so can't Bob really claim anything he wants? Hm, the other philosophers of physics in the room are nodding, so maybe the claims Bob is making about physics aren't so crazy, even though they sound crazy. (Like a good scientist, I will use proxy indicators if any are available.) But maybe these are the philosophers of physics who agree with Bob, and others disagree. How am I supposed to tell?

The next talk is by Carol. This will hopefully be better, because it's on biology, which I understand better. Carol is telling us about recent discoveries and experiments using biomarkers. They combine observational studies, computer simulations and in vitro and in vivo experiments. But as the talk goes on, I don't think I agree with Carol's approach. In the question time, when I try to object, though, Carol just

---

[30] What happens to the assassins is not usually recorded, but Billy and Suzy were born and brought up in the causality literature, unlike the other characters, who are merely regular visitors. (We include Glennan's toilet by special request—with thanks!) Billy and Suzy escape unscathed in the chaos, claim their money as the president is safely dead, and continue their violent careers by becoming a bomber and a fighter pilot in World War III (Hall, 2004).

[31] Or: Did the squirrel save the president's life? Who is responsible for Merlin's absence? Did the secret service kill the president?

keeps saying that it's complex, and insists, 'It just doesn't work that way'. I don't know what to say, since I don't know the specific case Carol's talking about, but I'm making a more general point about methodology. Do the other philosophers of biology look puzzled? Hm, it's difficult to tell. I've now seen two talks, and both Bob and Carol have inferred from one or two cases—cases they clearly know very well and care about a lot—to quite general claims. And Carol talked about a lot of entities, so they seem to contradict each other! I really don't know what to believe, and I can't find my supervisor. The speakers are clearly too important to bother with my questions, which must be dumb as everyone else in the room understands perfectly well. I wonder if they'll have some more of those nice chocolate biscuits for the coffee break?

Anyone attempting to make sense of the philosophical literature without training—and even with training!—can be forgiven for wondering what the examples so ubiquitous in the causality literature are intended to show. Indeed, it might be difficult for those working within the literature to grasp how alienating the use of examples in the literature can be to those trying to use it, whether they are academics from other fields or even philosophy students. The rather extreme parodies we offer above are an attempt to illustrate this alienation.

Nevertheless, using such examples is very important to philosophical dialectic. In this chapter, we will try to help make the philosophical literature more intelligible to those outwith it, by examining how examples can be used badly, and how they can be used well. Our book is concerned with causality in the sciences, and in the chapters in Part II we primarily used examples from the sciences themselves, for instance epidemiological studies on cancer or other diseases, biological investigations into the mechanisms of metabolism, etc. We call examples that actively engage with the scientific practice 'scientific examples'. There are also 'toy examples', which we occasionally used in Part II. These include causal stories drawn from everyday life, simplified scientific cases, or altogether fictitious stories where we seek to imagine causes and effects.

We will try to show that *both* toy examples and real scientific examples can be used well, and *both* can also be used badly. The idea is that examples are not good or bad per se—their purpose and the context in which they are produced also count, as we shall explain. There is one very general feature of their context which is relevant. Even if the manoeuvres in a particular paper seem odd, they are performed in the context of a whole literature, and they are assessed by the whole community. What the philosophical community achieves collectively might well be more nuanced, and far more careful, than a particular paper, whether written or spoken, appears to be. Sometimes one scholar, with one contribution, tries out a new idea, and then other people join the debate refining, improving, polishing and of course criticizing the idea. Each contribution cannot be fully considered in isolation any more, as there is already a network of ideas being created. For example, one group of philosophers begins discussing the mechanisms of protein synthesis, other philosophers gradually become expert on them, and these mechanisms rapidly become standard cases for PhD students to study. Part II discusses many other such episodes. Modern academia, organized around social networks and numerous conferences, is very much a collective enterprise, and ideas are created and assessed by a community. In other words, philosophers are not 'Leibnizian monads'![32]

20.2 Toy examples or scientific examples?

So, if you want to work on causality, or read work on causality, you should think about what kind of examples interest you, which will help you argue for or critically evaluate a point of view. Do you want to think about, or use, everyday, simple examples, or complex scientific ones?

20.2.1 Examples impeding the progress of debate

There is no simple answer to this question because, in the philosophical enterprise, both toy examples and scientific examples can fail, in the ways we parodied in the introduction to this chapter.

Complex scientific examples can impede debate in two ways. In the first way, they can be used to prevent criticism of a view being developed, by retreating from more obviously philosophical claims, into claims directly about a scientific case. This can be a tempting move if an opponent is not familiar with the case. Inevitably, given the need for a certain amount of expertise to assess scientific examples, there are limits to how an audience or readers unfamiliar with a case under discussion can engage with the paper. Over time, this problem is ameliorated, as others are quite free to examine the case, or consult others already expert in the case. But if debate is to be fruitful, both the scientific examples and the philosophical claims need to be made intelligible to those unfamiliar with the case. This, of course, takes skill and constant exercise, and it is easier for some cases than others.

The second way is in an over-quick induction from a small number of examples—or from a single example—to a more general claim than seems warranted. For instance, if all examples are taken from the health sciences, the scope of the claim will range over the health sciences, but not beyond. An account based on examples, or 'case studies' as scientific examples are generally called in philosophy of science, is only entitled to the scope of claim warranted by the examples used (see Table 20.2 near the end of the chapter).

Toy examples can also impede debate in two ways. The first way is when they are used without sensitivity to how much weight they can bear. They are often said to illuminate our 'intuitions' about causality. But what intuitions? Whose intuitions? And how secure are those intuitions? The debate seems to show that our intuitions about causality cannot bear a great deal of weight—just look at how many different accounts, examples and counterexamples our intuitions have produced! Philosophers have been able to stick to their personal intuitions even in the face of a great deal of opposition.

---

[32] In the philosophy of Leibniz, monads are 'self-contained', independent, unique substances that don't have interactions with the external world.

Toy examples used in this way only establish something if at least the academics working on the topic agree that those examples are a clear case of causality, and ...they usually don't! Even so, it is likely that the examples of causality we come across in everyday life, or in simplified scientific cases, or in imaginary situations, are systematically unrepresentative of all causes, as they are all drawn from an artificial, engineered environment that might be quite unlike the natural world.

The second way toy examples can impede debate is in beginning to drive problem selection. What philosophical problems we think it worth trying to solve is an important judgement. When an account is developed, anomalies are typically identified using some example or other. For instance, suppose a lecturer presents an account of process tracing (see chapter 11) and insists on the importance of tracking 'real' physical stuff that is likely to play a (causal) role in the process. Then he or she introduces the problem of absences, which is held to undermine the whole process-tracing approach, by asking the students to consider the following scenario. You go away on holiday, and your neighbour agrees to water your plants while you are away. But your neighbour forgets, and when you come home, your plants are dead. What causes the death of your pants? Is it your neighbour's failure to water your plants? The Queen also failed to water your plants. Did her failure cause their death? What kind of 'real' physical stuff is involved in *not* watering a plant? A toy example like this one can be extremely useful, but solving the problem of absences can become assimilated to the aim of dealing with the *single problematic example* identified, rather than refining the process-tracing account to explore whether we should consider that toy example a genuine causal story or an example of something else (for instance of moral or legal responsibility). These two strategies are not the same. It can seem, especially to outsiders, as if the fundamental problem of causality in a particular literature, in this case process-tracing, has become 'The plant wasn't watered and died. What is the cause?'. This appearance is, of course, misleading. But regularly pulling back from the example used to describe the anomaly to think of the anomaly itself as the real problem to be solved is important.

If we're honest, in the philosophical literature we've all been guilty of slipping into bad use of examples at some time or other. Under pressure, it is easy to adapt an offered counterexample to suit your view, without really addressing the underlying problem.

20.2.2 Examples enhancing progress in debate

However, examples are also used positively. Where do accounts of causality come from in the first place? Well, from looking to cases of causality, thinking about those cases, and wondering what causality is in those cases, how we find out about those kinds of causes, what we use such causal knowledge for, and what we mean by causality when we think about those cases. So examples can be used to suggest new accounts, and to suggest interesting developments in accounts.

We discuss the use of examples in the rest of this section, with respect to four particular papers: Hall (2004), Schaffer (2000), Machamer et al. (2000), and Salmon (1977). Table 20.1 (at the end of this section) summarizes the comparison of these papers according to six aspects: problem selection, aim, examples used, constraints on theorizing, intuitions and scope of claims. This comparison is an exercise you can and should do with other papers, but we use these because all four are clear cases of very important work on causality. Indeed, we take all four to be important to the research projects we are personally pursuing now. They have all been influential, but the research in the papers was conducted, and the papers were written, in very different styles. One thing that the four papers together clearly illustrate is how important both toy examples and scientific examples have been and still are to the philosophical debate on causality.

Hall and Schaffer. The first and most obvious point about examples is that they need to be drawn from the domain of interest. We return to this point in section 20.3 below. If you are interested in a folk concept of causality, used in daily life, then look at everyday causal claims, and of course search the philosophical literature for examples of everyday causes that other philosophers have considered important, and look into the psychology literature, specifically about causal learning, or into the recent development of experimental philosophy. Your work will be better, however, the more sensitive you are to the possibility that the examples you have drawn on are tailor-made to the problem you are addressing, and the more open you are to alternative examples to alleviate this possible worry by making sure your examples are representative of your domain.

Your job is to design an account that helps illuminate the thinking of other scholars and that propels others to build on it, for instance exploring it in other fields or explaining how it relates to other accounts, concepts, methods, etc. (Floridi, 2011a). This is a collective enterprise, which has generated important work, of which our examples here are Schaffer and Hall. Both mention scientific examples, but do not investigate cases in any depth, focusing instead primarily on more everyday examples to make their arguments.

What Hall does, for instance, is characterize an anomaly for existing approaches to causality, approaches that attempt to capture all our causal concepts in a single account. He uses a very simple example to present that anomaly, in a story about Billy and Suzy throwing stones at a bottle. Billy's stone smashes the bottle, while Suzy's stone flies through the shards of glass immediately behind Billy's. Hall argues that this case shows that Billy's throw produces the smashing, but the smashing does not depend on Billy's throw, as Suzy's throw would have smashed the bottle without Billy's throw. Based on this and on further arguments, Hall concludes that we have in fact two, not one, concepts of causality, dependence and production, that do not always align. This, the original story of Billy and Suzy, is a simple story that is easy to communicate, to teach and even to put in funding applications.

If you are interested in the domain of causality in daily life, examples like this can be of interest to you. Ask yourself: what is this example supposed to show? If it is supposed to show something about our everyday concept of cause, then the artificial, imaginary setting does not impede grasping the real problem. Even so, the case of Billy and Suzy only succeeds because it does capture a real problem. It captures the *structure* of the anomaly that is the real problem, a structure that appears in many cases, not only for the fictional Billy and Suzy!

But if we are interested in wider domains, such as understanding causality in the sciences, Billy and Suzy can still be of interest if the distinction Hall offers has application in domains beyond the one it originated in. We can ask questions like: Does biology track dependence, production, both or none? Does physics? Or epidemiology? Concerns about *methods* and *evidence* generating or tracking production and dependence are exactly what lie behind accounts such as evidential pluralism (see chapter 6).

Shaffer illustrates this kind of transition, working on a problem he claims applies in both everyday and scientific cases. Schaffer's achievement is to persuade the philosophical literature to take the problem of absences seriously. A trend of philosophical work was trying to explain causation by absences away as somehow not real cases of causality. Dowe (2000b), for one, argues that they are cases of *quasi*-causation. Schaffer (2000) marshalled many examples of cases of absence causality, amounting to a powerful objection to the explaining-away strategy. Schaffer is sensitive to the need for multiple examples really to push through this point. But the most reused case in this paper is the causing-death-by-shooting case, so much so that it is often now referred to as 'Schaffer's gun'. The idea is that the design of some guns is set up so that pulling the trigger releases a catch holding back the spring. Once released, the spring expanding fires the bullet. This mechanism allows one to cause death not by connecting something, but by disconnecting something—releasing the catch on the spring. This case is very powerful, because killing someone by shooting them is just one of our paradigm causes. This shows that we can never entirely disregard everyday causality. If a particular account implies that shooting someone doesn't count as causality, we seem to have gone very wrong (and our assassins should resign). As a matter of fact, many people have found this case persuasive.

Schaffer himself is interested in more than our everyday concept of causality, arguing that absences are also ubiquitous in science, and so the problem of absences cannot be ignored in science. Real scientific examples show us the full complexity of causality in the world. This is likely to be much more complex than in the more controlled world of artefacts, including the gun, since artefacts are deliberately designed to have simple and easy-to-manage causal relations. Schaffer's work, usefully summarized in the example of the gun, actually captures the structure of problems in scientific work. Absences pose a real conceptual challenge, and we sketch the prospects of an informational approach in chapter 13. But absences also pose methodological challenges, for instance in the design of knock-out experiments in genetic research, discussed in e.g. chapter 10. Absences also have a fundamental role in plenty of social and biological mechanisms, so we need to develop a language for mechanisms that accommodates this need—see chapter 12. In these ways Schaffer's gun is but the beginning of important work that seriously engages with scientific problems and practice. We can work out the details of scientific examples, in current scientific practice and in the history of science, in order to show that the problem of production and dependence, and the problem of absences, are scientific problems.

Machamer, Darden & Craver (MDC) and Salmon. Many philosophers are now interested in understanding concepts in the domain of science. Increasingly, philosophy of science looks to scientific practice, doing original work to uncover areas where ideas of causality are being or have been changed in science, or are under stress, or scientists are struggling to articulate new ideas. Machamer et al. (2000), for instance, illustrate this, showing detailed engagement with the practice of science, including the history of science, notably in biology. Traditionally, the philosophical literature has been primarily about physics—philosophy of biology is a comparatively newer subdiscipline in philosophy of science. Machamer et al. (2000) take there to be a story missing in the philosophical literature, about causal explanation using *mechanisms*. At the time they were writing, the process account of Salmon and Dowe, mainly using paradigmatic examples from physics, was *the* view under debate. But they looked to other areas of science to see what other concepts are at work, in particular how mechanistic explanation works in biology and neuroscience, to fill out that story. They make claims for mechanistic explanation only in the limited domain they examine, but subsequently there have been many attempts to examine mechanistic explanation in other domains, for instance in cognitive science. Machamer et al. (2000) use a certain amount of expertise about biology, but since the publication of that paper, many other philosophers have assessed their claims, in the light of an existing or newly acquired expertise concerning protein synthesis. Simultaneously, a parallel literature was developed about mechanisms in social science, and current work is trying to build bridges between these two strands (see chapter 12).

So the conceptualization of 'mechanism' offered by Machamer et al. (2000), originally largely based on examples such as protein synthesis, was but the beginning of a much broader project. We need to understand not only whether a given characterization grasps the essential features of mechanisms across different domains, we also have to develop accounts of related notions, for instance explanation or evidence, that contribute to illuminating the importance of mechanisms for scientific practice.

In drawing on science, scientific examples are an important source of *novelty* for philosophical debate. They drive change in the concept of causality and generate new material for philosophical debate, bringing novel examples to our attention. Scientific examples done well tap into this novelty, maybe even uncover new areas of it. In this way, examining science in depth can be very exciting for problem selection.

Problem selection is particularly interesting in Salmon (1977). Salmon does not engage in any detailed investigation of a scientific case in that paper. What is interesting is that nevertheless Salmon is engaging with science in a particular way, and that way lies in his problem selection. Salmon's aim is to illuminate the difference between causal processes and pseudo-processes, which he takes to be different according to the theory of relativity. Ultimately, his aim is to articulate what science—primarily physics—seems to be telling us about causality. Again, the problem of distinguishing causal from pseudo processes may not be confined to physics. So, again, a useful broader project is to investigate whether a particular account is tailored to one scientific domain, whether it can be extended to other domains or whether other accounts are better for the job.

It can be tempting to think that scientific examples, because they are unskewed by the peculiarities of our everyday lives, are the only examples of possible relevance to the philosophy of causality. However, this is not true in any simple sense. Scientific examples can be skewed in their turn. After all, we choose examples to further our own cause! This problem can be exacerbated by the fact that effective assessment of scientific examples as detailed case studies, or in problem selection, requires expertise. Here again, the collective philosophical literature can achieve far more than any single philosopher. Philosophy of science—just like science—is now definitely a collective enterprise, as only groups of scholars together have the necessary expertise. As with other philosophical fields—indeed, any humanities discipline—the final test is whether other philosophers, and of course possibly also academics from other disciplines, engage in a fruitful debate with a piece of work.

Our academic system is currently structured around peer reviewing and building an academic reputation. This is far from being a perfect system (for a critical assessment of the UK system, see e.g. Gillies (2008)). We may be pressed to think that publication in top journals and numbers of citations are good measures of academic achievement. But there should be more than that. Academics have to strive for high-quality pieces of work, where high quality means good description of a problem, good argumentation and explanation of why the problem and attempted solution is relevant and can help advance the debate. So peer review should aim to ensure that such high standards are met, rather than guarantee that peers *agree* and converge towards a mainstream view. Our primary goal, as academics, is to strive for high-quality research that peers recognize as valuable, as having the potential to contribute to the collective enterprise. Our primary goal, as peer reviewers, is to ensure that research follows high standards, not highly standardized ideas. This presupposes a good deal of intellectual honesty!

Looking to the complexity of science is very interesting. But then, why do so many serious philosophers of science also use toy examples? Well, when used well, toy examples also have their own unmatched virtue. They are the result of boiling down a problem so that it can be expressed in a quick and easy way, one that can prompt and inspire follow-up work in scientific domains. Many of the most famous toy examples in the causality literature can be read exactly like that: expressions of anomalies for one or more theories of causality, such as Hall's Billy and Suzy, and Schaffer's gun. Debate often then coalesces around such examples for reasons that have been recognized since Thomas Kuhn (1962). But these examples, remember, are useful in understanding causality in the sciences because they express anomalies that are not specific to Billy and Suzy or the gun. The examples are 'bas reliefs' of much more widespread, serious scientific problems. So long as this is the case, the community frequently needs ways to express these anomalies in a simple way just as much as it needs detailed scientific examples where the anomaly is at work.

Expressions of anomalies can also be given quickly with reference to a real scientific example, of course. Probably the most famous example is the EPR correlations, where we get instantaneous correlations in direction of spin between particles, prepared a certain way, that are so far apart in space that a causal relation seems impossible. But without more information, such examples are only intelligible to those with a certain expertise. For instance, in EPR we talk about correlations... but are those correlations the same as those that social scientists encounter and analyse? Until we explain these aspects, we may fail to understand each other. Toy examples are intelligible to anyone, even undergraduates meeting causality for the first time.

So long as the toy example being used really does represent the issue you want to talk about, they are irreplaceable. They allow academics to communicate effectively about their work, and also facilitate teaching. They can, however, seem to drive problem selection, even when they are not. However, we haven't met *any* philosopher who really thinks the fundamental problem of causality is about Billy and Suzy smashing bottles or who killed the plant! Instead, these are clean presentations of the structure of problems many philosophers recognize to be in need of a solution—in the domains of both everyday and scientific causes. The same holds for scientific anomalies. For more examples, flick through the 'Distinctions and warnings' sections at the end of the chapters in Part II.

In summary, toy examples and scientific examples have complementary virtues:

- Scientific examples: allow us to grasp the full complexity and novelty of cases of causality, and conceptual problems of causality, and can be particularly important for problem selection.
- Toy examples: allow us to capture recognized anomalies in a simple and communicable form.

Finally, note that today's scientific concepts might very well be tomorrow's everyday concepts, so understanding scientific concepts of causality is not distinct from the project of understanding everyday concepts. In turn, an account of causality based on science that implies that shooting someone doesn't count as killing them should also arouse suspicion. If a paradigm case doesn't count, have we made a bad mistake? The two projects are not entirely independent. Ultimately, *both* kinds of examples need to be used with sensitivity to how they might impede debate on one hand, and enhance it on the other hand. Recall that although in the causality literature there is significant stylistic variation on a spectrum between a general concentration on toy examples and serious work uncovering novel conceptualizations of causality in scientific work in progress, the literature as a whole achieves something no individual can achieve. Ultimately, toy examples and scientific examples are both extremely useful, used with a sensitivity to the potential problems. A healthy field of inquiry needs both.

20.3 Counterexamples

We have explained how both kinds of example can impede or enhance building an account of causality. We turn now to a particularly visible negative use of examples in philosophy: the ubiquitous counterexample. Counterexamples are negative because they are examples used to show that (other people's) accounts of causality are wrong or fail in some respect.

A counterexample is an example, a case of causality that—allegedly—does not fit the account currently being advocated, and so shows that account to be no good either at all or in some respects. Arguments over proposed counterexamples are popular in philosophy, particularly in seminars and at conferences. Students asked to critically assess an account might be forgiven for thinking: 'I produced a counterexample. Job done, right?' Counterexamples are undeniably important, so it is worth examining their purpose.

Just as for the examples previously discussed, counterexamples can be used well and used badly. They can be used well when they are sensitive to the purpose of the account, particularly to the scope of the account. Let's consider a range of possibilities. We begin with accounts having a very broad scope. Traditional philosophical accounts

TABLE 20.1 Different styles of philosophical theorizing about causality

	Problem selection	Intuition	Aim	Constraints on theorizing	Scope of claims	Examples
Hall (2004)	Spells out meaning of key concepts borrowed from counterfactuals literature, e.g. 'intrinsic', borrowed from Lewis.	Uses intuitions but realizes they conflict, and should not be regarded as 'data.'	Spells out elements of production-dependence intuitions and shows that they conflict. Uses that to solve philosophical problems.	Does occasionally mention use in ordinary language. Uses intuitions and counterexamples, but sensitive to possible failings.	Sufficiently general to capture cases of traditional counterfactual analysis. Does not attempt to capture all possible causes.	Forest fires and Billy & Suzy pre-emption and double prevention, laid out with neuron diagrams.
Schaffer (2000)	Understands absences in salient instances of causation.	Argues that salient instances of causation do match our intuitions about causation, but also says that some intuitions do not withstand philosophical scrutiny.	Shows that, pace Salmon, Dowe, Menzies and Armstrong, causation by disconnection is a genuine case of causation, with no physical connection.	Puts weight on matching our 'central connotations', 'paradigmatic cases' and 'theoretically salient cases' of causation.	Any world with causation by disconnection, which is presumably any world where omissions occupy these paradigmatic roles, and can bear that theoretical work.	Bomb, heart and gun, laid out with neuron diagrams. Some simplified scientific examples.

continued

TABLE 20.1 continued

	Characterize what mechanisms are, and what their role is in science.	Say that what is intelligible (about causality) is likely to be a product of the ontogenic and phylogenetic development of human beings.	Provide an account of explanation that is based on the description and discovery of mechanisms.	Dualistic metaphysics (entities and activities) for mechanisms is justified by different needs of biological cases.	This world, possibly just biochemistry, or even just protein synthesis and synapses.	Extended discussion of the mechanism of protein synthesis, including the history of its discovery; mechanisms of chemical transmission at synapses.
Machamer et al. (2000)						
Salmon (1977)	Distinguishes between genuine causal process and pseudo-processes, in line with relativity.	Not mentioned.	Develops 'mark method' as a criterion to distinguish causal processes from pseudo-processes.	The criterion succeeds in characterizing causal transmission without invoking anti-Humean 'powers' or 'necessary connections'.	This world, and other relativistic ones.	Rotating spot of light, possibly instantiated by pulsars.

were intended to capture *all* instances of causality.[33] Further, the aim was usually to capture necessary truths about causality, and this was conceived of as capturing anything that could possibly be causal. This is often thought of, following David Lewis (see chapter 9) as characterizing a concept that covered all causal relations in all possible worlds. For those unfamiliar with Lewis's theory, this can be thought of more informally as something like 'all the ways the world could have been'. This is an extraordinarily broad scope. Whether or not you are interested in such broad claims, for this kind of claim even a purely imaginary counterexample is enough to show that the proffered account is false. If you can imagine a case that other folk (particularly the other folk at the philosophy seminar) agree to be a possible case of causality, and it doesn't fit the account, then the account must be rejected. If Merlin casting a spell at Morgana is agreed to be a case of causality, but doesn't fit the account, the account has something wrong. But we can question the strategy of developing the broadest account possible. A corollary of the earlier discussion on the use of examples is precisely this: you may want to develop a more local account of causality, fit for a particular domain, and then explore possibilities for extending it into other domains.

This brings us to accounts with a restricted domain of application. In fact, most current philosophical work on causality does have some restriction on scope. Increasingly, philosophers are working on causality in, for example, particular case studies, as paradigmatic examples of one scientific field, say molecular biology, or analytical sociology, or experimental economics. This is at least partly due to the views of individual philosophers converging towards the idea that a widely general account is not likely to work. The upshot of this kind of work is an account that is intended to apply only to a particular scientific domain, sometimes in a particular time period, which of course raises the question about pluralism in causality (see chapter 23). There is also another feature of this work which can be concealed. Increasingly, philosophers offering such accounts are interested in attempting to capture the typical case of causality in that domain at that time. They no longer even seek to capture *all* such cases. Philosophers used to the traditional aim of an 'all possible worlds' account might see this kind of approach as an astonishing climb-down. But it turns out to be quite hard work to offer a successful account even of the typical case of causality in a particular scientific domain at a particular time. For this kind of account, only a real, typical case from the domain and time period in question counts as a relevant counterexample.

A great deal of frustration can be created in the literature when philosophers taking different approaches, due to different aims, misinterpret each other's accounts. One clear thing is that what counts as a counterexample to a philosophical account of causality depends entirely on what the *intended scope* of that account is. Table 20.2 schematically represents the relation between the scope of an account of causality and the range of relevant counterexamples. Notice the gradual changes as we move down the table.

First, as the scope of the account of causality narrows, fewer counterexamples are relevant. If you develop an account of causation in say, biochemistry, a counterexample based on economic modelling does not prove your account wrong, but rather

---

[33] One can develop analogous arguments for other notions, say probability, explanation, law, etc.

TABLE 20.2 The scope of philosophical questions and relevant counterexamples

	Accounts of causality		Counterexamples
Scope	Relevant questions	How many	Found in the literature
All possible worlds.	What does causation logically mean?	One logically possible example.	Witches casting spells; Angels protecting glasses.
Worlds close to the actual world.	What is causation metaphysically?	One metaphysically possible example.	World with reverse temporal direction; Salmon's moving spot of light.
This world.	What is causation in this world?	One or more real examples.	Causality and the kinetic theory of gases or quantum mechanics; Billy and Suzy throwing rocks at a bottle or bombing the enemy town.
Some region in this world.	What is causation in biochemistry, or physics?	A few real examples in the relevant domain; Typical not skewed examples.	Causality in protein synthesis mechanisms.
Some region of this world at some time.	What kind of causal explanation can we give of the economic crisis in 1929? Can we give the same kind of explanation of the economic crisis now?	A few real examples in the relevant domain at the relevant time; Typical not skewed examples.	Causality in the discovery of protein synthesis. Causality in systems biological approaches to protein synthesis.

raises the interesting question of whether your account may work elsewhere. Offering a counterexample that is inappropriate to the kind of account offered is frustrating—and refusal to consider a possible-worlds counterexample to an actual-world account is not a philosophical cop-out, but simply logically the right thing to do. So, if a 'Billy and Suzy story' is offered as a counterexample to your account of causation in biochemistry, you should accept the objection only to the extent that it captures the *structure* of a problem relevant to your domain. If it does, the toy example could be very useful. Alternatively, considering possible-worlds counterexamples like Merlin and Morgana to accounts intended to capture all possible worlds is simply logically the right thing to do, no matter how sceptical many philosophers of science might be about such an intended account.

Other changes become evident as we move down Table 20.2, which tend to be much more concealed, and so can impede debate. We have highlighted the second change, pointing out that increasingly the accounts towards the bottom of the table are intended to capture merely the typical case, rather than all cases. This means that even a counterexample in the relevant domain, so long as it is an odd, warped or 'pathological' example, can sensibly be rejected as irrelevant to the account offered. In such cases, you need to understand whether any counterexample offered is meant to illustrate a real problem for the account offered. Of course, this is not to say we can dismiss objections out of hand, but instead to set up clear rules for the game, so that everyone can play fairly.

The third change warrants emphasis. It is the time period an account is supposed to apply to. This is made explicit in the bottom row of Table 20.2, and is obvious when work makes specific reference to a time period being studied. However, some philosophers leave it implicit in how they work whether they are concerned with offering an account that is timeless, or whether an account that works just now will do. In general, philosophers working on questions such as 'what does causality logically mean?' or 'what is causality, metaphysically?' attempt to offer a timeless account. So relevant counterexamples can be drawn from any time at all, even far into the past or the distant future. An account offered of causality in this world might be intended to be similarly timeless, or it might intend to deal only with the concept we have now, and this restriction might not be explicitly stated. Counterexamples from the wrong time period will rightly be ignored by the proponent of the account, for instance because the methodological background is different, or because the body of 'accepted' knowledge is different in different time periods. In other words, the history of science and of ideas matters a lot for philosophical theorizing about causation. For those offering a time-sensitive account, examples do not have to be immutable. Consider our use of HIV as an example of a sufficient cause of AIDS in chapter 4. Recent good news suggests that there are cases of babies born HIV positive later testing as HIV negative. This suggests that, hopefully, HIV will not be sufficient for later development of full-blown AIDS for very much longer, and so HIV will cease to be an example of a sufficient cause.

It is perfectly possible to criticize accounts that do not attempt an account of the scope you prefer. You are authoritative about your aims, after all, and so you can still sensibly point out that an account doesn't fulfil your aims, and exclude it from your portfolio of possible accounts on that basis. But this is quite different from criticizing the account on its own terms—and to make that explicit is far more nuanced. The message of this section is to be sensitive to the goal of the philosophical account of causality that you are offering, and those accounts offered by other people, and critically assess them in the light of that goal. In other words, you have to be crystal clear about *your* questions, and about how we use or criticize other accounts to solve *that* question. An account can fail to respond to *your* question, and still be perfectly valuable for other questions.

There is a fourth change in Table 20.2 that mirrors the two kinds of examples we parodied at the beginning of the chapter. There is gradual change down the table from focusing on toy examples, which are unsophisticated from a scientific point of view, to focusing on increasingly detailed scientific examples. This is sociologically accurate. In general, those interested in accounts with wider scope spend more time on toy examples, and those interested in accounts with more targeted scientific scope spend more time on scientific examples. This is largely because those interested in more narrow accounts are intrigued by a new problem, or some kind of stress on causal concepts, or development in them, and this tends to come from the sciences. However, we have already explained how both scientific and toy examples can be useful, for any scope of claim, for those interested in scientific practices concerning causality, and in our everyday causal language.

In sum, use and misuse of examples can cause a great deal of confusion in the causality literature. However, we have explained the mistakes to avoid, both with toy examples and with scientific cases. We have given you a guide for using counterexamples effectively. We have also illustrated how both kinds of examples are best used to push debate forward. Go forth and find your example! Look for those that really intrigue you and worry you, because those will be the ones that you really enjoy working on. Avoid getting so fascinated by a particular example that you forget that one example is not the entire problem of causality; pull back regularly to think about what the real problem is, and all will be well.

CHAPTER 21
Truth or Models?

21.1 Two approaches to causal assessment

In chapter 9 we introduced the example of mesothelioma and exposure to asbestos. We now know how dangerous exposure to asbestos fibres is and how it can lead to asbestosis and lung cancer even with very long latency. Lack of proper protection led to real epidemics in Casale Monferrato (north-west Italy), site of the Eternit factory. Another well-known epidemic to do with asbestos exposure happened in Barking (London area) around the Cape Asbestos factory.

Do asbestos fibres cause mesothelioma and asbestosis? This is a question that ordinary people in these areas no doubt asked themselves, seeing so many of their relatives dying of these diseases. The question has also been asked by epidemiologists and medical doctors trying to understand the causes of the epidemics. The question was also asked by sociologists trying to understand how these local factories changed the social practices and health conditions of the population. The question has also been asked by lawyers and judges during trials in court.

The question whether asbestos fibres cause mesothelioma and asbestosis only *looks* the same for all these categories of people. There are in fact at least four different questions hiding behind the formulation 'Does C cause E?' Ordinary people would like to know whether it is true that their relatives died because of asbestos. Lawyers and judges also ask, to some extent, a question of truth, or at least of likelihood, in order to establish liability. Scientists look for models that explain the data on exposure and disease, via statistics and biomedical mechanisms, or via social practices. Broadly speaking, we have on one side a request for *truth*, and a request for *model validation* on the other side. Let us examine these in order.

A traditional concern of philosophers worrying about causality is to ask what the *truthmakers* of causal claims are, which means to ask what is it in the world that makes 'C causes E' true. In the scientific literature, you are much more likely to find discussion of *modelling*, and questions like 'what model should we use?' and 'should we trust our model?' Although these approaches are both discussing causality, and seem to have some kind of common concern about the relation between the world and our model, concepts or language, the approaches are so different that it can be extremely difficult to translate between them. In this chapter, we introduce the approaches, and compare them, to help promote accurate interpretation and useful dialogue.

With respect to causality, the primary aim of science is to establish what causes what, in a way that is strongly connected to the aims of causal explanation, prediction and control. The problem of establishing what causes what, and to what extent, is often called 'causal assessment'. Examples abound both in everyday parlance and in science:

Why was Alice late for class?
What causes the tides?
Why is delinquency higher in socially depressed areas?
How much should Bob exercise to minimize the chance of cardiovascular disease?

We ask questions like these in different forms, but ultimately we can rephrase them as stylized questions often used in the causality literature:

Does C cause E?
Is it true that C causes E?
How much does C cause E?

In this chapter, we tackle the following problem: how do, or should, we answer questions about causal assessment? An intuitive initial answer is that we need to establish whether any of the claims above are *true*. Another, more science-based strategy, is to establish whether we have enough *evidence* to trust such claims. These two strategies may seem thoroughly opposed to each other. In this chapter, we examine each, and their relationship. Ultimately, we will argue that both the truthmakers literature and the literature on models are relevant to the five scientific concerns about causality that have been the focus of this book—causal reasoning, inference, prediction, control and explanation—although in very different ways.

21.2 Causal assessment using models

Causal modelling in the sciences aims to perform a number of activities: to decide what causes what, to explain how something causes something else and to inform policy decisions, forecasting, etc. These are precisely the five scientific problems of causality we identified in chapter 1: inference, explanation, prediction, control and reasoning. These scientific activities can be reconducted to the quite general question of whether C causes E.

In the sciences, the answer to the question of whether C causes E is not simple or direct. Most of the time we also need to establish to what extent C causes E. Answering such questions requires careful modelling. Modelling, in turn, involves many stages and many approaches, as discussed in chapters 2 and 7. Causal assessment becomes, in causal modelling, a problem of *model validation*. We came across model validation already in chapter 2. The main aim is to decide whether, or to what extent, to trust what your model says about the causal structure of the system under investigation. Here, we go deeper into a couple of different aspects of model validation.

First, never forget the *background assumptions*. Models have many assumptions, some of which are made explicit, while others remain quite implicit. Some concern the causal structure, based on knowledge already available to the modeller; others instead concern deeper metaphysical assumptions. For instance, if Alice wants to study household expenditures in a Western developed country such as Belgium and applies the model developed by Bob's group for France, she is *assuming* that the two causal structures are similar in some relevant respects. When Alice analyses data from the past five years to assess whether and to what extent the recession has been affecting household expenditures, her metaphysical assumption is that causes precede their effects in time.

A useful and important exercise, when reading a scientific paper, is to consider the background of the phenomenon under study ... and also the background of the researchers! Does the choice of a specific model used in the study depend on embracing a particular school of thought? Economics, for one, is a field particularly sensitive to this issue. In these years of economic crisis we debate the causes of crisis and measures to get out of crisis. But what are the assumptions underlying these studies? What difference does it make for the results of a study or for a recommendation to embrace 'Keynesian' economics?

Another important aspect of model validation is *model relativity*. Model relativity means that whether, and to what extent, C causes E is in fact relative to the model being developed. Sometimes philosophers express this idea by saying that causal claims are contextual. This means, in a more scientific vocabulary, that the claim holds for the population of reference in the first place, and it is an *empirical* question whether it also holds elsewhere. This also means that the results are dependent on specific modelling choices, such as the variables included in the model, or the available indicators, and also on the data collected.

But to say that results are model relative doesn't mean that everything is relative and therefore we cannot establish any causal relations! There is still a fact of the matter about what holds and what does not hold. It is precisely by spelling out the background assumptions and the choices and strategies at different stages of the modelling process that we can decide what holds and what does not hold.

But what does 'hold' mean in the sentence above? Is it a synonym of 'true'? The philosophical literature has investigated the relation between models and reality, emphasizing different aspects. For instance, some authors conceive models as 'fictions' (Frigg, 2010), others as 'mediators' (Morgan and Morrison, 1999*b*) and still others as 'maps' (Giere, 2006), or as 'isolations' (Mäki, 1992). In different ways, these accounts try to make sense of the fact that, strictly speaking, models are all false, since their assumptions are not true but aim to approximate, idealize, simplify and abstract from a much more complex reality.

So to go back to the issue about causal assessment and modelling, the question is not so much whether models allow us to establish the *truth* of a causal claim, but rather whether they are of any *use* for a given problem—inference, explanation, prediction or control. The choice of a scientific problem, together with a model to tackle that problem, is a matter of choosing the appropriate level of abstraction, in the sense discussed in chapter 3.

This may look like an altogether different enterprise than the truthmakers question presented below. Later in this chapter we will face the question of whether the modelling and truthmakers enterprises are really in opposition.

21.3 Causal assessment identifying truthmakers

The primary aim of the truthmakers literature is fundamentally different. Here, the aim is to identify or characterize what it is in the world that makes causal claims true. The 'makes' here is a kind of metaphysical making, and the question is not at all to do with particular causal relationships, but to do with causings in general. The truthmakers literature, when it comes to causality, wonders whether probabilistic relationships make causal claims true, or is it natural laws, or wordly capacities? Consider a stock example in the philosophical and scientific literature: 'Aspirin relieves headaches'. Is this true because taking aspirin raises the chances of a headache cure, because there are natural laws by which aspirin affects pain, or because aspirin possesses a capacity to cure headaches? (You can of course fill in your favourite example and ask the same questions.)

In this book, we have met many candidate truthmakers of causal claims. The accounts of causality using capacities, probabilities, regularities and possible worlds can all be read as attempts to give the truthmakers for causal claims. They are often read in this way, within philosophy, although we have also offered alternative interpretations in the relevant chapters, that more directly relate to the five scientific problems.

In this chapter we will use these accounts as examples of stories about the truthmakers of causal claims. Certainly, if you know what the truthmakers of causal claims are, then that is going to be relevant to reasoning about those causal claims. And in turn this will feed through to reasoning about evidence for causal claims. In turn again, the evidence you gather will feed into limitations on your ability to control and predict the world.

For example, if the truthmakers of causal claims are capacities, then what we are reasoning about is abilities of individual objects, abilities they retain—to some extent—when moving from context to context. If aspirin has a stable capacity, it should be true that it relieves headache whether Alice takes aspirin today or tomorrow, or when she's in a different place. We try to find out what capacities are by studying objects of similar kinds, or in different circumstances, to find out what it is that they can or cannot do. But capacities can be hidden in some contexts. You have to do the right thing with aspirin to be able to observe it curing headache, such as take it with water rather than wine. This is relevant to how we seek evidence of the capacity we are looking for. Perhaps capacities have mechanistic explanations and we should seek those mechanisms. Understanding which contexts capacities won't manifest in is then important to prediction and control.

As an alternative example, if the truthmakers of causal claims are regularities, perhaps probabilistic regularities, then what we are reasoning about when we reason about causes are probabilistic relationships. This would suggest that we measure probabilities when we seek to find out about causes. Clearly, we want to examine populations so that we will detect the harder-to-find probabilistic relationships. Again, this will affect prediction and control.

The truthmakers of causal claims don't affect prediction, explanation and control in a simple direct way, but via structuring our reasoning about causes. In the truthmakers approach, the usual view is that there is *one* thing that the causal relation always is, and the aim is to find out what that thing is, with the view that this will make our reasoning about causality—and all our other practices with respect to causality—much more precise. Certainly, if there really is one thing that the causal relation always is, then it would be nice to know what! We don't seem to have grasped that yet, and that's why all the concepts in Part II are important for causality.

A second question is strongly related to the question of the truthmakers of causal claims. This is usually first interpreted as a concern about what makes causal relations obtain. But we also need to know what the relation actually relates—what are the *relata* of the causal relation (Schaffer, 2008; Moore, 2009)? There is an extensive literature on the relata of the causal relation. In accord with the usual approach to truthmakers, the usual aim is to establish, by metaphysical argument, what the one thing in the world that is always the relata of a causal relationship is. There are, in fact, not one but many candidates.

Objects. Hume speaks of one 'object' causing another. Although he probably meant objects quite vaguely, perhaps something like entities, which are a very natural candidate for causal relata. After all, the *ball* broke the window. In science, it makes sense to allow different 'objects' to be possible relata as, for example, a virus, which is an entity, but also exposure to the virus, which is an event (see below) can both be considered causes. People also seem to be important causes in legal and social contexts: *Bob* broke the window. However, most philosophers have sought something more specific to be the relata of the causal relation.

Events. Many philosophers have thought of events as the relata of the causal relation, particularly Donald Davidson and Jaegwon Kim. But notice that they have very different accounts of events! For Davidson, events are singular occurrences that occur once in space and time. However, they may be described in multiple ways. For example, a single event can be described as Alice writing on a piece of paper, Alice signing a contract, and Alice buying a house (Davidson, 1969). For Kim (1976), an event is the exemplification of a property at a time. Kim's account is finer-grained than Davidson's, since Alice instantiating the property of writing, and Alice instantiating the property of signing, which are the same event for Davidson, will be different events for Kim. While both events and objects are concrete and spatiotemporally located, events—on both accounts—are finer-grained than objects, thus allowing more information about the relata, and so about the causal relation. It is not merely Alice who buys the house, but particular features of Alice, such as her being a legal adult, which are properties of Alice necessary for her to be able to buy a house. It also makes sense to allow events as possible causal relata, because we sometimes model systems using state transitions and transition rules. These look a lot like transitions between static events in the system.

Processes. Other philosophers argue that causal relata are processes. For Chakravartty (2008), processes are systems of continuously manifesting relations between objects with causal powers. Processes are extended in time. Note that some if not all events are extended in time, too, such as WWII, which lasted several years. Chakravartty's point is that causal relata are always extended in time, even if that time is very brief, and they are always connected to each other. It makes sense to allow processes as causal relata, because sometimes we have a cause that is continuously varying, and continuously affecting your continuously varying effect of interest, and in that case it looks like you have a causal relationship between two processes. Events, as we have suggested, can also be useful causal relata, allowing a finer graining that can be useful. Often when modelling systems we can abstract time steps and model relations between them. This can be very useful because this yields a cleaner model. However, when you have a complex interacting system, you really have to work to create such an abstraction. It is worth remaining firmly aware of this abstraction, because it will limit the effectiveness of your model, and ultimately might make your model fail badly in its intended purpose. The fact that causal relata seem often to be dynamically extended processes explains why dynamic systems theory is so important—it allows modelling of simultaneous processes that affect each other.

Facts. A number of philosophers have tried to argue that causality is not a relation between events but rather between sets of *facts*. According to Mellor (1995), facts are actual states of affairs that can be expressed in sentences. It follows that if relata are facts, the focus will be on actual, single-case causal relations rather than generic ones. Facts, in Mellor's view, are also particular 'things', which may include people or particular events. So, Alice is a particular, and Alice buying a house is also a particular. Causation links particulars, and thus in a sense it links facts. But to decide what facts are is a matter of philosophical dispute on its own (see e.g. Mulligan and Correia (2013)). So the claim that causation links facts, or is a set of facts, rests on the specific concept of fact, which may collapse into other notions, such as events.

Variables. The final popular candidate for causal relata is variables. Woodward isn't explicit, but his view deals with relations between variables, as do Pearl and other scholars interested in causal models (see chapters 7 and 10). It makes sense to allow variables as causal relata because variables are so often the relata in causal models in science, particularly in quantitative causal analysis (see chapter 7). Further, variables are a particularly interesting possibility for causal relata, as they can vary over objects, events, processes and facts—i.e. *all* the other candidate relata.

So the focus of the truthmakers approach is the question: What *kind of thing* makes causal claims true? The answer is typically given in terms of the nature of the relata, i.e. C and E, and of the the nature of the relation itself, i.e. the 'thing' connecting C and E. The idea is that the right answer to this question will feed into causal reasoning, and thereby may affect the other scientific problems, e.g. causal inference and explanation, because presuppositions about the truthmakers will affect what you will look for—and how you will build the model you use to establish causal claims. This is very different from the modelling literature, which is concerned with how we establish causal claims, and focuses on model validation.

We have now introduced both the modelling literature, and the truthmakers literature. We will go on to compare them.

21.4 Truth or models?

There are undeniably some very important differences between the truthmakers approach and the modeling approach. Models are concerned with validation of causal claims. Once validated, these claims are explicitly model-relative, and they are local and not eternal, since they might easily change with more data and different models, and they are subject to background assumptions made while building the model that has been validated. Traditionally, the truthmakers literature about causality seeks to regiment, to figure out the One Truth about causal relata and the causal relation, and solve that problem for all time.

There is no doubt that if the relata and relation of causality were always one thing, that would be very nice to know. It would sharpen up our reasoning very nicely, and make our causal practices easier. However, one way to read the whole literature about both truthmakers of the causal relation, and causal relata, is as a very long and completely convincing argument that we should allow all of the above as possible relata and truthmakers of the causal relation! If each tradition has made a good case for allowing their favoured relata, then we can react by accepting them all.

Other than tradition, there is no particular reason the truthmakers approach cannot accept multiple causal relations, and multiple relata. Relatedly, there is no particular reason the 'truth' must be rigid. Correspondent (or correspondence) theories of truth hold the view that truth is a relation to reality; to facts out there. You may have come across formulations of this theory like this:

'The snow is white' is true if, and only if, the snow is white.

So we have a proposition, 'The snow is white', and if there are facts or things that correspond to what is claimed, then the proposition is true, and false otherwise. We'll know whether the proposition is in fact true or not by checking the world out there. There are a number of reasons to support or criticize the approach (for a comprehensive overview, see David (2009)). But prior to empirical investigation, why should we suppose that the reality we are interested in has a rigid constraint on what the causal relation and its relata must be? Indeed, the reasons for pluralism about causality we examine in chapter 23 suggest that we have evidence that it cannot be. Allying this with the arguments in the truthmakers literature, we suggest the release of the 'One Truth' straightjacket on the truthmakers approach. The resulting picture changes enormously!

With this change, the truthmakers literature and the modelling literature can be seen as addressing sets of complementary questions. Both concern a worldly relation: the relation of models to the world; the relation of our concepts/language to the world. There may very well be cases where you have some idea what the truthmakers of the causal claims you are investigating are in the particular system you are working on.

For example, you may know very well that you are investigating the capacities of a new protein. This will affect what you put in your model. On the other hand, you may have no idea what the truthmakers of causal claims about your system are, but you have some successful models. This means you can look at your model to get some idea of the truthmakers. So you can move in both directions. You might start from what you know about the causes in your system and begin making guesses about truthmakers, then constrain the rest of your model-building. Or you might start with some idea, say, that you're looking for a capacity of a protein, and choose variables and build a model on those grounds. In this way, the scientific and philosophical questions are complementary. The philosopher is traditionally trying to work out what the world is 'really' like. The scientist phrases this concern rather differently and asks: What should I measure? How should I model it? But if the philosopher wants to keep answers related to causal methodology, and the scientist builds in assumptions about the system itself, these approaches inform each other, rather than being in opposition.

21.4.1 Wait, what happened to truth?

A natural worry is that, having released the straightjacket, we have made all claims model-relative! Does this mean we abandon truth? We have abandoned truth only if truth must be monistic, rigid and eternal. If truth can be plural and diverse, there is no reason to think we have abandoned it.

The possibility of a non-rigid, plural and diverse concept of truth is discussed, for instance, by Floridi (2011b), who examines how all our interactions with the world are mediated by a particular level of abstraction (see chapter 3). What Floridi means by this is that we pay attention to some features of any system we are interacting with—whether we are describing it, modelling it, talking about it or even just shoving it—and ignore other features. Floridi says that a level of abstraction (LoA) is constituted by the set of variables that are attended to. These are chosen for the purpose of the interaction, and the interaction will only be successful if the variables are well chosen. This is just a very general way of saying what we have said about choosing what to model, according to what you want to do with the model you are building. For example, if you are modeling variables in a population to try to find out the causes of child mortality in order to implement policy programmes to reduce child mortality, you had better measure the right variables, or ultimately you will not manage to reduce child mortality.

Floridi offers an answer to the philosophical question about truth whereby truth can be genuinely complex, and diverse, and can depend on modelling practices. This is entirely coherent with a particular answer to the scientific question of what to model. That answer goes: measure whatever you can, whatever might be useful, and see how your models work, and go measure some other stuff based on what you learned. Find proxies if you can't measure what you really want to, and assess how well they work. Don't worry if it's a mess—that's how it goes!

What is important is that we cannot abandon all constraint. Science will be no good if we end up able to make any claim about models. The best way to understand this is to see how models are constrained. Models are constrained in many ways. One way is the background assumptions mentioned above. If our current best scientific theories exclude backwards causation, models are constrained by time direction. We need to find cause-variables that precede effect-variables. Models are also constrained by knowledge of mechanisms. For instance, we may make a causal relation involving both social factors and biological factors plausible by appealing to underlying mechanisms. More importantly, models are constrained by what we *do not* know. We take the scientific enterprise to be constantly striving to produce knowledge and reduce ignorance, but there are also socio-political-budgetary pressures that hold up progress in gaining causal knowledge.

So metaphysical questions about causality, which may arise in the form of some of the constraints alluded to in the previous paragraph, are not just legitimate but also vital (see chapter 22). But this does not mean that an accompanying causal metaphysics has to be monolithic, immutable or eternal. We advocate a form of 'liberalism' about relata: we want as much information as possible, and sticking to just *one* type of relatum will constrain our modelling procedures too much. This is perhaps why variables have become such popular relata in the most recent philosophy of science literature: variables can stand for processes, events, objects or whatever one can measure. This of course raises the issue of another important constraint on modelling: measurement. We cannot measure whatever we want. Some difficulties are technical (do we have an appropriate measurement apparatus?) but others are conceptual (how do we measure 'education', or 'motivation'?). Much research in the social sciences employs 'proxies' precisely to overcome such difficulties: scientists try to measure something close to the concept they are interested in, for instance 'class attendance' in the place of 'motivation'.

Even if we advocate liberalism with respect to relata, it doesn't follow that *anything* can be a cause. Once you establish the LoA, you can also decide what can count as a cause *in that context*. In some cases we need to identify the 'objects' that are causes (like identifying a mechanism, or a biological entity); in some others we need to identify the 'variables', or the 'events'. For instance, a biologist studying *E. coli* will rightly say that the *bacterium* causes gastroenteritis. The interest here is to identify the right 'object' responsible for the effect. But for an epidemiologist studying the *E. coli* outbreak in Germany in Spring 2011, causal relata will rather be variables recording exposure to the bacterium, or eating supposedly poisoned food, or hygiene measures such as washing hands, and so on. The epidemiologist will then have to work out their correlation with symptoms typical of gastroenteritis, and thus posit that strong, positive correlations will provide evidence for the 'right' cause of the outbreak. (It should turn out that this is presence of *E. coli* in a sufficiently large number of individuals in the sample.) But surely probabilities aren't enough, and in fact chapter 6 gives an overview of the different kinds of evidence that may enter into causal assessment; chapter 7 provides an overview of the main types of models to probe the data and chapter 9 explains how we can use counterfactual reasoning in causal assessment in different contexts.

The truthmaker strategy is legitimate in contexts where we wish to constrain our reasoning, e.g. if we want to be alert for particular problems. If we know already about *E. coli*'s behaviour, we might look for a specific capacity developed by one strain of the bacterium. So lab experiments and analyses should really focus on the capacity aspect and worry about the masking problem (see chapters 6 and 14) in the search for evidence. But a 'narrow-minded' truthmaker strategy may be unhelpful in scientific causal assessment. For instance, what does it mean to ask whether it is *true* that smoking causes cancer? We routinely ask such unspecific questions, to which science has no answer, precisely because science tries to establish specific answers, such as: What type of smoking? Cigarette? Pipe? Joints? How much? How long for? What type of cancer? In what people? . . .

In sum, the truthmakers strategy and the model validation strategy need not be seen in opposition to each other. They give different answers to different facets of causal assessment, and they can inform each other.

CHAPTER 22

Epistemology, Metaphysics, Method, Semantics, Use

22.1 Fragmented theorizing about causality

An important lesson taught in philosophy courses is about category mistakes. If Alice asks Bob whether he would like tea with milk or with sugar and lemon (according to the traditions of different countries) and Bob answers, 'Yes, please', Bob does reply to the question, but his reply doesn't help Alice understand what he would like to be served. That's a category mistake. Category mistakes can and do happen in the philosophical debate, including the causality debate. If Alice gives a talk about what productive causality is and Bob objects that the Bayesian nets formalism allows one to capture all the necessary dependencies, Alice and Bob are talking at cross purposes.

In this chapter we categorize philosophical questions about causality according to whether they tackle questions of (i) epistemology, (ii) metaphysics, (iii) method, (iv) semantics or (v) use. We comment on the importance of drawing conceptual boundaries between them, explain reasons for asking these questions separately and investigate what bearing they have on each other.

There is some overlap between these five philosophical questions and the five scientific problems we have been working with (inference, explanation, prediction, control and reasoning), reflecting the slightly different priorities of the philosophical and the scientific literatures. We will comment on this too. This chapter is primarily intended to help those who are beginning to study philosophy (particularly philosophy of causality) navigate through the huge and fast-growing causality literature in philosophy. It may also help those from other disciplines, particularly the sciences, who wish to make an intensive study of the philosophical literature, and need to understand how the philosophical framing of questions lines up with their concerns. Initially we concentrate on disentangling the five philosophical questions, and later we suggest how answers to questions posed in one area may influence answers to questions in posed another area. This means the five questions should not be treated as wholly independent, after all. We argue that in studying causality it is important to define the scope of the question asked, and therefore the scope of the answer. Once this has been clarified, a more fruitful debate is more likely, using relevant examples and counterexamples, and avoiding misdirected objections (see also chapter 20).

In the following, we discuss a possible categorization of the 'research domains' in the study of causality, based on five different kinds of questions you will see asked when you read the philosophical causality literature.

22.1.1 Epistemology

Epistemology studies human knowledge—how we get to know what we say we know, and how we should understand key concepts involved in our cognitive inquiries such as belief, knowledge, truth or evidence. The epistemology of causality asks how we *know* causal relations, what notion or notions guide causal reasoning, or what concepts of knowledge and evidence and so on underpin causal methods. Epistemology is more general than methodology, but the line between them is thin, as a method is based on several epistemological assumptions, although these may not always be explicit. The epistemology of causality also asks questions about our relation to the external world—what grants us epistemic access to reality, whether reality is detached from the epistemic agent, or whether the epistemic agent, in trying to find out about reality, interferes with it.

This philosophical question most naturally relates to the scientific problem of causal inference—how we know about causes.

22.1.2 Metaphysics

Metaphysics studies what there is—the fundamental nature of the social and natural world we study. Metaphysics is concerned with very general questions about the nature of our experience and the world. What is the basic stuff out there—entities, relations, or what? But it is also concerned with more local, specific questions. Are electrons real? Are genes? The metaphysics of causality asks: what *is* causality? What are the features that causes must have in order to be causes? What *are* causes? What kind of entities are causes? Conceptual analysis is a methodology often used in philosophy to 'dissect' the meaning, underpinning, content and use of concepts in a given context. It can be seen as part of the methodology of metaphysics in so far as it aims to 'fix' what is in the world, as the role-fillers of our concepts. (See the Canberra Plan in chapter 19.) Metaphysics of science usually starts from what science says about the world, while conceptual analysis usually proceeds from philosophical intuitions about what the world is like, but the goals are similar (see also chapter 19 and section 22.3 below).

This philosophical question is most distant from the five scientific problems, but we have shown in chapters such as 14 and 21 how metaphysical assumptions can affect our causal reasoning, and so shade our thinking about all five scientific problems.

22.1.3 Method

Methodology studies the development and implementation of procedures apt to address specific scientific problems, such as the ones presented in chapter 2. Methodology is the core business of the sciences, because it is through methods that they make discoveries, analyse data and so on, to come to conclusions. Philosophy also has a methodology, as different ways to tackle questions are possible. Methodology is constantly under scrutiny, and tends to be improved in its ability to deal with specific problems. So methodology is far from being monolithic. The methodology of causality asks how to *study* phenomena in order to disentangle, discover, establish, etc. causal relations. Scientists have developed and implemented methods that are tailored to specific scientific problems. For instance anthropologists may need different methods than, for example, computer scientists. We discuss evidence-gathering methods in chapter 2, and data-assesment methods, mainly in the quantitative tradition, in chapter 7.

This philosophical question is the most directly related to a particular scientific problem: the causal inference problem.

22.1.4 Semantics

Semantics is the study of meaning. It analyses the relation between words and other 'signifiers' such as numbers and what they denote, or stand for. In linguistics, semantics is closely related to other fields of investigation such as semiotics, lexicology or pragmatics. All this serves to understand human language and communication. Semantics can also be investigated specifically with respect to causality. For instance, we may ask what the *meaning* of causality is in its ordinary or in its scientific usage. This was the motivation of some approaches to causality that studied causal language and it also informs the Causality in the Sciences (CitS) approach, in so far as it is interested in understanding the scientific meaning and use of causal terms (see chapter 19). Conceptual analysis is also used here, to understand the concepts that lie behind the quirks of a particular language. We assess what our concept of causality really is, to help us understand what our causal language means. The semantics of causality can be used to help answer epistemological questions, and also to answer metaphysical questions, depending on whether conceptual analysis fixes the metaphysics or the epistemology of causation.

This philosophical question affects the scientific problem of causal reasoning, and so all five problems. For example, we show in chapter 9 how muddled assessment of counterfactual claims, which are important pieces of language used when talking and thinking about causes, can impede reasoning about causes.

22.1.5 Use

How do we *use* causal relations? Why do we bother establishing *causal* relations? Such questions are receiving increasing attention, thanks to Donald Gillies, who put forward an 'action-oriented' theory of causation in 2005 (see chapter 17), and to Nancy Cartwright, who has been pushing philosophers of causality to think about policy issues and search for a dialogue with practising scientists and policy-makers. This is probably the youngest of the domains presented here, but it is receiving increasing attention now, possibly aided by the pressure from funding bodies in favour of impactful research. Cartwright (2007) argues that our conceptual work on causality cannot afford to ignore how we *use* causes, or we'll get the analysis wrong. She points out that we often have more than one use for causes, and one of the things we need to do is think about whether the evidence we have for our causal conclusions justifies how we intend to use them (see chapter 6 for more detail). Philosophers are also increasingly concerned with why conceptual work itself is useful; why we are providing accounts and theories of causality; why this helps scientific practice and decisions. All this should foster a genuine interdisciplinary culture.

Questions about 'use' are not unrelated to methodological questions. Methodological concerns do include issues like what you want to use the results for, or what conclusions can be 'exported' from one study to another. But the emphasis on 'use' reminds us that we study the natural and social world for a reason, and we also conceptualize causal relations for a reason. Understanding our practices concerning causality does need to take into account the multiple uses we have for causes when we find them—see for instance chapter 6. The CitS approach is sensitive to this aspect of theorizing. This philosophical question relates to the scientific problems of explanation, prediction and control, which are distinct ways we try to use causes, once we find them.

It is because we aimed to present the philosophical literature on causality to those outwith the discipline that we recast the philosophical accounts of causality in terms of problems that scientists will more readily recognize. But it can be seen that answering the philosophical questions does help address the scientific problems, if you are prepared to hunt a bit in the philosophical literature.

22.2 Which question to answer when?

In presenting the views in Part II, we highlighted the five scientific problems that different accounts encounter and also suggested how these accounts, in spite of problems that have become known in the causality literature, can still be useful to address the scientific problems of inference, explanation, reasoning, prediction and control (chapter 1). This means that the views and accounts in Part II can be interpreted differently as answers to different philosophical questions in ways we will illustrate shortly. The variety of concepts and accounts of causality presented in Part II is evidence that causal theory is immensely rich and we suspect that no single account can exhaust, on its own, this diversity, assisting with all five scientific problems, in all domains. Thus, philosophical debate will be most useful not in proving some other account wrong, but in helping shape the borders of the account, allowing different pieces of the mosaic of causality to be developed to fit together and help explore all aspects of causality in the sciences (see chapters 20 and 24).

Let us give some examples of interpreting accounts as answers to different philosophical questions, and of the debates these differing interpretations have generated.

Counterfactual dependence. Lewis's work (chapter 9) is very much in the Canberra Plan tradition, so his analysis of counterfactuals arguably aims to give an account of what causality is, by clarifying a folk-psychological concept of causality. Lewis's strategy rests on the assumption that analysis of language illuminates ontological questions (see chapter 19). This means that counterfactuals are answers to metaphysical questions. Consider now Hall's two concepts of causation, that is dependence and production. If you accept Hall's distinction, then counterfactual dependence has trouble providing an account of *productive* causality. This means that if you think *only* productive causality is real, counterfactual dependence doesn't say what causality in the world is. Would that mean that there is no question counterfactuals successfully answer? You would have to say what your stance is, by clarifying what question you are interested in answering. Alternatively, despite the objections raised to counterfactual dependence, you may still want to take counterfactuals on board. Counterfactuals are widely used in ordinary, scientific, legal language, and so on, and one can suggest that they are good for reasoning, as a heuristic tool for the selection of hypotheses, and for giving an account of difference-making relations. In such cases counterfactuals are useful as part of an answer to epistemological and methodological questions.

Evidence. The 'Russo-Williamson Thesis' (chapter 6) was formulated as an *epistemological* thesis about causal assessment. Thus, in the mind of the proponents, the RWT answers an epistemological question. The RWT has a descriptive component and also a normative component. Some commentators have objected that in some cases in the history of medicine decisions have been made without knowledge of the underlying mechanisms. The question is therefore whether the RWT is also a thesis about decision-making, and thus is also an answer to a question of use. It is possible that for the RWT to guide decisions, amendments to the original formulation are needed. There is also a question about the RWT and its relation to the metaphysics of causality. You may want to interpret the RWT as a thesis about what causality *is*, in which case you would have to explain how dependence and production both constitute causation.

Manipulation. The manipulability account developed by Woodward (chapter 10) provides a conceptual analysis of causality and also methodological precepts. So one could locate the account simultaneously as an answer to semantic and methodological questions. Some commentators (see e.g. Strevens (2007) and Strevens (2008)), however, argued that the account can't answer both questions simultaneously, since the methodology that seems to follow from Woodward's conceptual analysis is too restricted to apply satisfactorily to important scientific fields where manipulations are not performed. This means that manipulationism could be construed as an answer to a more narrow methodological question, and a question remains about its answer to semantic questions.

We leave it to the reader, according to their interests, to think how the views and accounts discussed in Part II can be interpreted from different perspectives, as answers to epistemological, metaphysical, methodological and semantic questions, or questions of use. Generally, most accounts address a combination of these questions, rather than just one.

In light of this, it is good practice to explain where one's theorizing about causality stands with respect to these questions. Is the question being asked about the nature of causation? Or about causal reasoning? Or about decisions based on available causal knowledge? Muddling these questions can muddle accounts, and inhibit useful debates between proponents of accounts primarily concerned with different questions.

22.3 Which question interests me?

We said that it is *conceptually* useful to separate these approaches and to explain the different scope of philosophical questions and answers about causality. It is a matter of personal interest which question or questions of causality are of primary interest. Scientists are more likely to start with questions of epistemology and method, for the obvious reasons that led us to focus on the five scientific problems; policy-makers are likely to pay attention to questions of use; while many philosophers, for traditional reasons, prefer to begin with metaphysical questions. But perhaps these questions are not best treated as wholly independent. That it is useful to be clear about which question or questions are being addressed does not imply that theorizing about causality should remain entirely fragmented.

For example, suppose that the question that initially interests Alice is a question of semantics: what do we mean by our everyday use of causal terms? Should Alice treat this question as completely independent of all the other questions? Well, assuming that what we mean by our everyday notion of cause must successfully track worldly causes—at least to some extent—then an answer to the semantic question of what we mean by our causal terms will impose *some* constraints on an answer to the metaphysical question of what causes themselves are. Further, assuming our scientific notion of cause cannot be wholly disconnected from the everyday notion of cause—whether the scientific notion alters the everyday notion over time, or vice versa—then the answer to the semantic question will impose some constraints on answers to, say, questions of the epistemology of causality, of how we find causes, within science. Answers to what we use causes for, when we find them, should also impose some constraints on what we mean by our everyday notion of cause.

As an alternative example, consider Bob, who is primarily interested in questions of metaphysics. But if we have an answer to the question of what the nature of causality is, then surely that would impose some constraints on the answers to methodological questions, such as what should count as evidence of causes. Further, it seems that an answer to questions of the nature of causality should also impose some kinds of constraints on what we can use causes for. Ultimately, if an answer to what the nature of causality is imposes *no constraints whatsoever* on the methodology and epistemology of causality, on what our causal language means, or on what we can use causes for, many people will be in a bit of a puzzle as to why we should address metaphysical questions at all.

The constraints imposed by the answer to one question on answers to the others might be fairly minimal, if the relations between such questions are not very strong. But even such minimal constraints imply that it is not the best approach to ignore the flourishing literature on the other questions of causality within the causality literature itself. Whatever question is of primary interest to you, it is worth having a look at developments concerning the other questions within the literature.

Note that we are not saying there is any kind of fixed order of importance, or priority, among the various questions: either priority of justification or even merely of temporal priority. As a matter of fact, we presented them strictly in order of alphabetical priority! We hold that which question to begin with is a matter of personal interest. The methodological decision of where to start theorizing—in metaphysics over epistemology or in use over semantics—is decided by the researcher, according to their interests and goals. Choosing which to take as a primary focus is, we believe, a useful piece of advice for a graduate student who needs to narrow down the scope of their thesis. But it is also a useful piece of advice for a scientist who is interested in addressing causal issues. The question that follows will then be how different approaches can cohabit, that is whether we manage to develop approaches that complement each other, rather than conflict with each other.

22.4 Should we integrate the fragments?

Many philosophers will accept the minimal constraints we explain in the section above. The question of whether theorizing about causality should be even further integrated is, however, controversial. We hold that it should be treated as further integrated, and we explain here why. We begin with the difference between offering concepts of causality, and saying what causality is, but we will explain how the other questions above also become involved in the overall project of understanding causality.

There are two different ways of thinking about core questions about causality; two core goals of philosophical work:

> **Goal 1:** Conceptual design, or engineering, of concepts that are more suitable for science than previously existing, now stressed concepts.[34]
> **Goal 2:** Saying what causality is, according to science.

Meeting these goals *might* be considered as two largely separate activities. Since, presumably, our concepts track the world in some kind of way, we may have to accept that answers to the first question of what our causal concept is, and to the second question of what causality in the world is, must constrain each other in *some* way, as we explained in section 22.3. But this could be regarded as a pretty minimal constraint,

---

[34] We take conceptual design or engineering to extend more widely than conceptual analysis, in the way laid out extensively by Floridi (2011a).

leaving the two activities largely independent. After all, they do seem to be quite different, at least prima facie. On the one hand, the first question seems to be about the conceptual level, that is about our ideas, notions, concepts. In this sense, it is part of the legacy of the Canberra Plan (see chapter 1), albeit updated and with current practice in science put more centrally. On the other hand, the second question seems to be about the world, that is about what causality is as a worldly relation. Put this way, the two questions seem to be reasonably independent.

However, if we add in some considerations about the methodology of philosophy itself, the activities of answering the two questions begin to look more closely related. This can be seen by asking the following methodological questions—questions of *philosophical* methodology—for each activity:

> **Goal 1:** Conceptual design, or engineering, of concepts that are more suitable for science than previously existing, now stressed concepts.
> **Methodological question 1:** What kinds of *constraints* should we look to in order to inform our conceptual design?
> **Goal 2:** Saying what causality is, according to science.
> **Methodological question 2:** How do we go about *figuring out* what causality is, according to science?

Meeting the two goals could be independent if there was a way of improving our concepts of causality (answering the first question), that did not turn to questions about what causes are (answering the second question), and was independent of our scientific practices concerning causality; *and* if there was a way of finding out what causes in the world are that could proceed entirely independently of our concepts of causality, and our scientific practices concerning causality.

However, there may be no route to answering question 1 independently of answering question 2; and no route to answering question 2 independently of answering question 1. Suppose that we know about the deeply theoretical things in the world, such as causality, by looking at our best scientific theorizing, which involves our conceptualization of the world. In other words, our best scientific practices are pervaded with conceptualization. This means we have to look at our concepts *in practice*, in the empirical engagement of the sciences, and in the history of science, in order both to design better concepts, and to say what the nature of the world is, according to our current best knowledge. *This leads us to give the same answer to both methodological questions*:

> **Goal 1:** Conceptual design, or engineering, of concepts that are more suitable for science than previously existing, now stressed concepts.
> **Methodological question 1:** What kinds of *constraints* should we look to in order to inform our conceptual design?
> **Methodological answer:** We should look to scientific practice concerning causality, our epistemology and methodology, how we use causes when we find them, and how we use concepts of causality themselves.
> **Goal 2:** Saying what causality is, according to science.
> **Methodological question 2:** How do we go about *figuring out* what causality is, according to science?

Methodological answer: We should look to scientific practice concerning causality, our epistemology and methodology, how we use causes when we find them, and how we use concepts of causality themselves.

Looking at 'scientific practice' includes examining current science, the history of science, and other aspects of studying the sciences as explained in chapter 19. If both methodological questions get the same answer, then the two practices do not look so very different any more. Notice that the methodological answer brings in the other theory fragments; the questions of methodology, epistemology, semantics and use discussed earlier in this chapter.

In our view, when working in the CitS approach it is perfectly legitimate to pay more attention to one question at one time, and at other times to pay more attention to other questions, depending on the researchers' interests. The literature in the CitS tradition seems to agree on two things. First, science in practice is our best source of problems to solve. Second, science in practice is our best source for constraints on our theorizing about causes—whichever question we are trying to answer. On these two shared assumptions, the enterprises (1) and (2) come to much the same thing. There is a tendency for those who have more 'pluralistic' inclinations to think in terms of concepts (1), and for those having a more 'monistic' inclination to think in terms of what causality is (2), but there is nothing to force such a choice, for reasons we lay out in chapter 21.

One lingering objection to these arguments is that philosophical work concerns what is possible and what is necessary—what causality has to be, or what causality could be. Examining this world is quite different, telling us only what actually *is*—it tells us only what causality is here now, not what it has to be or could be. However, many now disagree with this sharp separation. The concepts of necessity and possibility, and any sense of necessity or possibility of other concepts, are themselves concepts formed in our evolutionary, personal developmental and cultural engagement with the actual world. No naturalist can believe otherwise. So anything we believe about possible and necessary causes must itself also come from interaction with this world.

Ultimately, then, we are suggesting that the enterprise of advancing our thinking about causality involves all the philosophical questions above: methodology, epistemology, metaphysics, semantics, use. It is a continuum, passing through epistemology, metaphysics, methodology, and semantics, which all feed questions about use of causal knowledge—see Figure 22.1.

The overall project of understanding causality is a large project; a project for a field as a whole, rather than for one individual researcher or small group. The causality literature is currently worrying about all the questions above, although it doesn't always manage to put answers to the different questions together effectively. We hope this chapter will help to foster and enhance collaborations between those working on different questions. Ultimately, we hope that the idea of a causal *mosaic* will show how a coherent picture can be constructed from the tiles of the Part II accounts, with attention to the five philosophical questions about causality, and the five scientific problems (see chapter 24). The mosaic, however, is to be dynamic, rather than static— to be something that moves with changing science, and the fast-moving literature

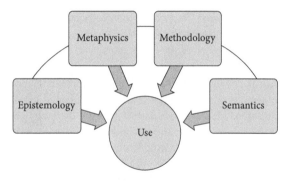

Fig 22.1 Relations between epistemology, metaphysics, methodology, semantics and use.

on causality. We merely try to give a temporary order and shape to the very many interesting ideas available now.

Note we are not here negating the message of the first half of the chapter. Confusion about which question is being asked and answered can easily impede debate. It is always worth thinking about which question or questions are the primary concern in any particular philosophical work. Nevertheless, we are saying, it is also worth thinking how the different answers inform each other, even if they do not inform each other in a simple way. The overall enterprise is fascinating.

In this book we talk about concepts, framing the chapters in Part II as introducing different concepts of causality. We think it is easier for scientists to grasp that enterprise, and see how it is both genuinely philosophical, and of interest to scientific concerns. We have noted answers to some of the five philosophical questions, with a light touch, in the chapters in Part II, and we encourage those who are interested in philosophical methodology to re-read those chapters with a view to the five questions introduced here.

PART IV

Conclusion: Towards a Causal Mosaic

Philosophers have worked towards developing theories of causation that would account for as many cases as possible and that would escape as many objections as possible. In the 15 chapters of Part II we discussed the main accounts, concepts, and ideas that have resulted, and discussed their limits and strengths. We want to suggest, however, that the idea of developing One True Causal Theory is not promising.

Philosophically minded scientists and scientifically oriented philosophers ought instead to engage in the enterprise of filling in the huge causal mosaic of empirical studies and theoretical investigations. In a mosaic, in order to see the whole picture, we have to put each element in place, painstakingly, tile after tile. This is what we have to do with respect to causation, too. Thus, there are no matters of principle to reject one or the other approach as such. The issue is rather to identify precisely the type of question we wish to ask, and the account, concept or method that best does the job we are interested in. In the following chapters, we take a stance on some important philosophical debates.

CHAPTER 23
Pluralism

23.1 If pluralism is the solution, what is the problem?

Part II presented various concepts and accounts developed and discussed in the philosophical and scientific literature. Part III explained that there are different traditions; different angles from which to approach causal questions. Are the various accounts we present and the different approaches we describe in opposition to each other? Or are they compatible? If they are in opposition, how can we choose the 'right' account, and the 'right' approach? If they are compatible, how do they go together? A number of the accounts discussed in Part II were intended by their advocates to be 'monistic' theories, that is, they were intended to be complete causal theories, all by themselves. To date, though, none of the accounts of Part II has been generally held to succeed in that ambition. We can find objections and counterexamples to show that any given account fails to capture at least some situations that we would consider causal, and we have presented the well-known ones.

In response to this, *pluralism* may be suggested as the solution: you don't have to choose *one* account, just take them all, or a carefully selected subset of accounts, to handle all the causal situations you need to think about. Pluralism may seem an easy way to avoid taking a stance about the 'right' account of causality. However, since many forms of pluralism exist, there is also a question of choosing the 'right' form of pluralism! It may turn out that several forms of pluralism are compatible and successful, thus giving the impression that we don't have to choose ... but in fact we do, because we still need to justify, to give an argument for, a form of pluralism, or indeed pluralism*s*.

In chapter 24 we defend *a form* of pluralism, but not as a way of avoiding choosing an account of causality. Nor do we advocate reaching for any account ad hoc. Our effort in Part II has been to show the variety and diversity of causal concepts and accounts, and show why each is useful. The sections 'Core ideas' and 'Distinctions and warnings' in each of the chapters of Part II were meant to show what problems a given account can solve and what problems it has more difficulty with. But the selection of accounts to put in the 'pluralistic mosaic' is not ad hoc. We hold that no single theory of causality currently available can meet the very varied needs of the sciences, and address the problems of causal reasoning, inference, explanation, prediction and control in all the diverse cases that we meet in the sciences. Instead, we suggest that causal concepts and accounts are placed into the mosaic in so far as they help deal with these scientific problems, set out in chapter 1, and discussed throughout the book.

We will begin our discussion of pluralism with examples of pluralistic and monistic stances defended in the literature, showing that these views are pluralistic or monistic about very different things. We will then make explicit what form or forms of pluralism we accept, and explain how this kind of pluralism contributes to the idea of a mosaic of causality (see chapter 24).

23.2 Various types of causing

This first form of pluralism concerns the diversity of causes we find in the world. Primitivist pluralism can be thought of as the view that we cannot give a general account of causation, because there are multiple types of 'primitive causings'. These are captured in our language. The idea is that pulling, pushing, blocking, sticking, breaking, binding, lifting, falling … and many others are all causal terms. Each such term has richer content than the bare word 'causes'. For example, 'X pushes Y' tells us not just that X causes Y, but how, and what aspect of Y, X affects. This position was famously held by Elizabeth Anscombe (1975). The richness of these causings is vital to causal explanation in at least some domains, one being biochemistry, which uses an extremely rich and varied language to describe its mechanisms. Anscombian pluralism also inspired current views such as those of Nancy Cartwright (2004).

Anscombe is sometimes criticized for not giving a general account of causality, but this is an unreasonable objection, as she argues that such an account is impossible. Anscombe held that there is no single property or group of properties that all instances of causing have in common. So no general account of cause can ever be given. One can merely characterize typical cases. In this view she follows the famous philosopher Wittgenstein, who persuaded the philosophical community that necessary and sufficient conditions for falling under a concept cannot always be given. Wittgenstein's famous example is the concept 'game'. The best way to see this is to try to give a property all games share, that no non-games share, and see that this is at the least extremely difficult. Some games have teams, some don't. Some are energetic, like football, but some are board games. Some are solitary. Not all have a winner or a loser, like skipping games, and so on.... You may object to Anscombe that you wish for a general account, but it is not reasonable to criticize her for failing to do something she considers impossible.

A form of pluralism similar to Anscombe's can also be found in the field of legal causation. Although legal texts do not always employ explicit causal terminology, what is meant when talking about assessing responsibility and liability for facts is nevertheless about cause-effect relations. Moore (2009, p5) explains:

Sometimes the liability rules of the criminal law and of torts do not use the word 'cause'. Sometimes they use what linguists call 'causative' verbs, those verbs that seem to require that there be some causal relation even if they do not use the word 'cause', such as verbs like 'kill', 'hit', 'penetrate', 'disfigure', 'abuse', etc. So, for example, liability doctrines that do not explicitly use the word 'cause' nonetheless define homicide (a wrongful act) as *killing*, not causing death; battery as *hitting*, not as causing contact; rape as *penetrating*, not as causing penetration; mayhem as *disfiguring*, not as causing disfigurement; child abuse as abusing a child, not as causing abuse, etc. On the surface at least, the law of torts and of crimes treats this second form of liability doctrines as equivalent to the first. That is, 'killing' is treated as equivalent to 'causing death', 'hitting' with 'causing contact', etc. If this equivalence is true, then whenever the law uses causative verbs it is requiring causation as a prerequisite to liability fully as much as when it uses the word 'cause'.

Aristotle famously developed the 'Doctrine of the four causes'. His theory can also be seen as setting out different types of causing. In the *Posterior Analytics* and also in the *Physics*, causes are the first principles—the explanatory factors—through which we explain something. In *Physics* (II3) and *Metaphysics* (V2), Aristotle presents his Doctrine, which is supposed to apply to physics and biology, but also to action and production, such as artistic production (for an overview of Aristotle's Doctrine, see e.g. Falcon (2012)). Four types of cause are distinguished: material, formal, efficient and final. The *material* cause is 'that out of which' something is done; the stock example is that bronze is the material cause of a statue. The *formal* cause is 'the account of what-it-is-to-be'. To continue with the example of the statue, the shape is its formal cause, i.e. what gives it the shape it has. The *efficient* cause is 'the primary source of the change or rest'. This is the meaning of cause closest to our modern concepts, specifically of productive causality. The *final* cause is 'the end, that for the sake of which a thing is done', namely the purpose of doing something. Aristotle thinks the four causes he identifies are the ones to use in giving an explanation for something. We use all four, as each grasps different aspects of the reason for something, although in his account final causes have explanatory priority.

23.3 Various concepts of causation

We have just seen that one form of pluralism concerns diverse types of causing in the world. A different view is pluralism about our concept of causation—how we think about causation. Causation is whatever it is in the world, but we think about it, that is we conceptualize it, in different ways. This is an idea that has been wandering around the literature for a long time, and it is not unrelated to the pluralism above. For instance Hart and Honoré (1959, p17) say that they accept

> [...] the possibility that the common notion of causation may have features which vary from context to context, that there may be different types of causal inquiry, and that there may not be a single concept of causation but rather a cluster of related concepts.

What are such concepts, then?

The most famous current conceptual pluralism was developed by Ned Hall (2004). Hall argues that we have two causal concepts: dependence and production. We examine this distinction in detail in chapters 19 and 20. In the broadest terms, dependence concerns a relation between cause and effect, without concern about what happens in between cause and effect, while production concerns the linking between cause and effect. Hall thinks that most cases of causation are cases of both production and dependence, but sometimes we get one without the other. In these cases we are happy to shift to reasoning just about production, or just about dependence.

Weber (2007) argues for a pluralistic stance within the social sciences. He says that social research is in need of a concept of causation at the population level, of a concept of probabilistic causality, and also a concept of causal mechanism.

Different forms of conceptual pluralism cut across each other. For example, for Hall, we get both production and dependence concepts in many different scientific domains. So the question is whether there is one concept of causation covering all scientific domains, whether different concepts are required for different scientific domains, or whether even within one scientific domain we need more than one concept of causation. Such issues have been discussed by, for example, Longworth (2010) and Godfrey-Smith (2010).

The other question is how pluralism about the concept of causality goes with pluralism about diverse causes. The answer is: many ways! It is perfectly possible to hold that there are as many causal concepts as Anscombian primitive causings. Alternatively, there could be production and dependence as concepts, along with diverse Anscombian worldly causings. An interesting view is monism about our *concept* of cause, allied with pluralism about diverse causings, and also other things. Williamson's epistemic theory is monistic about the concept of cause (Williamson, 2005), while Russo and Williamson (2007a) have argued that the epistemic theory still requires pluralism about causal inference, evidence and methods, which we will discuss shortly.

23.4 Various types of inferences

In chapter 18 we presented accounts that try to explicate causality through our inferential practices. These theories have an interesting common motivation. Williamson and Reiss explicitly say that they are unconvinced by any previously existing monistic theory, and it seems reasonable to infer that Beebee and Spohn, who take a similar approach, find them similarly uninspiring. Instead, these theories attempt to reject pluralism while respecting the real diversity that generates pluralism.

There are extraordinarily many things we seek in causal inference, and multiple methods we use, which Reiss and Williamson suggest inspires pluralism about *inferential practices*.[35] While Reiss focuses on what he calls the 'inferential base' and

---

[35] Williamson, as we explain in chapter 18, provides his own conceptually monistic theory, which is supposed to account for the variety of inferential practices. So while Williamson holds a monistic view about the concept of causation, he holds a pluralistic view about evidence and inference—see also chapter 6.

the 'inferential target'—i.e. what we base our inferences on and what we perform inferences for—Williamson illustrates how Bayesian net methodology can be used for different types of inferences. Both agree, however, that there is lots of diversity in the inferences we routinely draw. Some concern prediction, others explanation, or causal efficacy, or the effect of policy interventions, etc. (see chapter 6). The thought is, there is nevertheless a *unified job* that we use all these things for, and that is where we find any unity in the concept of causality. Seeking some kind of unified guidance for our actions is also our reason for worrying about causality. This is of course very close to Nancy Cartwright's worry that most accounts of causality neglect our *use* of causal relations, which we discuss in chapter 22.

This generates the idea that causality is what we use the concept for: to help our reasoning processes as agents who need to come to have beliefs about the world to navigate the world. In some ways this might accord very well with how science approaches its endeavours—in a highly pragmatic way, seeking whatever evidence can be found, trying to think of ways to test causal conclusions, inferring what we can, on the evidence available, and gradually modifying the picture of the world along the way. The epistemic and inferential approaches try to take account of the whole of this practice in their thinking about causality.

23.5 Various sources of evidence for causal relations

Chapter 6 shows that establishing a causal claim may mean very many different things, from finding the productive cause to establishing efficacy of treatment to establishing how *much* of a cause is needed to produce an effect. It also shows that in order to establish a causal claim we may need to seek evidence of different things. This has been maintained by Russo and Williamson in what is now called the 'Russo-Williamson Thesis' (RWT), which holds that in medicine, typically, we seek both evidence of difference-making (Hall's dependence) and of underlying mechanism to establish causal claims. The RWT is pluralistic about the *sources* of evidence. We also have multiple *methods* for gathering evidence, discussed in chapter 2, and for probing the data we have gathered, discussed in 7. This is yet another form of pluralism (see below).

This debate is about the *evidence* that support causal relations. This is related to how we gather evidence, which is a different, albeit related, issue (see the next section).

23.6 Various methods for causal inference

Pluralism about methods is closely related to pluralism about evidence. In chapter 6 we distinguish between evidence and evidence-gathering methods, and in chapter 7 we present various methods for causal analysis. What we present in those chapters does not exhaust the variety of available methods, but at least offers a primer on them. One reason to insist on the need for a variety of methods is that each of them has merits (as well as limits) in particular scientific contexts.

However, causal methods are not fragmented in the sense that they are unrelated and disconnected. After all, various methods for causal inference are still instantiations of a general *scientific approach* in the sense that they all instantiate our empiricist stance in studying the world. The distinction between, for example, experimental and observational methods in chapter 2 is more of a pedagogical distinction to illustrate different data-gathering and inference techniques, adjusted to the context of inquiry. But such a distinction should not entail the superiority of one method over the other.

This is a delicate point, and providing a full argument for such a claim would lead us far away from the main track. Nonetheless, this is the right moment to pause, and to reflect on an issue that goes beyond causality per se. We already made some relevant points in chapter 2. The point, boiled down, is to deconstruct the idea of a gold standard, or an 'in principle' superior method, based on the following considerations. A good, carefully planned, observational method is better than a bad, sloppily performed experimental study. This is also to say that quality is not an intrinsic property of studies, but something we should strive for, together with rigour, intellectual honesty, etc. As a consequence, studies ought not to be compared with, or evaluated against, a single golden method, but rather with other comparable methods. One reason this is needed is that we (scientists, ethicists, research policy officers, philosophers, the public ...) put boundaries on what it is practically or ethically possible to experiment on. The imposition of boundaries is partly a reaction to experiments that have been carried out in the past, that pushed forward our knowledge immensely, and yet we think they shouldn't have been done. Would you force people to smoke? Well, in the name of science, something close to that has been done. So declaring the 'in principle' superiority of the experimental method may open doors to science we don't want to have open, after all. Instead, we should work on the idea of *complementarity* of the experimental and observational method, because they both achieve a lot, and may achieve even more if used in a synergetic way.

John Goldthorpe (2001) expresses similar concerns in sociology—his own field—and explains:

> The other, contrasting reaction is that to be found most fully argued in the work of Lieberson (1985). This entails a straight rejection of the attempt to impose the experimental model (or, at any rate, that adopted in medical or agricultural research) onto sociology, on the grounds that this represents an undue 'scientism'—i.e. an undue regard for the form rather than the substance of scientific method—and with the implication, then, that sociologists have to find their own ways of thinking about causation, proper to the kinds of research that they can realistically carry out and the problems that they can realistically address.

This is an important idea. It suggests being 'libertarian' about methods, as we cannot impose methods from one field onto another, due to the need to assess their strengths and weaknesses in the particular context and inquiry. It also suggests that there is nevertheless a certain unity in scientific method, which we take to lie in methodological rigour, from the study set up until the interpretation and communication of results.

23.7 The pluralist mosaic

It is time to wrap up and explain where *we* stand with respect to the plurality of pluralisms currently available in the causality literature! After all, we are working towards creating a causal mosaic to be completed in chapter 24.

In our view, the various pluralisms currently held in the philosophical literature on causality are extremely important, and must be taken seriously. They have been irritants, like the Socratic gadfly, impeding the development of an easy monistic account of causality—and this is good. They have forced the causality literature to increasing sophistication, and better engagement with the real diversity of causings in the sciences, and also with the different kinds of problems of causality, the different things needing reasoned about, in the sciences. The most important lessons are:

Diversity. Anscombian primitivist pluralism really has a point about diversity of causes. This is what we experience in everyday life, but also looking at the diversity of things the sciences try to establish. Such a position captures the idea that it is very difficult to say what *all* worldly causes share in common.

Production and dependence/difference-making. The distinction between production and dependence (or difference-making), also stated in Hall's paper, is a major contribution to the literature. We suggest in chapter 20 that this is not because of unique features of Billy and Suzy throwing stones, but because distinguishing, or tracking, production and dependence is in fact a widespread problem in scientific practice. The distinction has generated important work, for instance on evidence for causal relations, and also helped us lay out the accounts and theories of Part II. Some accounts can be seen as aiming to provide an account of causal production—see e.g. chapters 11, 12 and 13—while others aim to provide an account of causal dependence or difference-making—see e.g. chapters 8, 9 and 10. These accounts can be seen as doing different jobs, capturing different causal concepts.

Evidence. In establishing causal relations, we need to appeal to different pieces of evidence, and each piece of evidence is itself multifaceted. The reason is that we need to establish different things about causal relations (see chapter 6), and one piece of evidence will rarely do the whole job on its own. Thus, while statistical methods help establish a robust difference-making relation, evidence of underlying mechanisms helps support the statistical results, particularly helping us infer to the single case, and to alternative populations. This conveys an idea slightly different from the usual 'triangulation of evidence', where we seek to probe the same result via different routes. The idea that multifaceted evidence conveys, instead, is that of *integration* of evidence. Just as reinforced concrete needs different materials to fare better with different types of stress, different pieces of evidence help support causal claims much better than single items, as each have their own strengths and weaknesses, in the context of inquiry.

Causal methods. There is a plurality of methods for causal inference; from observational to experimental, from qualitative to quantitative. Each of these comes with specific inference techniques, sometimes using very specific softwares. We need plurality here because research questions are multiple and diverse, and different scientific contexts may also require validation through different methods. Different data-processing methods, or evidence-gathering methods, also help produce the evidence sought to establish causal claims.

In view of these lessons, we think that Anscombe is right in saying that worldly causes are very diverse. This is indeed a serious impediment to saying what all causes share, particularly to giving any non-trivial, informative account of what all causes share. Any approach to causality also needs to admit the extensive methodological and evidential diversity we find in the sciences. This is a real problem, and it's not going to go away. Nevertheless, we can say something informative about groups of kinds of causes, just as we can say something general and informative about board games, and team games. So, for example, we take Woodward to have described precisely a concept that captures a broad range of worldly causes. So does the mechanisms account, and the other accounts we have described. In this view, so long as the limitations of each account are recognized, they can be used very effectively to sharpen up thinking about some causes: *all* the accounts have their place in the mosaic.

As a result, we think the theoretical resources that the philosophical literature can offer to anyone trying to think carefully about the sciences encompasses this: *all the developed accounts and theories in the literature*. From this point of view we offer a cheerful conceptual pluralism. We have described a variety of concepts of causality. We have explained what jobs they were designed to do, and their inspiration or the perspective from which they were developed. We have shown examples of scientific problems that we think they can usefully be applied to. We think, together, they can be used by scientists or other people thinking about science to sharpen up their thinking about causality, and better articulate distinctions that are important to science. We will illustrate this by looking at an extended case in chapter 24.

Note that while pluralism has caused great controversy in the philosophical literature, conceptual pluralism in this sense is a variety of pluralism that all academic philosophers working on causality might well be able to sign up to. All we are saying is that the full array of concepts of causality developed within philosophy could very well be of interest to practising scientists, *if* explained in terms of the scientific problems most familiar to them. This does not entail an 'anything goes' approach to philosophical accounts. To make an account useful to a practising scientist, you may have to re-frame the question, and perhaps alter the answer, while keeping the bulk of the original idea. We tried to do that with some of the accounts in Part II. In practice, any particular scientist would probably want to pick just a few to work with. But they need a library of concepts from which to pick, and that is what we have tried to offer in this book.

The mosaic suggests that causal notions have a role for different tasks such as inference, reasoning, prediction, explanation and control. In Part II we presented and explained 'traditional' concepts, and we also pushed for newly emerging notions in the literature, for instance information or variation. Even if these concepts fare well with a number of scientific problems, or with conceptualization, we don't aim to *replace* all other concepts. The idea of a mosaic developed in chapter 24 allies with that of conceptual engineering (Floridi, 2011a). Successful engineering requires a variety of materials, with different strengths and weaknesses! So let us repeat again: we hold the view that different concepts help in building causal knowledge, in virtue of their usefulness for specific tasks.

CHAPTER 24

The Causal Mosaic Under Construction: the Example of Exposomics

24.1 Making mosaics

A mosaic is an assembly of tiny tiles, all different. When the tiles are carefully composed, positioned and angled, they will make an image. Mosaics can be tiny personal pieces of craft, or vast and stunning public pieces of art, such as the mosaic on the ceiling of the Basilica di San Marco in Venice. In either case, every tile contributes to making the whole mosaic. In this book we have provided the materials to make a causal mosaic, to arrange and rearrange the different accounts of causality.

The accounts of causality. The accounts of causality developed in the literature are the tiles that can be used in building any causal mosaic. They are descibed in Part II, and there are plenty to choose from, stretching from counterfactuals and agency, to physical process, or INUS conditions. In chapter 1 we explained the five scientific problems of causality: explanation, prediction, control, inference and reasoning. We assessed all of the accounts of causality according to whether they do or do not help with one or more of the five scientific problems. It seems that no one account addresses all five problems successfully. Right from the start, we abandoned the idea that *one* account will on its own provide a full-blown causal theory, allowing us to do everything we might need to do, for all scientific domains. Instead, the various accounts—the tiles—each do *something* valuable.

The five philosophical questions. Part III examined different ways in which the tiles— the accounts—can be arranged, according to different perspectives or interests. In chapter 22 we explained the five philosophical questions: epistemology, metaphysics, methodology, semantics and use. We saw the resulting philosophical theory fragments constructed as answers to these questions. We explained why it is important first to clarify the scope of the question asked, then to understand what purposes the different accounts of causality serve, and finally to examine how they can be used together. One tile may be better placed in epistemology, and another in metaphysics; or the same tile may need to be used in different ways to address different questions. This means that many of the Part II accounts can be seen as complementary, addressing different questions, rather than as competing answers to a single question. If we are clear about what question we are asking, it is more likely that we will find complementary notions, from the other fragments, to help us complete our account, and solve our problem or problems. This accords with the idea that the philosophical questions, and the theory fragments built in response to them, are distinct but not independent.

Diversity and pluralism. The idea of using the different accounts in a complementary way may sound good, but it is not simple. In chapter 23 of Part IV we explained some of the challenges for recomposing the accounts of causality—challenges for making a mosaic out of these materials. These were the challenges of diversity and pluralism. Worldly causes seem to be very different; we also have different sources of evidence; and the methods of the sciences themselves are also pretty diverse. What form of pluralism suits a given scientific context is at least partly an empirical question. We need to study the field, its particular problems and methods, and work out what tiles and what fragments we need, so that we can start to build up a mosaic.

The brief recap on accounts (tiles), questions (fragments) and diversity serves as a reminder that there are a lot of materials available in the literature on causality. So how do we put them together? Most of the accounts address more than one of the five scientific problems, and most can address more than one of the five philosophical questions. Further, the different accounts—the tiles—also interrelate. For example, some of them form into the difference-making or dependence family, others into the production family. Alternatively, some are primarily conceptual, some primarily epistemological or methodological—or single accounts have alternative conceptual or methodological interpretations.

This is why we illustrate how we think this complex philosophical literature can be used by scientists and others trying to think about causality with the idea of making a mosaic, because a mosaic is precisely a whole made out of diverse pieces. We acknowledge a debt to Carl Craver's useful metaphor in writing of the 'mosaic unity of neuroscience' in Craver (2007). Russo (2009) also talks about the 'mosaic of causality', of which the epistemology of causal modelling is but one tile. In this chapter we extend these useful metaphors and link them to both philosophical and scientific practice. We will demonstrate how the exercise can be done using the example for exposomics, an emerging field of science that we will explain. Note that we do not think we are giving the final answer for exposomics, nor for the philosophy of causality—we suggest just *one* useful arrangement of the tiles for exposomics.

24.2 Preparing materials for the exposomics mosaic

24.2.1 Exposomics, or the science of exposure

Exposomics, or the science of exposure, is an emerging area of research in the biomedical field. It draws on biology, epidemiology, environmental science, statistics, bioinformatics, and information and communication technologies (ICTs). The field is new, proposing innovative and challenging methods to study exposure to environmental factors and their effects on several diseases. Exposomics faces many challenges at the conceptual and methodological level and also in making policy recommendations. The hope is that interactions between exposomics scientists and Causality in the Sciences (CitS) philosophers will help address these challenges.[36]

To understand what exposomics science tries to achieve, we need to take a step back to more traditional studies on environmental exposure and disease. This has been done by epidemiologists, who study the distribution and variation of exposure and disease in populations. Traditional epidemiology, and more precisely environmental epidemiology, managed to establish links between environmental factors and classes of diseases. They found strong and stable correlations between categories of factors (such as air pollution, chemicals, etc.) and diseases (such as cancer, allergies, etc.). *How* exactly environmental factors lead to disease is much less understood, and the methods used so far will not illuminate that question. This calls for a change in the methods used in epidemiology, which is precisely what is happening in exposomics. Scientists think that we need to study exposure and disease at the molecular level, in order to understand the molecular basis of life and disease.

To do this, the question of the connection between exposure and disease has to be translated to the molecular level: How do we track changes at the micro-molecular level due to levels of chemicals in, for example, air or water? The answer to this question lies, according to exposomic scientists, in the study of *biomarkers*. 'Biomarker' means biological marker. A biomarker is a characteristic which is objectively measured and which indicates normal biological processes, or pathogenic processes, or a pharmacologic reaction to a therapeutic intervention. Where do we look for those markers? There are various candidates, for instance metabolites in blood, proteins or features of gene expression. These can be detected and measured, and so tell us something about what's going on in the body.

Carrying on the study of exposure and disease at the molecular level is, in a sense, the effect of a much bigger change—a change in the concept of *exposure*. One aspect of exposure is familiar to us: there is stuff 'out there' we have contact with and to which our body responds in some way. Exposomic scientists call this the 'external exposome'. A second aspect of exposure is the novelty in the approach: there is stuff that happens

---

[36] At the time of writing, there is a major FP7 project studying environmental exposure and its effects on several diseases: <http://www.exposomicsproject.eu>. The project is coordinated by Prof. Paolo Vineis at Imperial College, London. We are extremely grateful to Prof. Vineis and collaborators for the opportunity to work and discuss with them.

inside the body, once we have contact with the stuff out there. But what happens inside depends not just on the stuff out there, but also on the internal environment that our body creates. Exposomics scientists call this the 'internal exposome'. Now, to understand how, say, pollution leads to allergies, we need to understand both aspects: the internal and the external, i.e. the total exposome (Wild, 2005; Wild, 2009; Wild, 2011; Rappaport and Smith, 2010).

In practice, how can we study the internal and the external exposome? For the external, that is relatively easy, provided that we can make accurate measurements of the environmental factors. The internal is more complicated, and this is where new statistical methods and new technologies come to the rescue.

After exposure, scientists need to collect bio-samples and analyse them, looking for relevant biomarkers of exposure. But what biomarkers? That is precisely the problem. Exposomics is seeking help from biological theory *and at the same time* is helping refine biological theory (Vineis and Chadeau-Hyam, 2011). Exposomic scientists also look for biomarkers that indicate the presence of a disease. And finally they look for the biomarkers that are in between, the biomarkers that indicate that some clinical change within the body has occurred or is occurring. Exposomics scientists are creating new cohorts (i.e. groups of people) to carry out these studies, and are also using data available from previous studies, such as the EPIC cohort (European Prospective Investigation into Cancer and Nutrition, <http://epic.iarc.fr>).

Once the data are in, the issue is how to analyse them. This is where a novel 'meeting-in-the-middle' methodology enters the scene. Exposomic scientists try to match the biomarkers that are most correlated with the exposure with those that are most correlated with the disease, hoping to find a sensible overlap—in the middle—and also hit on those biomarkers that may indicate early clinical changes; the very first indicators of the onset of disease. This requires highly sophisticated statistical modelling and network theory approaches.

But why are exposomics scientists bothering with all this? Well, in a sense this is a response to genomics. Studies of the genome achieved a lot, but still less than what was hoped for (Manolio et al., 2009). In particular, we hoped to gain much more insight into disease mechanisms than we actually did. Exposomics scientists are trying a new venture, in quite a difficult field. If, as it seems, disease is not all in our genes, we need to study how our body interacts and reacts to stuff that may well be responsible for a number of diseases. This is the line of argument of exposomics scientists, who have been quite successful in gaining the support of European funding bodies. The hope is to get a much deeper understanding of disease mechanisms and also to advise policy-makers about public health matters.

This gives a quick overview of what exposomics science is, what its aims are, and its position with respect to traditional epidemiology and to genomics. We will now see in some more detail what scientific problems are at stake. In this chapter we illustrate the CitS approach by showing how some of the different concepts and methods discussed in Part II play a role in a scientific project dealing with many causal issues. Exposomics enables us to touch simultaneously on epistemological, methodological and metaphysical issues, and also on issues concerning how we use causal knowledge. It also helps us show how the distinction between difference-making and production can be useful, how probabilistic and variational issues come into play, and so on.

Our aim here is to illustrate how the resources we have laid out in the book can be marshalled to help think about the issues. We lay this out as a series of questions to address. Precisely because these questions will have different answers for different purposes, different people might find different tiles more useful. This is why we think having a guide to them *all* is helpful. We make no suggestion that the illustration we provide here is unique. It is also a timely philosophy. Exposomics is research-in-progress. In a few years, the situation will have changed and the mosaic of causal concepts useful to thinking about it may well also have changed. Here we show how to do the exercise; we don't say that this is the one mosaic for exposomics, nor the one mosaic for science looking for causes more generally. Since this is one of the cases we both work on, this means we also give hints about what we think would be exciting research to do in the CitS spirit. We are carrying out some of these projects ourselves, but there is so much going on that we hope many other people join the venture.

24.2.2 Question 1: What scientific problems would it be useful to address?

Thinking through the five scientific problems of causality can help identify distinct issues for any scientific case. What are the scientists trying to do? Is there more than one aim, and do they conflict?

Inference. Is there a causal relation between X and Y? Does X cause Y? What are the causes of Y? How much of X causes how much of Y?

Basically, causal inference concerns finding out about causes. In exposomics, we want to know what the environmental causes of disease are. We want to find links between environmental factors and diseases. We also want to establish links between exposure and some biomarkers and diseases and some biomarkers, so that ultimately we can link up the whole chain from environmental exposure to disease. We need to think about how we can establish such correlations. Where can we get the data? (For exposomics, technologies help a lot, but produce big data sets, which are their own challenge.)

Explanation. How do we (causally) explain phenomena? To what extent is a phenomenon explained by statistical analyses? What level of explanation is appropriate for different audiences, such as fellow specialists, the public or jurors in a court?

We often want to know not just what happened or will happen, but how it happened, and why. This is causal explanation. In exposomics, the explanations offered by our background knowledge are very important. We are trying to probe an area where little is known, but we depend very heavily on what we already know about the human body, and about environmental causes of disease, found in traditional epidemiology. A lot of our knowledge of the body takes the form of known mechanisms, which explain, for example, how cells make proteins, or how our immune system works. This is important to help us make sense of the many correlations we will find in big data sets. Correlations alone won't do. We need a plausible biochemical story of what happens after exposure and inside the body in order to believe we are starting to make progress on understanding disease causation by environmental factors—at the molecular level.

Prediction. What will happen next, and how do we find out what will happen next? How accurate are our predictions about the evolution of given population characteristics (e.g. mortality, morbidity, ...)? What does a physical theory predict will happen in a given experimental setting?

Prediction is simply working out what will happen. In exposomics, we want to know things like what will happen to populations if current levels of environmental contaminants remain the same, or continue to rise. We also want to know whether we can predict disease given knowledge of biomarkers. If we can better predict disease trends, thanks to biomarkers, this will help us decide what public health actions we should take.

Control. How do we control variables that may confound a relation between two other variables? In what cases is it appropriate to control for confounders? How do we control the world or an experimental setting? How do we find out about that? This extends all the way from interfering with a system to get a slightly different result, to the construction of entirely new causal systems, such as smartphones, or GPS, or a political system.

Control is going beyond prediction, to alter what will happen. In exposomics, our core concern is how to prevent disease. But we also want to know how we can understand the pathways from exposure to disease, and thus better control possible confounders, and get a 'cleaner' link or pathway from exposure to disease.

Reasoning. What reasoning underlies the construction and evaluation of scientific models? What conceptualization of causation underpins causal methods? How do we reason about all aspects of causality? How can we sharpen up that reasoning, making it more precise and so more effective?

Causal reasoning is our broadest scientific problem, as it concerns all the ways we think about causality in science, explicitly and implicitly. In exposomics, we need to think in terms of causes that are hard to find and that are fragile, i.e. whose actions or capacities are easy to disrupt. We need to adjust our rationale for our methods accordingly, because looking only for correlations isn't enough, and we don't have a well-developed theory of the causation of disease by interactions of genes and environmental factors within the body to help us much either. How should we model the interrelations between biomarkers?

Note that under causal reasoning is where we worry about relations between the other four problems. In exposomics, we have to think about what existing explanations are available and what causal inference methods are available, decide how they fall short, and design the methods of the project to meet the needs of the current state of the science. It is here that it becomes useful to ask the most obviously philosophical questions; the conceptual questions. For exposomics, it is useful to wonder: how can we conceptualize the link from exposure to disease? How does the macro-environmental level connect to the micro-molecular level? How am I thinking about causality so that these links and connections make sense? Does that thinking about causality accord with my methods?

24.2.3 Question 2: What do the scientists want to know or do, and what problems of diversity do they face?

As we explained in chapter 22, philosophical work on causality in the spirit of CitS can begin at any point; from science or from philosophy. So we offer these questions merely as an illustration of how one might choose to approach a scientific case study. This question is useful to ask, to get a sense of the scope of the research you are looking at, and the scale of the problem understanding it poses.

Fragility. In exposomics, scientists are looking for fragile relations; relations which disappear in some contexts. This problem is important to exposomics, but also affects every field where populations are heterogenous, where complex mechanisms exist, where there are homeostatic mechanisms that change and adapt in response to multiple stimuli, and so on. So the question is: how do we isolate such fragile relations? How can we establish that they are causal given their fragility? What are the right methods to circumscribe them? Experiments? Observations? Simulations? What?

Diverse research groups. Exposomics science is not the result of one single brain. Research *groups* are essential. Exposomics projects usually involve consortia of several institutions, and within each institution there are several groups (epidemiologists, statisticians, etc.). But this is not unique to exposomics. Another example is the discovery of the Higgs boson, which has been possible thanks to numerous research groups working on different experiments. Much of current research needs synergies between different groups and approaches.

Diverse fields of science. Exposomics is (molecular) epidemiology, but it is also medicine, and biomedicine, and biochemical medicine; and medicine using knowledge from nuclear physics, and from statistics, and from sociology, etc. Understanding, predicting and preventing disease is a complex enterprise where each of these fields contributes something vital. Again, this is not a problem only for exposomics, but is a feature of many important collaborations nowadays.

Diverse technologies. The embeddedness of technology in modern medicine goes far beyond looking through a microscope. We don't 'see better' using omics technology; we collect data about stuff that we wouldn't see at all otherwise! Actually, we don't 'see' even using these technologies. We are able to identify signals that then need to be interpreted in order to find something in there. The technologies used to generate, collect and analyse data in exposomics (GPS and sensors, omics technologies, network-based statistics software, etc.) change the landscape of what we can observe and study and of how we understand and conceptualize diseases. How should we conceive of the relation between science and technology in the face of techno-science? What questions does technology pose for the practice and use of medicine? Again, this does not only arise in exposomics. Think for instance of diagnostic tools, that allow us to see *possible* tumours very early, and intervene. While this is generally seen as an achievement and is praised, some controversies are also arising, for instance the 'slow medicine' movements.[37] New technological tools have the power to enhance the production and analysis of data, and also to 'create' new phenomena, by making diseases appear too early or even when they are merely a possibility. How we should react in response to these technological tools is quite a difficult issue: think about Angelina Jolie undergoing mastectomy to avoid cancer development. Is this ethical? Is this action justified by the evidence available?

Data overload. Big data is a problem for exposomics, as for many other sciences. Having a lot of data, massive datasets, is not a solution per se. Big data creates a big problem of analysing data, of using them, or re-using them for other research questions. In exposomics, for instance, they are re-using samples and data from other cohorts. Maintaining their quality is a serious challenge (Leonelli, 2009; Leonelli, 2014; Illari and Floridi, 2014).

Significant diversity in evidence and data-probing methods. One idea in exposomics is that to establish how pollutants induce changes in the body at the molecular level we use different methods: analyse the exposure (chemistry tools), analyse samples (again chemistry *and* nuclear physics *and* omics tools, etc.), and analyse data with statistics and other similar tools (e.g. network theory, calibration methods, etc.). Another idea is that data should be available for re-analysis by different people with different tools. Making all this work effectively can be ferociously difficult, and also touches on political issues about accessibility of data, who funds research, who owns the data and so on.

Clearly there are some significant challenges of diversity in understanding exposomics.

24.2.4 Question 3: What philosophical questions would it be useful to address?

Asking the five philosophical questions can be a useful way to push beyond the surface, to uncover assumptions that may be underlying the research methodologies or questions.

Epistemological issues. What do the Exposomics scientists want to know? What kinds of assumptions about scientific knowledge might they be making? Would the assumptions they are making have a more general bearing on causal epistemology?

Exposomics scientists want to understand how environmental exposure is linked to disease. In that sense, their epistemological question is fairly simple. But answering it is rendered complex by surrounding knowledge. They know a lot about the system, and so many mechanisms within the body are assumed, as are some known

---

[37] See e.g. <http://blogs.bmj.com/bmj/2012/12/17/richard-smith-the-case-for-slow-medicine/>, accessed 23rd July 2013, and McCullough (2009), who initiated the 'slow medicine' movement.

mechanisms for environmental causes of disease, such as radiation damaging DNA, and DNA methylation due to smoking. But a lot is still unknown, and in the middle of that exposomics scientists are looking for many small causes, with small effects, and large interaction effects. They expect widely different factors to be causes, including e.g. social and chemical factors. Their use of new omics technologies allows a comprehensive approach, but generates vast amounts of data that must all be stored, maintained and processed. Naturally, this is strongly linked to the scientific problem of causal inference. But exposomics is also a fascinating lesson in how increasing knowledge happens in such a complicated way, particularly when we are taking the first tentative steps in an attempt to push back the current boundaries of knowledge.

Metaphysical issues. Are there assumptions about the nature of the domain being investigated? What are scientists assuming about the causal relations they are seeking?

In exposomics, there are certainly assumptions, based on the known mechanisms described above. Nevertheless, scientists assume that there is something to track in the middle of the vast uncharted territory, whereby particular environmental factors cause disease. Even though it can be hard to get such factors to show up—hard to find the fragile correlations in the middle of the many misleading correlations likely to exist in large datasets—they assume the causes are there, and can be found. They are interested in how to characterize them.

Methodological issues. What methodology or methodologies are being used? What do the scientists say about why they chose as they did? Are they using a novel methodology? What are the background assumptions of any models or other standard techniques?

As a reaction to the novel methodological challenges of exposomics, the scientists have articulated a new 'meeting-in-the-middle' methodology. Broadly, they are trying to track biomarkers of exposure, and biomarkers of early disease onset, and match up biomarkers that correlate with both in the middle, so that ultimately they can track the entire evolution of disease using biomarkers. This new methodology raises interesting issues about the relations between background knowledge and new causal discoveries. What makes us think that the meeting-in-the-middle methodology and use of omics technologies will really advance understanding of disease mechanisms? How do we—and funders—know that this is not mere hype?

Semantic issues. Do the scientists seem to be using causal language in a standard way? Are they innovating? How? What does it mean that macro-environmental factors cause micro-molecular changes in the body? What kind of worldly causation would that be?

The main reason we are personally so drawn to studying exposomics is that the scientists do seem to be innovating. They have explicitly created a new word, the 'exposome', by analogy with the 'genome', to capture the system-wide nature of the interactions they expect to see. They are then trying to think about causality from within the exposome, while simultaneously struggling with some serious methodological challenges. One thing they do, when talking about causality, is frequently talk about signalling. Do they think this has something to do with the meaning of causality?

Is it a way to point to one notion that could give an account for causality (see below) or is it a way of avoiding causal talk (like sometimes using the language of determinants instead of causes in biomedical sciences)?

Issues of use. What are the scientists intending to use their results for? What are other people intending to use their results for? How can results be communicated to research funders and policy-makers so that actions can be taken? How can the public use the results? How is fine-grained molecular knowledge of disease mechanisms going to be used for policy purposes?

Clearly the work is ultimately intended to inform health-care recommendations. But it is not obvious how to use exposomics results for policy. For instance, if exposomics results mainly concern relations at the *molecular* level, how can we inform governments about interventions at a *social* level, i.e. trying to change patterns of behaviour, or banning certain kinds of pollutants? How do results about 'molecules' translate into actions about 'people' and 'behaviours'?

Again, asking all these questions about a domain of science is an interesting study, and helps us to explore aspects of the work that may otherwise not occur to us. We have argued in chapter 22 that these questions cannot be addressed wholly in isolation. Nevertheless, addressing them all simultaneously is a lot of work. We also suggested that we have to identify which question or questions are of most interest to us, in light of our goals and interests, and keep that question or questions clear in our work.

24.3 Building the exposomics mosaic

We are now in a position to begin building the mosaic itself, beginning by selecting the tiles for the mosaic.

24.3.1 Question 4: What accounts might be useful?

Which of the five scientific problems are most of interest to us? The summary tables of the appendix can be used to identify accounts that might be useful on that basis. Which of the theory fragments do we want to concentrate on? We can use the summary tables, and the examples 'Core ideas' and 'Distinctions and warnings' of each chapter in Part II to track down accounts that touch on the scientific problems and philosophical questions we have identified. We selected the following tiles to help us with our thinking about exposomics:

Levels. The 'exposome' is a new (causal) concept put forward by epidemiologists in order to redefine the causal context in which causal relations at different levels take place, so chapter 5 on levels of causation may be useful. The issue of the levels here concerns more than the relation between the generic and the single-case level. It also concerns the integration of factors of different natures (social and biological) into the same explanatory framework. So, what understanding of 'levels' helps illuminate concerns in exposomics science?

Evidence. In this study the interplay between evidence of difference-making and evidence of mechanisms is crucial to establish causal relations successfully, so chapter 6 could be interesting. Biomarkers of disease are supposed to make a difference to the probability of disease, but this probability raising needs to be substantiated by a plausible underlying mechanism. How are the scientists searching for each?

Production: mechanisms. Exposomics scientists talk about mechanisms, examined in chapter 12, and clearly know a lot about mechanisms such as biochemical mechanisms in the cell and mechanisms such as the immune system in the body. How are they thinking about these things? Note that exposomics scientists also conceptualize the evolution of biomarkers as a process, so how should we think about this with respect to the distinction between difference-making and mechanisms? Is 'mechanism' the right way to think about causal linking in exposomics?

Production: information. As well as talking about processes, exposomic scientists talk about 'signal detection'. Since they are looking for something underlying their data—a kind of linking—it looks like they are seeking productive causality, and they associate this idea with the idea of picking up signals. So it looks like productive causality, in the minds of exposomics scientists, is associated with the concept of information, discussed in chapter 13. So, information, as a production account, should be investigated. Would it suit exposomics? Why? Could it help solve problems elsewhere too?

Capacity. What does it mean that pollutants have the capacity to induce changes in the body at the molecular level? How can we find out exactly what this capacity is? When it is activated, what is its threshold to induce changes? Also, is the predictive power of a biomarker due to its own capacity, or to the capacity of some entity it is a proxy for? Here, the examination of capacities in chapter 14 is of interest, particularly the idea of the masking of capacities and the resulting difficulty of getting evidence of capacities when they are sensitive to context.

Depending on what you want to know, and what you already know, there could be material in many other chapters that is useful to you. For example, a background grasp of probabilistic approaches to causality is given in chapter 8; more about what powerful techniques are available for probing data can be found in chapter 7; and a discussion of how diverse sources of evidence might be thought of as contributing to causal inferences is available in chapter 18.

24.3.2 Question 5: How can we put together these resources to help us in our thinking?

Philosophical concepts can't make science easy—*nothing can do that*. What these accounts of causality can do is to sharpen up thinking. This in turn can help the actual practice being done to be done with more clarity, aiding communication among scientists themselves, and outside science and academia more generally.

Now we have accumulated the resources available, we really have to begin choosing tiles for our mosaic, according to which questions are of most interest to our project.

Exposomics is such a rich case study that there are many questions of both scientific and philosophical interest. We will just have to pick some to illustrate the process of building a mosaic. Suppose we take the process of discovery as of primary interest, at least to start with. Suppose we are interested in how the conceptualization of causes can help with the struggle to build a new methodology apt for exposomics. Then we're interested in the methodological challenges. We are interested in how semantics and metaphysics (philosophical questions) might help with constructing a novel methodology (philosophical question and scientific problem) for this case of exposomics. In turn, of course we are interested in how the construction of that novel methodology might inform semantic and metaphysical questions.

We can use the chapters in Part II, beginning with the core ideas and distinctions and warnings, as guides to place the tiles we have identified in our mosaic:

Levels resources. We can watch out for mistakes translating between the philosophical and scientific literatures. Inference between the population level and the single-case level is frequently a problem, and we can be alert for mistakes. Often, we can have evidence of one level, and want to know about the other, but be unable to make the inference directly. There are also concerns about different kinds of measurements, integrating them into single model or single explanation, and so on.

Difference-making. This helps us see the difference between worrying about the relation between cause and effect variables, in isolation from worrying about what happens in between. In exposomics, as in any other case which generates a large data set, there is a problem with finding too many correlations, and requiring some means of isolating those of interest.

Mechanisms. The idea of entities and activities organized to produce a phenomenon, initially in an attempt to explain the phenomenon, might be of interest. Can activities and entities and their organization be found? Can evidence of such things help us? Can they help answer philosophical questions? Can they help with the scientific problems? The problems with mechanisms—properly understanding their context and organization—do seem to arise here. Context shifts make activities and entities difficult to detect in exposomics, and the organization of many mechanisms here are so complex we are not even sure how to *begin* describing them.

Information. This might offer an interesting way to conceptualize linking in exposomics. Partly, it is interesting that a causal link could be something so thin, so apparently intangible. From the problems discussed in chapter 13, we might be wary that thinking of the link as informational might not be so very informative. On the other hand, it is reasonable to think of information as something that can work with the methodology of exposomics—detect at point x, detect at point y, detect at point z, match up the chain, infer that there's a link right through the system.

Capacities. These help conceptualize something that is stable, at a particular locality, in a particular context. Capacities are useful if you can find them. But they are only useful if they can be found in spite of context shifts, or if they are stable enough that they only change with known context shifts. What are the known capacities in exposomics? What are their limits?

What we end up with is a mosaic showing the methodological problems faced by exposomics scientists, where the tiles of the mosaic have given us the language to express these problems very clearly, yielding a deep understanding of them. We can say how these factors relate; how known mechanisms and capacities structure the problem of exposomics, giving us the background knowledge against which much finer-grained causal links are sought. In this way we can clear the ground, to show the importance of the construal of some kind of linking to causal reasoning, and so to addressing the methodological problems in exposomics. In view of the innovation within the science, and the absence of existing language in the philosophical debate, it can now be no surprise that the scientists are reaching for new language to express this. Here philosophical work can feed directly back into the scientific enterprise by helping provide new concepts, designing a conceptual apparatus that can help support the science.

Note that we are not attempting to say how to write a philosophical paper, but to indicate how to use the philosophical literature on causality to understand a scientific problem and to address it philosophically. This, however, is what it is to do philosophy of science in the CitS style. Our hope is that in creating your own mosaic from the materials we have provided in this book, you end up with something adapted to your questions, your problems, allowing you to move forward with them, and communicate them to others to get any help you need. Not the least of your achievement is to have *identified and refined your questions and problems*.

24.3.3 How many mosaics?

There could be indefinitely many such mosaics. If we choose different questions as our starting point then, consequently, we collect different tiles, focus on different philosophical questions, and design a different mosaic. Philosophers with different backgrounds might naturally start from different perspectives, focusing on different questions from each other. Likewise, scientists with different backgrounds may approach the same problem differently. And, clearly, philosophers have taken a different perspective on causality, by focusing on different questions than scientists usually do.

The building of this kind of mosaic might not only be useful for a single person—whether philosopher or scientist—but also for a research group. For example, exposomics researchers benefit from building a common mosaic, to make their language more precise, and facilitate their communication and their training of postdocs and doctoral researchers. This 'mosaic of brains' has been discussed in the literature as 'distributed understanding' (Leonelli, 2014), and also using resources from social epistemology, such as in Andersen and Wagenknecht (2013), Fagan (2012) and Beaver (2001).

The *mosaic*, and the understanding of exposomics science that we have developed along with it, is something we have made. This does not mean that there is no reality out there. There is. Reality, to echo Floridi (2008), is a *resource* for knowledge and we interact with that reality in various ways: using evidence-gathering methods through epistemological notions. We don't merely passively imitate, which sometimes seems to be all that is meant by 'represent', causal relations. It takes work to construct our knowledge about causal relations: gathering data, probing the data using advanced methods—and then of course using that knowledge to build things, like social or health policies, which in turn generate more causal knowledge. All this is very active; not at all passive. Exposomics science shows even more detail. We know there is something there, something important to us, but it is very difficult to find. We construct technology, construct research teams, control circumstances so that what we want to find is discoverable. Even after that, data processing is still necessary to find anything. There is a reality out there that we can hit and act upon, study, model and understand, using biomarkers, and we can conceptualize disease causation using processes or other notions. But we have to interfere a great deal with the system and constantly design or re-design concepts in order to find out. Ultimately we get causal knowledge by a very sophisticated interaction between us and the world, using reality as a resource to construct our causal knowledge.

To conclude, in this chapter, as in the book generally, we have focused on explaining the philosophical literature on causality to people outwith that literature. From that point of view, we have used this chapter to illustrate how anyone can use the resources of the causality literature to think better about science. But, particularly in Part III, we also examined issues about how philosophy can successfully engage with science. This chapter can also be read as the finale to chapters 20 and 22, in that it shows in an extended way how science can be a rich resource for the development of philosophical problems.

APPENDIX

Accounts, Concepts and Methods: Summary Tables

A.1 THE SCIENTIFIC PROBLEMS OF CAUSALITY

In Chapter 1 we identify five scientific problems for causality. These are: inference, explanation, prediction, control and reasoning. In Part II, we discuss several accounts and concepts and assess them against these problems.

Inference	Does C cause E? To what extent?
Explanation	How does C cause or prevent E?
Prediction	What can we expect if C does (or does not) occur?
Control	What factors should we hold fixed to understand better the relation between C and E? More generally, how do we control the world or an experimental setting?
Reasoning	What considerations enter into establishing whether / how / to what extent C causes E?

A.2 THE PHILOSOPHICAL QUESTIONS ABOUT CAUSALITY

Chapter 22 identifies different philosophical questions that can be asked about causality. These concern: epistemology, metaphysics, methodology, semantics and use. In Part II, we explain that different accounts and concepts can be given one or more such emphases.

Epistemology	How do we know that / whether / how C causes E?
Metaphysics	What are causal relata? What is causation?
Methodology	How can we establish that / whether / how C causes E in scientific, ordinary, legal, …, contexts?
Semantics	What does it mean that C causes E? What inferences are justified?
Use	What actions and decisions can be taken knowing that / the extent to which / how C causes E?

A.3 THE ACCOUNTS: HOW THEY FARE WITH SCIENTIFIC PROBLEMS

Part II presents different accounts and concepts. This table summarizes their core ideas and then ticks the boxes of the scientific problems they primarily address. 'Core ideas' and 'Distinctions and warnings' at the end of each chapter also help track accounts with respect to the scientific problems.

Account	Core Idea	Inference	Explanation	Prediction	Control	Reasoning
INUS conditions	Causes are an Insufficient, but Non-redundant part of an Unnecessary but Sufficient condition	✓				✓
Pie charts in epidemiology	C is just a part of the complex cause/mechanism bringing about E		✓			✓
Levels of causation—philosophical	Type / population level 'C causes E': about properties. Token / individual level 'C causes E': about instantiations of the type properties	✓				✓
Levels of causation—scientific	'C causes E' is inferred either from aggregate variables or from individual level variables	✓				✓
Levels of causation—legal	General causal claims are used for causal attribution at the singular level	✓				✓
Causality and evidence	Various sources and types of evidence are needed to support a causal claim	✓	✓	✓	✓	✓
Causal methods	Causal methods help establish different sorts of causal claims	✓	✓	✓	✓	✓
Difference-making—probabilistic causality	The occurrence of C alters the chances of occurrence of E	✓		✓		
Difference-making—counterfactuals	C causes E means: if C hadn't happened, E wouldn't have happened either	✓	✓			✓
Difference-making—manipulation and invariance	C causes E if wiggling C makes E wiggle, and the relation between C and E is stable enough	✓	✓		✓	
Production—processes	C causes E if there is a (physics) process linking C to E	✓	✓	✓		

Account	Core Idea	Inference	Explanation	Prediction	Control	Reasoning
Production—mechanisms	C causes E if there is a mechanism linking C to E	✓	✓			✓
Production—information	C causes E if there is information transmission from C to E	✓		✓		✓
Capacities, powers, dispositions	Whether 'C causes E' is true depends on whether there is a CPD that makes the claim true	✓	✓			
Regularity (Hume)	C causes E means: instantiations of E regularly follow instantiations of C	✓		✓		✓
Variation	To establish whether C causes E we study how variations in C are related to variations in E	✓	✓	✓		✓
Causality and action (Gillies)	We can use C to produce or prevent E	✓			✓	
Causality and action (Price)	C causes E if there is a recipe for agents to bring about E by bringing about C			✓	✓	✓
Causality and inference	Causality is tied to several inferential practices, both to establish that C causes E and to take action once we know that C causes E	✓	✓	✓	✓	✓

A.4 THE ACCOUNTS: HOW THEY FARE WITH PHILOSOPHICAL QUESTIONS

Part II presents different accounts and concepts. This table summarizes their core ideas and then ticks the boxes of the philosophical questions they particularly provide an answer to. 'Core ideas' and 'Distinctions and warnings' at the end of each chapter, and particularly chapters 22 and 23, also help track which accounts address which philosophical questions.

Account	Core Idea	Epistemology	Metaphysics	Methodology	Semantics	Use
INUS conditions	Causes are Insufficient, but Non-redundant part of an Unnecessary but Sufficient condition	✓	✓	✓	✓	✓
Pie charts in epidemiology	C is just a part of the complex cause/mechanism bringing about E		✓	✓		✓
Levels of causation—philosophical	Type / population level 'C causes E': about properties. Token / individual level 'C causes E': about instantiations of the type properties		✓		✓	✓
Levels of causation—scientific	'C causes E' is inferred either from aggregate variables or from individual level variables	✓		✓		✓
Levels of causation—legal	General causal claims are used for causal attribution at the singular level	✓	✓			✓
Causality and evidence	Various sources and types of evidence are needed to support a causal claim	✓		✓		✓
Causal methods	Causal methods help establish different sorts of causal claims			✓		✓
Difference-making—probabilistic causality	The occurrence of C alters the chances of occurrence of E		✓	✓		✓
Difference-making—counterfactuals	C causes E means: if C hadn't happened, E wouldn't have happened either		✓		✓	✓
Difference-making—manipulation and invariance	C causes E if wiggling C makes E wiggle, and the relation between C and E is stable enough	✓	✓	✓	✓	✓
Production—processes	C causes E if there is a (physics) process linking C to E		✓	✓		✓

Account	Core Idea	Epistemology	Metaphysics	Methodology	Semantics	Use
Production—mechanisms	C causes E if there is a mechanism linking C to E	✓	✓	✓		✓
Production—information	C causes E if there is information transmission from C to E	✓	✓	✓	✓	
Capacities, powers, dispositions	Whether 'C causes E' is true depends on whether there is a CPD that makes the claim true		✓			
Regularity (Hume)	C causes E means: instantiations of E regularly follow instantiations of C	✓	✓			
Variation	To establish whether C causes E we study how variations in C are related to variations in E	✓		✓		✓
Causality and action (Gillies)	We can use C to produce or prevent E					✓
Causality and action (Price)	C causes E if there is a recipe for agents to bring about E by bringing about C		✓	✓	✓	
Causality and inference	Causality is tied to several inferential practices, both to establish that C causes E and to take action once we know that C causes E	✓	✓	✓	✓	✓

REFERENCES

Agresti, Alan (1996). *An introduction to categorical data analysis*. Wiley, New York.

Alexander, Joshua (2012). *Experimental Philosophy: An Introduction*. Polity, Cambridge.

Andersen, Hanne and Wagenknecht, Susann (2013). 'Epistemic dependence in interdisciplinary groups'. *Synthese*, 190(11), 1881–98.

Anderson, John (1938). 'The problem of causality'. *Australasian Journal of Psychology and Philosophy*, XVI, 127–42.

Anscombe, G.E.M. (1975). 'Causality and determination'. In *E. Sosa, (ed.) Causation and Conditionals*, pp. 63–81. Oxford University Press.

Armstrong, D.M., Martin, C.B., and Place, U.T. (1996). *Dispositions: A Debate*. Routledge, London.

Arntzenius, Frank (2008). 'Reichenbach's common cause principle'. In *The Stanford Encyclopedia of Philosophy* (Winter 2008 edn) (ed. E. N. Zalta). Available at: <http://www.plato.stanford.edu/archives/win2008/entries/physicsRpcc/>.

Aronson, Jerrold (1971). 'On the grammar of "cause"'. *Synthese*, 22, 414–430.

Bandyoapdhyay, Prasanta S., Nelson, Davin, Greenwood, Mark, Brittan, Gordon, and Berwald, Jesse (2011). 'The logic of Simpson's paradox'. *Synthese*, 181, 185–208.

Bandyopadhyay, Prasanta S. and Cherry, Steve (2011). 'Elementary probability and statistics: A primer'. In *Handbook of the Philosophy of Statistics* (ed. P. S. Bandyoapdhyay and M. Forster). Elsevier, North Holland.

Barndorff-Nielsen, Ole (1978). *Information and Exponential Families in Statistical Theory*. John Wiley & Sons, Chichester.

Baumgartner, Michael (2008). 'Regularity theories reassessed'. *Philosophia*, 36, 327–54.

Baumgartner, Michael (2009). 'Interventionist causal exclusion and non-reductive physicalism'. *International Studies in the Philosophy of Science*, 23(2), 161–78.

Beaver, Donald Deb. (2001). 'Reflections on scientific collaboration (and its study): Past, present, and future'. *Scientometrics*, 52(3), 365–77.

Bechtel, William (2006). *Discovering Cell Mechanisms: the Creation of Modern Cell Biology*. Cambridge University Press, Cambridge.

Bechtel, William (2007a). 'Biological mechanisms: organized to maintain autonomy'. In *Systems Biology* (ed. F. Boogerd, F. J. Bruggeman, J.-H. Hofmeyr, and H. V. Westerhoff), pp. 269–302. Elsevier, Amsterdam.

Bechtel, William (2007b). 'Reducing psychology while maintaining its autonomy via mechanistic explanations'. In *The Matter of the Mind*. Blackwell.

Bechtel, William (2008). *Mental Mechanisms: Philosophical Perspectives On Cognitive Neuroscience*. Routledge, Oxford.

Bechtel, William (2009). 'Looking down, around, and up: Mechanistic explanation in psychology'. *Philosophical Psychology*, 22, 543–64.

Bechtel, William (2010). 'The downs and ups of mechanistic research: Circadian rhythm research as an exemplar'. *Erkenntnis*, 73(3), 313–28.

Bechtel, William (forthcoming). 'What is psychological explanation?' In *Routledge Companion to Philosophy of Psychology* (ed. P. Calvo and J. Symons). Routledge, London.

Bechtel, William and Abrahamsen, Adele (2005). 'Explanation: A mechanist alternative'. *Studies in the History and Philosophy of the Biological and Biomedical Sciences*, 36, 421–41.

Bechtel, William and Abrahamsen, Adele (2008). 'From reduction back to higher levels'. In *Proceedings of the 30th Annual Conference of the Cognitive Science Society* (ed. B. C. Love, K. McRae, and V. M. Sloutsky), pp. 559–64. Cognitive Science Society.

Bechtel, William and Abrahamsen, Adele (2009). 'Decomposing, recomposing, and situating circadian mechanisms: Three tasks in developing mechanistic explanations'. In *Reduction and Elimination in Philosophy of Mind and Philosophy of Neuroscience* (ed. H. Leitgeb and A. Hieke), pp. 173–86. Ontos Verlag, Frankfurt.

Bechtel, William and Abrahamsen, Adele (2010). 'Dynamic mechanistic explanation: Computational modeling of circadian rhythms as an exemplar for cognitive science'. *Studies in History and Philosophy of Science*, 1, 321–33.

Bechtel, William and Richardson, Robert (2010). *Discovering complexity*. MIT Press, Cambridge, Massachussets.

Bechtel, William and Wright, Cory (2009). 'What is psychological explanation?' In *The Routledge companion to philosophy of psychology* (ed. P. Calvo and J. Symons), pp. 113–30. Routledge, London.

Beebee, Helen (2007). 'Hume on causation: The projectivist interpretation'. In *Causation, Physics, and the Constitution of Reality: Russell's Republic Revisited* (ed. H. Price and R. Corry), pp. 224–49. Clarendon Press, Oxford.

Bell, Graham (2008). *Selection: The Mechanism of Evolution* (Second edn). Oxford University Press, Oxford.

Bertram, Lars and Tanzi, Rudolph E. (2009). 'Genome-wide association studies in Alzheimer's disease'. *Human Molecular Genetics*, 18(R2), R137–R145.

Bickel, P.J., Hammel, E.A., and O'Connell, J.W. (1975). 'Sex bias in graduate admissions: Data from Berkeley'. *Science*, 187(4175), 398–404.

Bird, Alexander (1998). 'Dispositions and antidotes'. *The Philosophical Quarterly*, 48, 227–34.

Bird, Alexander (2005a). 'The dispositionalist conception of laws'. *Foundations of Science*, 10, 353–70.

Bird, Alexander (2005b). 'Laws and essences'. *Ratio*, 18, 437–61.

Bird, Alexander (2007). *Nature's Metaphysics: Laws and Properties*. Oxford University Press, Oxford.

Blackburn, Simon (2008). *How to Read Hume*. Granta Books, London.

Bogen, Jim (2005). 'Regularities and causality; generalizations and causal explanations'. *Studies in the History and Philosophy of Biological and Biomedical Sciences*, 36, 397–420.

Bogen, Jim (2008). 'Causally productive activities'. *Studies in History and Philosophy of Science*, 39, 112–23.

Bogen, Jim and Machamer, Peter (2011). 'Mechanistic information and causal continuity'. In *Causality in the Sciences* (ed. P. M. Illari, F. Russo, and J. Williamson), pp. 845–864. Oxford University Press, Oxford.

Boniolo, Giovanni, Faraldo, Rossella, and Saggion, Antonio (2011). 'Explicating the notion of 'causation': The role of extensive quantities'. In *Causality in the Sciences* (ed. P. M. Illari, F. Russo, and J. Williamson), pp. 502–5. Oxford University Press.

Boniolo, Giovanni and Vidali, Paolo (1999). *Filosofia della scienza*. Bruno Mondadori.

Boudon, Raymond (1967). *L'Analyse Mathématique des Faits Sociaux*. Plon, Paris.

Bradford Hill, Austin (1965). 'The environment of disease: Association or causation?' *Proceedings of the Royal Society of Medicine*, 58, 295–300.

Brading, Katherine and Castellani, Elena (2013). 'Symmetry and symmetry breaking'. In *The Stanford Encyclopedia of Philosophy* (Spring 2013 edn) (ed. E. N. Zalta). Available at: <http://plato.stanford.edu/archives/spr2013/entries/symmetry-breaking/>.

Brand, J. E. and Xie, Y. (2007). 'Identification and estimation of causal effects with time-varying treatments and time-varying outcomes'. *Sociological Methodology*, 37, 393–434.

Brandom, Robert (2000). *Articulating Reasons: An Introduction to Inferentialism*. Harvard University Press, Cambridge, Massachusetts.

Broadbent, Alex (2011). 'Inferring causation in epidemiology: mechanisms, black boxes, and contrasts'. In *Causality in the Sciences* (ed. P. M. Illari, F. Russo, and J. Williamson), pp. 45–69. Oxford University Press, Oxford.

Brunswik, E. (1955). 'Represenative design and probabilistic theory in a functional psychology'. *Psychological Review*, 62, 193–217.

Bukodi, Erzsébet and Goldthorpe, John H. (2011). 'Social class returns to higher education: chances of access to the professional and managerial salariat for men in three British birth cohorts'. *Longitudinal and Life Course Studies*, 2(2), 1–17.

Bukodi, Erzsébet and Goldthorpe, John H. (2012). 'Response: Causes, classes and cases'. *Longitudinal and Life Course Studies*, 3, 292–6.

Caldwell, J.C. (1979). 'Education as a factor in mortality decline: An examination of Nigerian data'. *Population Studies*, 33(3), 395–413.

Campaner, Raffaella (2011). 'Understanding mechanisms in the health sciences'. *Theoretical Medicine and Bioethics*, 32, 5–17.

Campaner, Raffaella and Galavotti, Maria Carla (2012). 'Evidence and the assessment of causal relations in the health sciences'. *International Studies in the Philosophy of Science*, 26(1), 27–45.

Campbell, D. T. and Stanley, J. C. (1963). *Experimental and Quasi-Experimental Designs for Research*. Rand McNally, Chicago.

Cardano, Mario (2009). Ethnography and reflexivity. Notes on the construction of objectivity in ethnographic research. Technical Report 1, NetPaper del Dipartimento di Scienze Sociali.

Cartwright, Nancy (1979). 'Causal laws and effective strategies'. *Noûs*, 13, 419–37.

Cartwright, Nancy (1989). *Nature's Capacities and their Measurement*. Clarendon Press, Oxford.

Cartwright, Nancy (1995). 'Ceteris paribus laws and socio-economic machines.' *The Monist*, 78, 276–94.

Cartwright, Nancy (1999). *The Dappled World: A Study of the Boundaries of Science*. Cambridge University Press, Cambridge.

Cartwright, Nancy (2002). 'Against modularity, the causal Markov condition, and any link between the two: Comments on Hausman and Woodward'. *British Journal for the Philosophy of Science*, 53, 411–53.

Cartwright, Nancy (2004). 'Causation: One word, many things'. *Philosophy of Science*, 71, 805–19.

Cartwright, Nancy (2007). *Hunting causes and using them*. Cambridge University Press, Cambridge.

Cartwright, Nancy (2011). 'Evidence, external validity and explanatory relevance'. In *The Philosophy of Science Matters: The Philosophy of Peter Achinstein* (ed. G. Morgan), pp. 15–28. Oxford University Press, New York.

Casini, Lorenzo, Illari, Phyllis, Russo, Federica, and Williamson, Jon (2011). 'Models for prediction, explanation, and control: Recursive Bayesian networks'. *Theoria*, 26, 5–33.

Chakravartty, Anjan (2008). *A Metaphysics for Scientific Realism. Knowing the Unobservable*. Cambridge University Press, Cambridge.

Chalmers, David, Manley, David, and Wasserman, Ryan (ed.) (2009). *Metametaphysics: New Essays on the Foundations of Ontology*. Oxford University Press, Oxford.

Chao, Hsiang-Ke (2009). *Representation and Structure in Economics. The Methodology of Econometric Models of the Consumption Function*. Routledge.

Chisholm, Roderick M. (1946). 'The contrary-to-fact conditional'. *Mind*, 55, 289–307.

Chisholm, Roderick M. (1955). 'Law statements and counterfactual inference'. *Analysis*, 15, 97–105.

Choi, Sungho and Fara, Michael (2012). 'Dispositions'. In *The Stanford Encyclopaedia of Philosophy* (Spring 2014 edn) (ed. E. N. Zalta). Available at: <http://plato.stanford.edu/archives/spr2014/entries/dispositions/>.

Christensen, L. B. (2001). *Experimental Methodology* (8th edn). Allyn & Bacon, Needham Heights, Massachusetts.

Clarke, Brendan (2011). *Causality in Medicine with Particular Reference to the Viral Causation of Cancers*. Ph.D. thesis, UCL.

Clarke, Brendan, Gillies, Donald, Illari, Phyllis, Russo, Federica, and Williamson, Jon (2013). 'The evidence that evidence-based medicine omits'. *Preventive Medicine*, 57(6), 745–747.

Clarke, Brendan, Gillies, Donald, Illari, Phyllis, Russo, Federica, and Williamson, Jon (2014). 'Mechanisms and the evidence hierarchy'. *Topoi*, **online first**, DOI:10.1007/s11245-013-9220-9.

Collier, John (1999). 'Causation is the transfer of information'. In *Causation, Natural Laws, and Explanation* (ed. H. Sankey), pp. 215–63. Kluwer, Dordrecht.

Collier, John (2011). 'Information, causation and computation'. In *Information and Computation: Essays on Scientific and Philosophical Understanding of Foundations of Information and Computation* (ed. G. D. Crnkovic and M. Burgin). World Scientific, Singapore.

Collingwood, R. G. (1938). 'On the so-called idea of causation'. *Proceedings of the Aristotelian Society (New Series)*, 38, 82–112.

Collingwood, R. G. (1940). *An Essay in Metaphysics*. Oxford University Press, Oxford.

Cook, T. D. and Campbell, D. T. (1979). *Quasi-Experimentation. Design and Analysis Issues for Field Settings*. Rand MacNally, Chicago.

Cornfield, Jerome, Haenszel, William, Hammond, E. Cuyler, Lilienfeld, Abraham M., Shimkin, Michael B., and Wynder, Ernst L. (2009). 'Smoking and lung cancer: Recent evidence and a discussion of some questions'. *International Journal of Epidemiology*, 38(5), 1175–91.

Courgeau, Daniel (1994). 'Du group à l'individu: l'exemple des comportements migratoires.' *Population*, 1, 7–26.

Courgeau, Daniel (ed.) (2003). *Methodology and Epistemology of Multilevel Analysis. Approaches from Different Social Sciences*. Kluwer, Dordrecht.

Courgeau, D. (2004). 'Probabilité, démographie et sciences sociales'. *Mathematics and Social Sciences*, 167, 27–50.

Courgeau, Daniel (2007). *Multilevel Synthesis: From the Group to the Individual*. Springer, Dordrecht.

Courgeau, Daniel (2012). *Probability and Social Science. Methodological Relationships between the two Approaches*. Springer.

Craver, Carl (2001). 'Role functions, mechanisms and hierarchy'. *Philosophy of Science*, 68: 1(1), 53–74.

Craver, Carl (2007). *Explaining the Brain*. Clarendon Press, Oxford.

Craver, Carl and Bechtel, William (2007). 'Top-down causation without top-down causes'. *Biology and Philosophy*, 22, 547–63.

Crockett, Seth D. and Keeffe, Emmet B. (2005). 'Natural history and treatment of hepatitis B virus and hepatitis C virus coinfection'. *Annals of Clinical Microbiology and Antimicrobials*, 4(13), 4–13.

Cummins, Robert (1975). 'Functional analysis'. *Journal of Philosophy*, 72, 741–65.

Darden, Lindley (1998). 'Anomaly-driven theory redesign: computational philosophy of science experiments'. In *The Digital Phoenix: How Computers are Changing Philosophy* (ed. T. W. Bynum and J. H. Moor), pp. 62–78. Blackwell Publishers, New York.

Darden, Lindley (2002). 'Strategies for discovering mechanisms: Schema instantiation, modular subassembly, forward/backward chaining'. *Philosophy of Science*, 69, S354–S365.

Darden, Lindley (2006a). *Reasoning in Biological Discoveries*. Cambridge University Press, Cambridge.

Darden, Lindley (2006b). 'Strategies for discovering mechanisms: construction, evaluation, revision'. In *Reasoning in biological discoveries* (ed. L. Darden), pp. 271–308. Cambridge University Press, Cambridge.

Darden, Lindley (2008). 'Thinking again about biological mechanisms'. *Philosophy of Science*, 75(5), 958–69.

Darden, Lindley and Craver, Carl (2002). 'Strategies in the interfield discovery of the mechanism of protein synthesis'. *Studies in the History and Philosophy of the Biological and Biomedical Sciences*, 33(1), 1–28.

David, Marian (2009). 'The correspondence theory of truth'. In *The Stanford Encyclopedia of Philosophy* (Fall 2009 edn) (ed. E. N. Zalta). Available at: <http://plato.stanford.edu/archives/fall2009/entries/truth-correspondence/>.

Davidson, Donald (1969). 'The individuation of events'. In *Essays in honor of Carl G. Hempel* (ed. N. Rescher), pp. 216–34. Reidel.

Dawid, A. Philip (2001). 'Causal inference without counterfactuals'. In *Foundations of Bayesianism* (ed. D. Corfield and J. Williamson), pp. 37–74. Kluwer, Dordrecht.

Dawid, A. Philip (2007). 'Counterfactuals, hypotheticals and potential responses: a philosophical examination of statistical causality'. In *Causality and Probability in the Sciences* (ed. F. Russo and J. Williamson), pp. 503–32. College Publications.

Demeulenaere, Pierre (ed.) (2011). *Analytical Sociology and Social Mechanisms*. Cambridge University Press, Cambridge.

Di Nardo, J. (2008). 'Natural experiments and quasi-natural experiments'. In *The New Palgrave Dictionary of Economics* (Second edn) (ed. S. N. Durlauf and L. E. Blume). Palgrave Macmillan.

Doll, Richard and Hill, Bradford (1950). 'Smoking and carcinoma of the lung'. *British Medical Journal*, 2(4682), 739–48.

Doll, Richard and Peto, Richard (1976). 'Mortality in relation to smoking: 20 years' observations on male British doctors'. *British Medical Journal*, 2, 1525–36.

Douven, Igor (2011). 'Abduction'. In *The Stanford Encyclopedia of Philosophy* (Spring 2011 edn) (ed. E. N. Zalta). Available at: <http://plato.stanford.edu/archives/spr2011/entries/abduction/>.

Dowe, Phil (1992). 'Wesley Salmon's process theory of causality and the conserved quantity theory'. *Philosophy of Science*, 59(2), 195–216.

Dowe, Phil (2000a). 'Causality and explanation: review of Salmon'. *British Journal for the Philosophy of Science*, 51, 165–74.

Dowe, Phil (2000b). *Physical causation*. Cambridge University Press, Cambridge.

Dowe, Phil (2004). 'Causes are physically connected to their effects: why preventers and omissions are not causes'. In *Contemporary debates in Philosophy of Science* (ed. C. Hitchcock). Blackwell, Oxford.

Dowe, Phil (2008). 'Causal processes'. In *The Stanford Encyclopedia of Philosophy* (Fall 2008 edn) (ed. E. N. Zalta). Available at: <http://plato.stanford.edu/archives/fall2008/entries/causation-process/>.

Driessen, Geert (2002). 'School composition and achievement in primary education: a large-scale multilevel approach'. *Studies in Educational Evaluation*, 28, 347–68.

Drouet, Isabelle (2009). 'Is determinism more favorable than indeterminism for the causal Markov condition?' *Philosophy of Science*, 76(5), 662–75.

Drouet, Isabelle (2011). 'Propensities and conditional probabilities'. *International Journal of Approximate Reasoning*, 52, 153–65.

Dupré, John (2012). *Processes of life. Essays in the philosophy of biology*. Oxford University Press.

Dupré, John (2013). 'Living causes'. In *Aristotelian Society Supplementary Volume*, Volume 87, pp. 19–37. John Wiley and Sons.

Durkheim, Emile (1895/1912). *Les Règles de la Méthode Sociologique*. Libraire Félix Arcan, Paris.

Durkheim, Emile (1897/1960). *Le Suicide*. Presses Universitaires de France.

Eells, Ellery (1991). *Probabilistic causality*. Cambridge University Press, Cambridge.

Engle, Robert F., Hendry, David F., and Richard, Jean-Francois (1983). 'Exogeneity'. *Econometrica*, 51(2), 277–304.

Fagan, Melinda (2012). 'Collective scientific knowledge'. *Philosophy Compass*, 7(12), 821–31.

Fair, David (1979). 'Causation and the flow of energy'. *Erkenntnis*, 14(3), 219–50.

Falcon, Andrea (2012). 'Aristotle on causality'. In *The Stanford Encyclopedia of Philosophy* (Winter 2012 edn) (ed. E. N. Zalta). Available at: <http://plato.stanford.edu/archives/win2012/entries/aristotle-causality/>.

Fennell, Damien (2011). 'The error term and its interpretation in structural models in econometrics'. In *Causality in the Sciences* (ed. P. M. Illari, F. Russo, and J. Williamson), pp. 361–78. Oxford University Press, Oxford.

Ferzan, Kimberly Kessler (2011). 'The unsolved mysteries of Causation and Responsibility'. *Rutgers Law Journal*, 42, 347–75.

Fisher, Ronald A. (1925). *Statistical Methods for Research Workers*. Oliver & Boyd, Edinburgh.

Fisher, Ronald A. (1935). *The Design of Experiments* (1st edn). Oliver & Boyd, Edinburgh.

Fitelson, Branden and Hitchcock, Christopher (2011). 'Probabilistic measures of causal strength'. In *Causality in the sciences* (ed. P. Illari, F. Russo, and J. Williamson), pp. 600–27. Oxford University Press, Oxford.

Florens, Jean-Pierre and Mouchart, Michel (1985). 'Conditioning in dynamic models'. *Journal of Time Series Analysis*, 53(1), 15–35.

Florens, Jean-Pierre, Mouchart, Michel, and Rolin, Jean-Marie (1993). 'Noncausality and marginalization of Markov processes'. *Econometric Theory*, 9, 241–62.

Floridi, Luciano (2008). 'A defence of informational structural realism'. *Synthese*, 161(2), 219–53.

Floridi, Luciano (2011a). 'A defence of constructionism: philosophy as conceptual engineering'. *Metaphilosophy*, 42(3), 282–304.

Floridi, Luciano (2011b). *The philosophy of Information*. Oxford University Press, Oxford.

Flowerdew, Robin and Al-Hamad, Alaa (2004). 'The relationship between marriage, divorce and migration in a British data set'. *Journal of Ethnic and Migration Studies*, 30(2), 339–51.

Franklin, Allan (1981). 'What makes a "good" experiment'. *British Journal for the Philosophy of Science*, 32, 367–77.

Franklin, Allan (1986). *The Neglect of Experiment*. Cambridge University Press, Cambridge.

Franklin, Allan (2012). 'Experiment in physics'. In *The Stanford Encyclopedia of Philosophy* (Winter 2012 edn) (ed. E. N. Zalta). Available at: <http://plato.stanford.edu/archives/win2012/entries/physics-experiment/>.

Freedman, David A. (2005a). 'Linear statistical models for causation: A critical review'. In *Encyclopedia of Statistics in Behavioral Science*. John Wiley & Sons.

Freedman, David A. (2005b). *Statistical Models: Theory and Practice*. Cambridge University Press, Cambridge.

Freedman, David A. (2006). 'Statistical models for causation: what inferential leverage do they provide?' *Evaluation Review*, 30, 691–713.

Frigg, Roman (2010). 'Models and fiction'. *Synthese*, 172, 251–68.

Frigg, Roman and Hartmann, Stephan (2012). 'Models in science'. In *The Stanford Encyclopedia of Philosophy* (Fall 2012 edn) (ed. E. N. Zalta). Available at: <http://plato.stanford.edu/archives/fall2012/entries/models-science/>.

Frisch, Mathias (2014). 'Physics and the human face of causation'. *Topoi*, **online first**, DOI:10.1007/s11245-013-9172-0.

Frisch, R. ([1938] 1995). 'Autonomy of economic relations'. In *The Foundations of Econometric Analysis* (ed. D. F. Hendry and M. S. Morgan), Chapter 37, pp. 407–26. Cambridge University Press, Cambridge.

Gasking, Douglas (1955). 'Causation and recipes'. *Mind*, 64(256), 479–87.

Gaumé, Catherine and Wunsch, Guillaume (2010). 'Self-rated health in the Baltic countries, 1994–1999'. *European Journal of Population*, 26(4), 435–57.

Gerring, John (2005). 'Causation. A unified framework for the social sciences'. *Journal of Theoretical Politics*, 17(2), 163–98.

Gerstenberg, Tobias and Lagnado, David A. (2010). 'Spreading the blame: the allocation of responsibility amongst multiple agents'. *Cognition*, 115, 166–71.

Giere, Ronald (1979). *Understanding Scientific Reasoning*. Holt, Rinehart, and Wiston.

Giere, Ronald (2006). *Scientific Perspectivism*. The University of Chicago Press, Chicago.

Gillies, Donald (2000). *Philosophical Theories of Probability*. Routledge, London and New York.

Gillies, Donald (2005a). 'An action-related theory of causality'. *British Journal for the Philosophy of Science*, 56, 823–42.

Gillies, Donald (2005b). 'Hempelian and Kuhnian approaches in the philosophy of medicine: the Semmelweis case'. *Studies in History and Philosophy of Biological and Biomedical Sciences*, 36, 159–81.

Gillies, Donald (2008). *How Should Research be Organised?* College Publications.

Gillies, Donald (2011). 'The Russo-Williamson Thesis and the question of whether smoking causes heart disease'. In *Causality in the Sciences* (ed. P. M. Illari, F. Russo, and J. Williamson), pp. 110–25. Oxford University Press, Oxford.

Gillies, Donald and Sudbury, Aidan (2013). 'Should causal models always be Markovian? The case of multi-causal forks in medicine'. *European Journal for Philosophy of Science*, 3(3), 275–308.

Glennan, Stuart (1996). 'Mechanisms and the nature of causation'. *Erkenntnis*, 44, 49–71.

Glennan, Stuart (1997). 'Capacities, universality, and singularity'. *Philosophy of Science*, 64(4), 605–26.

Glennan, Stuart (2002a). 'Contextual unanimity and the units of selection problem'. *Philosophy of Science*, 69(1), 118–37.

Glennan, Stuart (2002b). 'Rethinking mechanistic explanation'. *Philosophy of Science*, 69, S342–S353.

Glennan, Stuart (2005). 'Modeling mechanisms'. *Studies in the History and Philosophy of Biology and Biomedical Sciences*, 36, 443–64.

Glennan, Stuart (2008). 'Mechanisms'. In *Routledge Companion to the Philosophy of Science* (ed. S. Psillos and M. Curd). Routledge, Oxford.

Glennan, Stuart (2009a). 'Mechanisms'. In *The Oxford Handbook of Causation* (ed. H. Beebee, C. Hitchcock, and P. Menzies), pp. 315–25. Oxford University Press, Oxford.

Glennan, Stuart (2009b). 'Productivity, relevance and natural selection'. *Biology and Philosophy*, 24(3), 325–39.

Glennan, Stuart (2010a). 'Ephemeral mechanisms and historical explanation'. *Erkenntnis*, 72(2), 251–66.

Glennan, Stuart (2010b). 'Mechanisms, causes, and the layered model of the world'. *Philosophy and Phenomenological Research*, LXXI, 362–81.

Glennan, Stuart (2011). 'Singular and general causal relations: a mechanist perspective'. In *Causality in the Sciences* (ed. P. M. Illari, F. Russo, and J. Williamson), pp. 789–817. Oxford University Press, Oxford.

Glynn, Luke (2013). 'Of miracles and interventions'. *Erkenntnis*, 78, 43–64.

Godfrey-Smith, Peter (2010). 'Causal pluralism'. In *Oxford Handbook of Causation* (ed. H. Beebee, C. Hitchcock, and P. Menzies), pp. 326–37. Oxford University Press, Oxford.

Goldstein, Harvey (1987). *Multilevel Models in Educational and Social Research*. Griffin.

Goldstein, Harvey (2003). *Multilevel Statistical Models*, Volume Kendall's Library of Statistics (n. 3). Arnold, London.

Goldthorpe, John H. (2001). 'Causation, statistics, and sociology'. *European Sociological Review*, 17(1), 1–20.

Good, I. J. (1961a). 'A causal calculus I'. *British Journal for the Philosophy of Science*, 11, 305–18. Errata vol 13 pg. 88.

Good, I. J. (1961b). 'A causal calculus II'. *British Journal for the Philosophy of Science*, 12, 43–51. Errata vol 13 pg. 88.

Gopnik, Alison, Glymour, Clark, Sobel, David M., Schulz, Laura E., Kushnir, Tamar, and Danks, David (2004). 'A theory of causal learning in children: causal maps and Bayes nets'. *Psychological Review*, 111(1), 3–32.

Gopnik, Alison and Schulz, Laura (ed.) (2007). *Causal Learning: Psychology, Philosophy, and Computation*. Oxford University Press, Oxford.

Greenland, Sander (2000). 'Principles of multilevel modelling'. *International Journal of Epidemiology*, 29, 158–67.

Guala, Francesco (2005). *The Methodology of Experimental Economics*. Cambridge University Press, Cambridge.

Guth, Alan (1998). *The Inflationary Universe*. Basic Books.

Haavelmo, Trygve (1944, July). 'The probability approach in econometrics'. *Econometrica*, 12(Supplement), iii–115.

Hacking, Ian (1983). *Representing and Intervening*. Cambridge University Press, Cambridge.

Hall, Ned (2004). 'Two concepts of causation'. In *Causation and Counterfactuals* (ed. L. Paul, E. Hall, and J. Collins), pp. 225–76. MIT Press, Cambridge, Massachusetts.

Harré, Rom (1981). *Great Scientific Experiments: Twenty Experiments that Changed our View of the World*. Phaidon, Oxford.

Hart, H. L. A. and Honoré, T. (1959). *Causation in the Law*. Clarendon Press, Oxford.

Hartmann, Stephan (1996). 'The world as a process: Simulations in the natural and social sciences'. In *Modelling and Simulation in the Social Sciences from the Philosophy of Science Point of View* (ed. R. Hegselmann, U. Mueller, and K. G. Troitzsch), pp. 77–100. Springer.

Hausman, Daniel M. and Woodward, James (1999). 'Independence, invariance, and the causal Markov condition'. *British Journal for the Philosophy of Science*, 50, 521–83.

Hausman, Daniel M. and Woodward, James (2002). 'Modularity and the causal Markov condition: A restatement'. *British Journal for the Philosophy of Science*, 55, 147–61.

Heckman, James J. (2008). 'Econometric causality'. *International Statistical Review*, 76(1), 1–27.

Hedström, Peter and Swedberg, Richard (ed.) (1998). *Social Mechanisms: An Analytical Approach to Social Theory*. Cambridge University Press, Cambridge.

Hedström, Peter and Ylikoski, Petri (2010). 'Causal mechanisms in the social sciences'. *The Annual Review of Sociology*, 36, 49–67.

Hellevik, O. (1984). *Introduction to Causal Analysis*. Allen & Unwin, London.

Hempel, Carl G. (1965). *Aspects of Scientific Explanation*. Free Press, New York.

Hesslow, G. (1976). 'Discussion: two notes on the probabilistic approach to causality'. *Philosophy of Science*, 43, 290–92.

Hitchcock, Christopher (2001). 'The intransitivity of causation revealed in equations and graphs'. *The Journal of Philosophy*, 98(6), 273–99.

Hitchcock, Christopher (2011). 'Probabilistic causation'. In *The Stanford Encyclopedia of Philosophy* (Winter 2011 edn) (ed. E. N. Zalta). Available at: <http://plato.stanford.edu/archives/win2011/entries/causation-probabilistic/>.

Hitchcock, Christopher and Knobe, Joshua (2009). 'Cause and norm'. *Journal of Philosophy*, 106(11), 587–612.

Hoerl, Christoph, McCormack, Teresa, and Beck, Sarah R. (ed.) (2011). *Understanding Counterfactuals, Understanding Causation: Issues in Philosophy and Psychology*. Oxford University Press, Oxford.

Hofer-Szabó, Gàbor, Rédei, Miklós, and Szabó, László E. (2013). *The Principle of the Common Cause*. Cambridge University Press, Cambridge.

Holland, Paul (1986). 'Statistics and causal inference'. *Journal of the American Statistical Association*, 81, 945–60.

Holland, Paul (1988). 'Comment: causal mechanism or causal effect: Which is best for statistical science?' *Statistical Science*, 3, 186–88.

Hood, W. C. and Koopmans, T. C. (ed.) (1953). *Studies in Econometric Methods*. Cowles Foundation Monograph 14. John Wiley & Sons, New York.

Hoover, K. (2001). *Causality in Macroeconomics*. Cambridge University Press, New York.

Howick, Jeremy (2011a). 'Exposing the vanities—and a qualified defence—of mechanistic evidence in clinical decision-making'. *Philosophy of Science*, 78(5), 926–40. Proceedings of the Biennial PSA 2010.

Howick, Jeremy (2011b). *The Philosophy of Evidence-Based Medicine*. BMJ Books, Oxford.

Hume, David (1777). *Enquiries Concerning the Human Understanding and Concerning the Principles of Morals (edited by L. A. Selby-Bigge)* (1902, 2nd edn). Clarendon Press, Oxford.

Humphreys, Paul (1985). 'Why propensities cannot be probabilities'. *The Philosophical Review*, 94(4), 557–70.

Humphreys, Paul (1990). *The Chances of Explanation*. Princeton University Press, Princeton.

Humphreys, Paul (2004). 'Some considerations on conditional chances'. *The British Journal for the Philosophy of Science*, 55(3), 667–80.

Huntington, Samuel P. (1996). *The Clash of Civilization and the Remaking of World Order*. Simon and Schuster, New York.

Hutchinson, J. F. (2001). 'The biology and evolution of HIV'. *Annual Review of Anthropology*, 30, 85–108.

IARC (2006). Preamble to the IARC monographs. International Agency for Research on Cancer, <http://monographs.iarc.fr/ENG/Preamble/index.php>.

Illari, Phyllis (2013). 'Mechanistic explanation: integrating the ontic and epistemic'. *Erkenntnis*, 78, 237–55.

Illari, Phyllis and Floridi, Luciano (ed.) (2014). *The Philosophy of Information Quality*. Springer.

Illari, Phyllis and Russo, Federica (2014). 'Information channels and biomarkers of disease'. *Topoi*, **Online first**, DOI:10.1007/s11245-013-9228-1.

Illari, Phyllis McKay (2011a). 'Mechanistic evidence: disambiguating the Russo-Williamson thesis'. *International Studies in the Philosophy of Science*, 25(2), 139–157.

Illari, Phyllis McKay (2011b). 'Why theories of causality need production: an information-transmission account'. *Philosophy and Technology*, 24(2), 95–114.

Illari, Phyllis McKay and Williamson, Jon (2010). 'Function and organization: comparing the mechanisms of protein synthesis and natural selection'. *Studies in the History and Philosophy of the Biological and Biomedical Sciences*, **41**, 279–91.

Illari, Phyllis McKay and Williamson, Jon (2011). 'Mechanisms are real and local'. In *Causality in the Sciences* (ed. P. M. Illari, F. Russo, and J. Williamson), pp. 818–44. Oxford University Press, Oxford.

Illari, Phyllis McKay and Williamson, Jon (2012). 'What is a mechanism?: Thinking about mechanisms across the sciences'. *European Journal of the Philosophy of Science*, **2**, 119–35.

Jackson, Frank (1998). *From Metaphysics to Ethics*. Clarendon Press, Oxford.

Jadad, Alejandro R. and Enkin, Murray W. (2007). *Randomised Controlled Trials: Questions, Answers and Musings* (Second edn). BMJ Books, Oxford.

Jimenez-Buedo, Maria and Miller, Luis M. (2009). 'Experiments in the social sciences: the relationship between external and internal validity'. In *SPSP 2009: Society for Philosophy of Science in Practice (Minnesota, June 18-20, 2009)(PhilSci Archive)*.

Jimenez-Buedo, Maria and Miller, Luis Miguel (2010). 'Why a trade-off? The relationship between the external and internal validity of experiments'. *Theoria*, **69**, 301–21.

Joffe, Michael (2011). 'Causality and evidence discovery in epidemiology'. In *Explanation, prediction, and confirmation. New trends and old ones reconsidered.* (ed. D. Dieks, W. J. Gonzalez, S. Hartmann, T. Uebel, and M. Weber), pp. 153–66. Springer.

Joffe, Michael (2013). 'The concept of causation in biology'. *Erkenntnis*, **78**, 179–97.

Joffe, Michael, Gambhir, M., Chadeau-Hyam, Marc, and Vineis, Paolo (2012). 'Causal diagrams in systems epidemiology'. *Emerging themes in epidemiology*, **9**(1).

Joyce, James M. (2002). 'Levi on causal decision theory and the possibility of predicting one's own actions'. *Philosophical Studies*, **110**, 69–102.

Katok, Anatole and Hasselblatt, Boris (1995). *Introduction to the Modern Theory of Dynamical Systems*. Cambridge University Press, New York.

Killoran, A. and Kelly, M. P. (ed.) (2010). *Evidence-Based Public Health. Effectiveness and Efficiency*. Oxford University Press, New York.

Kim, J. (1976). 'Events as property exemplifications'. In *Action Theory* (ed. M. Brand and D. Walton), pp. 159–77. Reidel.

Kincaid, Harold (2011). 'Causal modeling, mechanism and probability in epidemiology'. In *Causality in the Sciences* (ed. P. M. Illari, F. Russo, and J. Williamson), pp. 70–90. Oxford University Press, Oxford.

Kistler, Max and Gnassounou, Bruno (ed.) (2007). *Dispositions and Causal Powers*. Ashgate, Hampshire, UK.

Kittel, Bernhard (2006). 'A crazy methodology? On the limits of macro-quantitative social science'. *International Sociology*, **21**(5), 647–77.

Kleinberg, Samantha (2013). *Causality, Probability, and Time*. Cambridge University Press, New York.

Kolmogorov, A. N. (1965). 'Three approaches to the quantitative definition of information'. *Problems of Information and Transmission*, **1**(1), 1–7.

Kolmogorov, A. N. (1983). 'Combinatorial foundations of information theory and the calculus of probabilities'. *Russian Mathematical Surveys*, **38**(4), 27–36.

Koopmans, T. C. (ed.) (1950). *Statistical Inference in Dynamic Economic Models*. Cowles Foundation Monograph 10. John Wiley, New York.

Korb, Kevin B. and Nicholson, Ann E. (2003). *Bayesian Artificial Intelligence*. Chapman and Hall / CRC Press, London.

Kuhn, Thomas S. (1962). *The Structure of Scientific Revolutions*. University of Chicago Press, Chicago.

La Caze, Adam (2011). 'The role of basic science in evidence-based medicine'. *Biology and Philosophy*, 26(1), 81–98.

Ladyman, James and Ross, Don (2007). *Every thing Must Go*. Oxford University Press. With David Spurrett and John Collier.

LaFollette, Hugh and Schanks, Niall (1996). *Brute Science: Dilemmas of Animal Experimentation*. Routledge, New York.

Lagnado, David A. and Channon, S. (2008). 'Judgments of cause and blame: The influence of intentionality and foreseeability'. *Cognition*, 108, 754–70.

Lagnado, David A. and Sloman, Steven (2004). 'The advantage of timely intervention'. *Journal of Experimental Psychology: Learning, Memory and Cognition*, 30(4), 856–76.

Laudisa, Federico (1999). *Causalità. Storia di un Modello di Conoscenza*. Carocci.

Leonelli, Sabina (2009). 'On the locality of data and claims about phenomena'. *Philosophy of Science*, 76, 737–49.

Leonelli, Sabina (2014). 'Data interpretation in the digital age'. *Perspectives on Science*, **in press**. Special issue edited by Henk de Regt and Wendy Parker.

Leuridan, Bert (2012). 'Three problems for the mutual manipulability account of constitutive relevance in mechanisms'. *The British Journal for the Philosophy of Science*, 63(2), 399–427.

Lewis, David (2004). 'Void and object'. In *Causation and Counterfactuals* (ed. J. Collins, N. Hall and L. A. Paul), pp. 227–90. MIT Press, Cambridge, Massachusetts.

Lewis, David K. (1973). *Counterfactuals*. Blackwell, Oxford.

Lewis, David K. (1983). *Philosophical Papers*, Volume 1. Oxford University Press, Oxford.

Lewis, David K. (2000). 'Causation as influence'. *Journal of Philosophy*, 97(4), 182–97.

Lieberson, S. (1985). *Making It Count*. University of California Press, Berkeley, California.

Lioy, Paul J. and Rappaport, Stephen (2011). 'Exposure science and the exposome: an opportunity for coherence in the environmental health sciences'. *Environmental Health Perspectives*, 119(11), a466–a467.

Lipton, Robert and Ødegaard, Terje (2005). 'Causal thinking and causal languange in epidemiology'. *Epidemiological Perspectives and Innovations*, 2, 8. <http://www.epi-perspectives.com/content/2/1/8>.

Little, D. (1991). *Varieties of Social Explanation: An Introduction to the Philosophy of Social Science*. Westview Press, Boulder.

Little, D. (1998). *Microfoundations, Method and Causation. On the Philosophy of the Social Sciences*. Transaction Publishers, New Brunswick, N.J.

Little, D. (2004). 'Causal mechanisms'. In *Encyclopedia of Social Sciences* (ed. M. Lewis-Beck, A. Bryman, & T. Liao), pp. 100–1. Sage.

Little, Daniel (2006). 'Levels of the social'. In *The Philosophy of Anthropology and Sociology* (ed. M. Risjord and S. Turner), pp. 343–71. Elsevier, Amsterdam.

Little, Daniel (2011). 'Causal mechanism in the social realm'. In *Causality in the Sciences* (ed. P. M. Illari, F. Russo, and J. Williamson), pp. 273–95. Oxford University Press, Oxford.

Longino, Helen (2013). 'The social dimensions of scientific knowledge'. In *The Stanford Encyclopedia of Philosophy* (Spring 2013 edn) (ed. E. N. Zalta). Available at: <http://plato.stanford.edu/archives/spr2013/entries/scientific-knowledge-social/>.

Longworth, Francis (2010). 'Cartwright's causal pluralism: a critique and an alternative'. *Analysis*, 70(2), 310–18.

Lucas, R. E. (1976). 'Econometric policy evaluation'. In *The Phillips Curve and Labor Markets* (ed. K. Brunner and A. Meltzer), Volume 1 of *Carnegie-Rochester Conference Series on Public Policy*, pp. 161–68. North-Holland, Amsterdam.

Mach, Ernst (1905). *Erkenntnis und Irrtum. Skizzenzur Psycologie der Forschung.* J.A. Barth, Leipzig.

Machamer, Peter (2004). 'Activities and causation: the metaphysics and epistemology of mechanisms'. *International Studies in the Philosophy of Science*, 18: 1, 27–39.

Machamer, Peter and Bogen, Jim (2011). 'Mechanistic information and causal continuity'. In *Causality in the Sciences* (ed. P. M. Illari, F. Russo, and J. Williamson). Oxford University Press, Oxford.

Machamer, Peter, Darden, Lindley, and Craver, Carl (2000). 'Thinking about mechanisms'. *Philosophy of Science*, 67, 1–25.

Mackie, J. L. (1974). *The Cement of the Universe. A study on Causation.* Oxford University Press, Oxford.

Mäki, Uskali (1992). 'On the method of isolation in economics'. In *Idealization IV: Intelligibility in Science* (ed. C. Dilworth), Poznan Studies in the Philosophy of the Sciences and the Humanities, pp. 319–54. Rodopi.

Malinas, Gary and Bigelow, John (2012). 'Simpson's paradox'. In *The Stanford Encyclopedia of Philosophy* (Fall 2012 edn) (ed. E. N. Zalta). Available at: <http://plato.stanford.edu/archives/fall2012/entries/paradox-simpson/>.

Malinowski, Bronislaw (1935). *Coral Gardens and Their Magic.* American Book Co.

Mandel, David R. (2011). 'Mental simulation and the nexus of causal and counterfactual explanation'. In *Understanding Counterfactuals, Understanding Causation* (ed. C. Hoerl, T. McCormack, and S. Beck). Oxford University Press, Oxford.

Manolio, Teri A., Collins, Francis S., Cox, Nancy J., Goldstein, David B., Hindorf, Lucia A., Hunte, David J., McCarthy, Mark I., Ramos, Erin M., Cardon, Lon R., Chakravarti, Aravinda, Cho, Judy H., Guttmachera, Alan E., Kong, Augustine, Kruglyak, Leonid, Mardis, Elaine, Rotimi, Charles N., Slatkin, Montgomery, Valle, David, Whittemore, Alice S., Boehnke, Michael, Clark, Andrew G., Eichler, Evan E., Gibson, Greg, Haine, Jonathan L., Mackay, Trudy F. C., McCarroll, Steven A., and Visscher, Peter M. (2009). 'Finding the missing heritability of complex diseases'. *Nature*, 461, 474–753.

Marmodoro, Anna (2013). 'Potentiality in Aristotle's metaphysics'. In *The Handbook of Potentiality* (ed. K. Engelhard and M. Quante). Springer.

Marschak, J. (1942). 'Economic interdependence and statistical analysis'. In *Studies in mathematical economics and econometrics. In memory of Henry Schultz* (ed. O. Lange, F. McIntyre, and T. O. Yntema), pp. 135–50. University of Chicago Press, Chicago.

Martin, C. B. (1994). 'Dispositions and conditionals'. *The Philosophical Quarterly*, 44, 1–8.

Mayntz, Renate (2002). 'Mechanisms and the analysis of social macro-phenomena'. *Philosophy of the Social Sciences*, 34, 237–59.

McCormack, Teresa, Frosch, Caren, and Burns, Patrick (2011). 'The relationship between children's causal and counterfactual judgements'. In *Understanding Counterfactuals, Understanding Causation* (ed. C. Hoerl, T. McCormack, and S. Beck). Oxford University Press, Oxford.

McCullough, Dennis (2009, January). *My Mother, Your Mother: Embracing "Slow Medicine," the Compassionate Approach to Caring for Your Aging Loved Ones* (Reprint edn). Harper Perennial.

McLaren, Lindsay, Ghali, Laura M., Lorenzetti, Diane, and Rock, Melanie (2006). 'Out of context? Translating evidence from the North Karelia project over place and time'. *Health Education Research*, 22(3), 414–24.

Mellor, D. H. (1995). *The facts of Causation*. Routledge, Oxford.

Menzies, Peter (2007). 'Causation in context'. In *Russell's Republic* (ed. H. Price and R. Corry), pp. 191–223. Clarendon Press, Oxford.

Menzies, Peter (2011). 'The role of counterfactual dependence in causal judgements'. In *Understanding Counterfactuals, Understanding Causation* (ed. C. Hoerl, T. McCormack, and S. Beck). Oxford University Press, Oxford.

Menzies, Peter (2014). 'Counterfactual theories of causation'. In *The Stanford Encyclopedia of Philosophy* (Spring 2014 edn) (ed. E. N. Zalta). Available at: <http://plato.stanford.edu/archives/spr2014/entries/causation-counterfactual/>.

Menzies, Peter and Price, Huw (1993). 'Causation as a secondary quality'. *British Journal for the Philosophy of Science*, 44(2), 187–203.

Merton, Robert K. (1964). *Social Theory and Social Structure*. The Free Press.

Mettler, F.A. Jr, Huda, W., Yoshizumi, T.T., and Mahesh, M. (2008). 'Effective doses in radiology and diagnostic nuclear medicine: a catalog'. *Radiology*, 248(1), 254–63.

Mill, John Stuart (1843). *A System of Logic, Ratiocinative and Inductive: Being a Connected View of the Principles of Evidence and the Methods of Scientific Investigation* (Seventh (1868) edn). Longmans, Green, Reader, and Dyer, London.

Millikan, Ruth Garrett (1984). *Language, Thought, and Other Biological Categories*. MIT Press, Cambridge, Massachusetts.

Millikan, Ruth Garrett (1989). 'In defense of proper functions'. *Philosophy of Science*, 56, 288–302.

Mitchell, Sandra D. (2003). *Biological Complexity and Integrative Pluralism*. Cambridge University Press, Cambridge.

Moneta, Alessio (2007). 'Mediating beetween causes and probability: the use of graphical models in econometrics'. In *Causality and probability in the sciences* (ed. F. Russo and J. Williamson), pp. 109–30. College Publications.

Moneta, Alessio and Russo, Federica (2014). 'Causal models and evidential pluralism in econometrics'. *Journal of Economic Methodology*, 21(1), 54–76.

Moore, Michael S. (2009). *Causation and Responsibility. An Essay in Law, Morals, and Metaphysics*. Oxford University Press, Oxford.

Moore, Michael S. (2011). 'Causation revisited'. *Rutgers Law Journal*, 42, 451–509.

Morgan, Mary and Morrison, Margaret (1999a). 'Models as mediating instruments'. In *Models as Mediators. Perspectives on Natural and Social Science* (ed. M. Morgan and M. Morrison). Cambridge University Press, Cambridge.

Morgan, Mary and Morrison, Margaret (ed.) (1999b). *Models as Mediators*. Cambridge University Press, Cambridge.

Morgan, S. L. and Winship, C. (2007). *Counterfactuals and Causal Inference*. Cambridge University Press, Cambridge.

Mouchart, Michel and Russo, Federica (2011). 'Causal explanation: recursive decompositions and mechanisms'. In *Causality in the sciences* (ed. P. M. Illari, F. Russo, and J. Williamson), pp. 317–37. Oxford University Press, Oxford.

Mouchart, Michel, Russo, Federica, and Wunsch, Guillaume (2010). 'Inferring causal relations by modelling structures'. *Statistica*, LXX(4), 411–32.

Mulligan, Kevin and Correia, Fabrice (2013). 'Facts'. In *The Stanford Encyclopedia of Philosophy* (Spring 2013 edn) (ed. E. N. Zalta). Available from <http://plato.stanford.edu/archives/spr2013/entries/facts/>.

Mumford, Stephen and Anjum, Rani Lill (2011). *Getting Causes from Powers*. Oxford University Press, Oxford.

Neapolitan, Richard E. (1990). *Probabilistic Reasoning in Expert Systems: Theory and Algorithms*. Wiley, New York.

NICE (2012). *The Guidelines Manual*. National Institute for Health and Care Excellence, London. Available from: <www.nice.org.uk>.

Nolan, Daniel (2011). 'Modal fictionalism'. In *The Stanford Encyclopedia of Philosophy* (ed. E. N. Zalta). Available from <http://plato.stanford.edu/archives/win2011/entries/fictionalism-modal/>.

Norris, P. and Inglehart, R. (2003). 'Islam and the West: Testing the 'clash of civilization' thesis'. *Comparative Sociology*, 1, 235–65.

Norton, John D. (2003). 'Causation as folk science'. *Philosophers' Imprint*, 3(4), 1–22.

OCEBM Levels of Evidence Working Group (2011). The Oxford 2011 levels of evidence. Oxford Centre for Evidence-Based Medicine, <http://www.cebm.net/index.aspx?o=5653>.

Oertelt-Prigione, Sabine and Regitz-Zagrosek, Vera (ed.) (2012). *Sex and Gender Aspects in Clinical Medicine*. Springer.

Osimani, Barbara (2013). 'Until RCT-proven? On the asymmetry of evidence requirements for risk assessment'. *Journal of evaluation of clinical practice*, 19, 454–62.

Osimani, Barbara (2014). 'Hunting side effects and explaining them: should we reverse evidence hierarchies upside down?' *Topoi*, **online first**, DOI:10.1007/s11245-013-9194-7.

Paneth, Nigel (2004). 'Assessing the contributions of John Snow to epidemiology. 150 years after removal of the broad street pump handle'. *Epidemiology*, 15(5), 514–16.

Parascandola, M. and Weed, D.L. (2001). 'Causation in epidemiology'. *Journal of Epidemiology and Community Health*, 55, 905–12.

Parsons, Craig A. and Smeeding, Timothy M. (ed.) (2006). *Immigration and the Transformation of Europe*. Cambridge University Press, Cambridge.

Pearl, Judea (1988). *Probabilistic Reasoning in Intelligent Systems: Networks of Plausible Inference*. Morgan Kaufmann, San Mateo CA.

Pearl, Judea (2000). *Causality: Models, Reasoning, and Inference*. Cambridge University Press, Cambridge.

Pearl, Judea (2009). *Causality: Models, Reasoning, and Inference* (Second edn). Cambridge University Press, Cambridge.

Pearl, Judea (2011). 'The structural theory of causality'. In *Causality in the sciences* (ed. P. M. Illari, F. Russo, and J. Williamson), pp. 697–727. Oxford University Press, Oxford.

Pearson, Karl (1911). *The Grammar of Science*. Adam & Charles Black, London.

Peirce, C. S. and Jastrow, J. (1885). 'On small differences in sensation'. *Memoirs of the National Academy of Sciences*, **3**, 73–83.

Pettifor, Ann (2003). *The Real World Economic Outlook*. Palgrave Macmillan, Hampshire, UK.

Popper, Karl (1963). *Conjectures and Refutations: The Growth of Scientific Knowledge*. Routledge, London.

Price, Huw (1996). *Time's Arrow and Archimedes' point: New Directions for the Physics of Time*. Oxford University Press, New York.

Price, Huw (2001). 'Causation in the special sciences: the case for pragmatism'. In *Stochastic Causality* (ed. M. C. Galavotti, P. Suppes, and D. Costantini), pp. 103–21. CSLI Publications, Stanford, California.

Psillos, S. (2002). *Causation and Explanation*. Acumen Publishing, Chesham.

Psillos, S. (2004). 'A glimpse of the secret connexion: harmonising mechanisms with counterfactuals'. *Perspectives on Science*, **12**(3), 288–319.

Psillos, Stathis (2009). 'Causation and regularity'. In *Oxford Handbook of Causation* (ed. H. Beebee, P. Menzies, and C. Hitchcock), pp. 131–57. Oxford University Press, Oxford.

Pundik, Amit (2007). 'Can one deny both causation by omission and causal pluralism? The case of legal causation'. In *Causality and Probability in the Sciences* (ed. F. Russo and J. Williamson), pp. 379–412. College Publications.

Puska, Pekka, Vartiaine, Erkki, Laatikainen, Tiina, Jousilahti, Pekka, and Paavola, Meri (ed.) (2009). *The North Karelia Project: from North Karelia to national action*. National Institute for Health and Welfare.

Rappaport, Stephen (2011). 'Implications of the exposome for exposure science'. *Journal of Exposure Science and Environmental Epidemiology*, **21**, 5–9.

Rappaport, Stephen M. and Smith, Martyn T. (2010). 'Environment and disease risks'. *Science*, **330**, 460–61.

Reichenbach, Hans (1956). *The Direction of Time*. University of California Press, California.

Reichenbach, Hans (1957). *The Philosophy of Space and Time*. Dover.

Reiss, Julian (2007). *Error in Economics: Towards a More Evidence-Based Methodology*. Routledge, London.

Reiss, Julian (2012a). 'Causation in the sciences: an inferentialist account'. *Studies in History and Philosophy of Biological and Biomedical Sciences*, **43**, 769–77.

Reiss, Julian (2012b). 'Counterfactuals'. In *The Oxford Handbook of Philosophy of Social Science* (ed. H. Kincaid), pp. 154–83. Oxford University Press, Oxford.

Rescher, Nicholas (2007). *Conditionals*. The MIT Press, Cambridge, Massachusetts.

Reutlinger, Alexander, Schurz, Gerhard, and Hüttemann, Andreas (2011). 'Ceteris paribus laws'. In *The Stanford Encyclopedia of Philosophy* (Spring 2011 edn) (ed. E. N. Zalta). Available at: <http://plato.stanford.edu/archives/spr2011/entries/ceteris-paribus/>.

Robinson, W. S. (1950). 'Ecological correlations and the behavior of individuals'. *American Sociological Review*, **15**, 351–7.

Roessler, Johannes (2011). 'Perceptual causality, counterfactuals, and special causal concepts'. In *Understanding counterfactuals, understanding causation* (ed. C. Hoerl, T. McCormack, and S. Beck), pp. 75–89. Oxford University Press, Oxford.

Rosen, Deborah (1978). 'In defence of a probabilistic theory of causality'. *Philosophy of Science*, 45, 604–13.

Rosen, Gideon (2011). 'Causation, counterfactual dependence and culpability: moral philosophy in Michael Moore's Causation and Responsibility'. *Rutgers Law Journal*, 42, 405–34.

Ross, Don and Spurrett, David (2004). 'What to say to a skeptical metaphysician: a defense manual for cognitive and behavioral scientists'. *Behavioral and Brain Sciences*, 27(5), 603–47.

Rothman, Kenneth J. (1976). 'Causes'. *American Journal of Epidemiology*, 104(6), 587–92.

Rothman, Kenneth J. and Greenland, Sander (2005). 'Causation and causal inference in epidemiology'. *American Journal of Public Health*, 95(S1), S144–S150.

Rothman, Kenneth J., Greenland, Sander, and Lash, Timothy L. (2008). *Modern Epidemiology*. Wolters Kluwer/Lippincott Williams and Wilkins.

Rubin, Donald (1974). 'Estimating causal effects of treatments in randomized and non randomized studies'. *Journal of Educational Psychology*, 66, 688–701.

Rubin, Donald (1986). 'Which ifs have causal answers'. *Journal of the American Statistical Association*, 81, 961–2.

Russell, Bertrand (1913). 'On the notion of cause'. *Proceedings of the Aristotelian Society*, 13, 1–26.

Russell, Bertrand (1948). *Human Knowledge: Its Scope and Limits*. Simon and Schuster.

Russo, F. (2006). 'The rationale of variation in methodological and evidential pluralism'. *Philosophica*, 77, 97–124.

Russo, Federica (2007). Frequency-driven probabilities in quantitative causal analysis. *Philosophical Writings*, 32, 32–49.

Russo, Federica (2009). *Causality and Causal Modelling in the Social Sciences. Measuring Variations*. Methodos Series. Springer, New York.

Russo, Federica (2011a). 'Correlational data, causal hypotheses, and validity'. *Journal for General Philosophy of Science*, 42(1), 85–107.

Russo, Federica (2011b). 'Explaining causal modelling. Or, what a causal model ought to explain'. In *New Essays in Logic and Philosophy of Science* (ed. M. D'Agostino, G. Giorello, F. Laudisa, T. Pievani, and C. Sinigaglia), pp. 347–62. College Publications.

Russo, Federica (2012). 'Public health policy, evidence, and causation. Lessons from the studies on obesity'. *Medicine, Health Care, and Philosophy*, 15, 141–51.

Russo, Federica (2014). 'What invariance is and how to test for it'. *International Studies in Philosophy of Science*. In press.

Russo, Federica and Williamson, Jon (2007a). 'Interpreting causality in the health sciences'. *International Studies in Philosophy of Science*, 21(2), 157–70.

Russo, Federica and Williamson, Jon (2007b). 'Interpreting probability in causal models for cancer'. In *Causality and Probability in the Sciences* (ed. F. Russo and J. Williamson), Texts in Philosophy Series, pp. 217–42. College Publications.

Russo, Federica and Williamson, Jon (2011a). 'Epistemic causality in medicine'. *History and Philosophy of the Life Sciences*, 33, 563–82.

Russo, Federica and Williamson, Jon (2011b). 'Generic vs. single-case causal knowledge. The case of autopsy'. *European Journal for Philosophy of Science*, 1(1), 47–69.

Russo, Federica and Williamson, Jon (2012). 'Envirogenomarkers. The interplay between difference-making and mechanisms'. *Medicine Studies*, 3, 249–62.

Russo, Federica, Wunsch, Guillaume, and Mouchart, Michel (2011). 'Inferring causality through counterfactuals in observational studies. Some epistemological issues'. *Bulletin of Sociological Methodology*, 111, 43–64.

Ruzzene, Attilia (2012). 'Meccanismi nelle scienze sociali'. *APhEx*, 5.

Salmon, Wesley C. (1984). *Scientific Explanation and the Causal Structure of the World*. Princeton University Press, Princeton.

Salmon, Wesley C. (1977). 'An "At-At" theory of causal influence'. *Philosophy of Science*, 44(2), 215–24.

Salmon, Wesley C. (1980a). 'Causality: production and propagation'. In *Causation* (ed. E. Sosa and M. Tooley), pp. 154–71. Oxford University Press, Oxford.

Salmon, Wesley C. (1980b). 'Probabilistic causality'. In *Causality and Explanation*, pp. 208–32. Oxford University Press (1988), Oxford.

Salmon, Wesley C. (1989). *Four Decades of Scientific Explanation*. University of Pittsburgh Press, Pittsburgh.

Salmon, Wesley C. (1994). 'Causality without counterfactuals'. *Philosophy of Science*, 61, 297–312.

Salmon, Wesley C. (1997). 'Causality and explanation: a reply to two critiques'. *Philosophy of Science*, 64(3), 461–77.

Salmon, Wesley C. (1998a). *Causality and Explanation*. Oxford University Press, Oxford.

Salmon, Wesley C. (1998b). 'Scientific explanation: three basic conceptions'. In *Causality and Explanation*, pp. 320–32. OUP.

Samaniego, Fernanda (2013). 'Causality and intervention in the Spin-Echo experiments'. *Theoria*, 28(3), 477–97.

San Pedro, Iñaki and Suárez, Mauricio (2009). 'Reichenbach's common cause principle and indeterminism: A review'. In *Philosophical Essays in Physics and Biology* (ed. J. G. Recio), pp. 223–50. Georg Olms Verlag.

Schaffer, Jonathan (2000). 'Causation by disconnection'. *Philosophy of Science*, 67(2), 285–300.

Schaffer, Jonathan (2004). 'Causes need not be physically connected to their effects: the case for negative causation'. In *Contemporary debates in Philosophy of Science* (ed. C. Hitchcock). Blackwell, Oxford.

Schaffer, Jonathan (2008). 'The metaphysics of causation'. In *The Stanford Encyclopedia of Philosophy* (Fall 2008 edn) (ed. E. N. Zalta). Available at: <http://plato.stanford.edu/archives/fall2008/entries/causation-metaphysics/>.

Schaffer, Jonathan (2012). 'Disconnection and responsibility'. *Legal Theory*, 18(4), 399–435.

Shannon, Claude (1948). 'A mathematical theory of communication'. *The Bell System Technical Journal*, 27, 379–423 and 623–56.

Shannon, Claude and Weaver, Warren (1949). *The Mathematical Theory of Communication*. University of Illinois Press (1964), Urbana.

Simpson, Edward H. (1951). 'The interpretation of interaction in contingency tables'. *Journal of the Royal Statistical Society*, Ser. B 13, 238–41.

Sloman, Steven (2005). *Causal Models: How People Think About the World and its Alternatives*. Oxford University Press, Oxford.

Sober, Elliott (1984). 'Two concepts of cause'. *PSA: Proceedings of the Biennial Meeting of the Philosophy of Science Association*, 2, 405–24.

Sober, Elliott (1986). 'Causal factors, causal inference, causal explanation'. *Proceedings of the Aristotelian Society, Supplementary Volumes*, 60, 97–113.

Sober, E. (2001). 'Venetian sea levels, British bread prices, and the principle of the common cause'. *The British Journal for the Philosophy of Science*, 52(2), 331–46.

Sober, Elliott and Wilson, David Sloan (1998). *Unto Others: The Evolution and Psychology of Unselfish Behavior*. Harvard University Press, Harvard.

Solomon, Miriam (2011). 'Just a paradigm: evidence-based medicine in epistemological context'. *European Journal for Philosophy of Science*, 1, 451–66.

Solomon, Phyllis, Cavanaugh, Mary M., and Draine, Jeffrey (2009). *Randomized Controlled Trials: Design and Implementation for Community-Based Psychosocial Interventions*. Oxford University Press, Oxford.

Soros, George (1994). *The Alchemy of Finance. Reading the Mind of the Market* (2nd edn). John Wiley.

Soros, George (1998). *The Crisis of Global Capitalism: Open Society Endangered*. Public Affairs.

Soros, George (2008). *The New Paradigm for Financial Markets: The Credit Crisis of 2008 and What it Means*. Public Affairs.

Spirtes, Peter, Glymour, Clark, and Scheines, Richard (1993). *Causation, Prediction, and Search* (Second (2000) edn). MIT Press, Cambridge Massachusetts.

Spohn, Wolfgang (1983). 'Eine Theorie der Kausalität'. Habilitationsschrift, München.

Spohn, Wolfgang (1986). The representation of Popper measures. *Topoi*, 5, 69–74.

Spohn, Wolfgang (1994). 'On the properties of conditional independence'. In *Patrick Suppes: Scientific Philosopher. Vol I: Probability and Probabilistic Causality* (ed. P. Humphreys), pp. 173–94. Kluwer.

Spohn, Wolfgang (2002). 'Laws, ceteris paribus conditions, and the dynamics of belief'. *Erkenntnis*, 57, 373–94.

Spohn, Wolfgang (2006). 'Causation: an alternative'. *British Journal for the Philosophy of Science*, 57(1), 93–119.

Spohn, Wolfgang (2012). *The Laws of Belief. Ranking Theory and its Philosophical Applications*. Oxford University Press, Oxford.

Stalnaker, R. (1968). 'A theory of conditionals'. *Studies in Logical Theory: American Philosophical Quarterly Monograph Series*, 2, 98–122.

Stalnaker, R. (1975). 'Indicative conditionals'. *Philosophia*, 5(3), 269–86.

Steel, Daniel (2008). *Across the Boundaries. Extrapolation in Biology and Social Science*. Oxford University Press, Oxford.

Stegenga, Jacob (2011). 'Is meta-analysis the platinum standard of evidence?' *Studies in History and Philosophy of Biological and Biomedical Sciences*, 42, 497–507.

Stella, Federico (2003). *Giustizia e Modernità. La Protezione dell'Innocente e la Tutela delle Vittime*. Giuffrè.

Strevens, Michael (2007). 'Essay review of Woodward, Making Things Happen'. *Philosophy and Phenomenological Research*, 74, 233–49.

Strevens, Michael (2008). 'Comments on Woodward, Making Things Happen'. *Philosophy and Phenomenological Research*, 77(1), 171–92.

Stroud, Barry (1981). *Hume*. Routledge, London.

Suárez, Mauricio (2007). 'Causal inference in quantum mechanics. A reassessment'. In *Causality and Probability in the Sciences* (ed. F. Russo and J. Williamson), pp. 65–106. College Publications.

Suárez, Mauricio (ed.) (2011). *Probabilities, Causes, and Propensities in Physics*. Springer.

Suárez, Mauricio and San Pedro, Iñaki (2007). 'EPR, robustness and the causal Markov conditions'. Technical Report PP/04/07, Centre for Philosophy of Natural and Social Science. LSE Philosophy Papers.

Suppes, Patrick (1970). *A Probabilistic Theory of Causality*. North-Holland, Amsterdam.

Sytsma, Justin, Livengood, Jonathan, and Rose, David (2012). 'Two types of typicality: rethinking the role of statistical typicality in ordinary causal attributions'. *Studies in History and Philosophy of Biological and Biomedical Sciences*, 43(4), 814–20.

Tenenbaum, Joshua B. and Griffiths, Thomas L. (2001). 'Structure learning in human causal induction'. In *Advances in Neural Information Processing Systems* (ed. T. Leen, T. Dietterich, and V. Tresp), Volume 13, pp. 59–65. MIT Press, Cambridge Massachusetts.

Thagard, Paul (1998). 'Explaining disease: correlations, causes, and mechanisms'. *Minds and Machines*, 8, 61–78.

Uttal, William R. (2001). *The New Phrenology: the Limits of Localizing Cognitive Processes in the Brain*. MIT, Cambridge, Massachusetts.

van Fraassen, Bas (1980). *The Scientific Image*. Oxford University Press, Oxford.

Vandenbroucke, Jan (2009). Commentary: '"Smoking and lung cancer"—the embryogenesis of modern epidemiology'. *International Journal of Epidemiology*, 38(5), 1193–6.

Vineis, Paolo (1998). 'Epidemiology between social and natural sciences'. *Journal of Epidemiology and Community Health*, 52, 616–17.

Vineis, Paolo (2003). 'Causality in epidemiology'. *Social and Preventive Medicine*, 48, 80–7.

Vineis, Paolo, Alavanja, M., Buffler, P., Fontham, E., Franceschi, S., Gao, Y. T., Gupta, P. C., Hackshaw, A., Matos, E., Samet, J., Sitas, F., Smith, J., Stayner, L., Straif, K., Thun, M. J., Wichmann, H. E., Wu, A. H., Zaridze, D., Peto, R., and Doll, R. (2004). 'Tobacco and cancer: Recent epidemiological evidence'. *Journal of the National Cancer Institute*, 96(2), 99–106.

Vineis, Paolo and Chadeau-Hyam, Marc (2011). 'Integrating biomarkers into molecular epidemiological studies'. *Current Opinion in Oncology*, 23(1), 100–5.

Vineis, Paolo, Khan, Aneire E., Vlaanderen, Jelle, and Vermeulen, Roel (2009). 'The impact of new research technologies on our understanding of environmental causes of disease: the concept of clinical vulnerability'. *Environmental Health*, 8(54).

Vineis, Paolo and Perera, Frederica (2007). 'Molecular epidemiology and biomarkers in etiologic cancer research: the new in light of the old'. *Cancer Epidemiology, Biomarkers & Prevention*, 16(10), 1954–65.

von Neumann, John (1955). *Mathematical Foundations of Quantum Mechanics*. Princeton University Press, Princeton.

von Wright, G. (1971). *Explanation and Understanding*. Cornell University Press, Ithaca, New York.

von Wright, G. (1973/1993). 'On the logic and epistemology of the causal relation'. In *Causation* (ed. E. Sosa and M. Tooley), pp. 105–24. Oxford University Press, Oxford.

von Wright, G. (1975). *Causality and Determinism*. Columbia University Press, New York.

Wagner, E. H. (1982). 'The North Karelia Project: What it tells us about the prevention of cardiovascular disease'. *American Journal of Public Health*, 72(1), 51–3.

Waters, C. Kenneth (2007). 'Causes that make a difference'. *The Journal of Philosophy*, CIV, 551–79.

Weber, Erik (2007). 'Conceptual tools for causal analysis in the social sciences'. In *Causality and Probability in the Sciences* (ed. F. Russo and J. Williamson), pp. 197–213. College Publications.

Weber, Erik and de Vreese, Leen (2012). 'Causation in perspective: are all causal claims equally warranted?' *Philosophica*, 84, 123–48.

Weber, Marcel (2012). 'Experiment in biology'. In *The Stanford Encyclopedia of Philosophy* (Spring 2012 edn) (ed. E. N. Zalta). Available at: <http://plato.stanford.edu/archives/spr2012/entries/biology-experiment/>.

White, C. (1990). 'Research on smoking and lung cancer: a landmark in the history of chronic disease epidemiology'. *Yale Journal of Biology and Medicine*, 63(1), 29–46.

Wild, Christopher P. (2012). 'The exposome: from concept to utility'. *International Journal of Epidemiology*, 41(1), 24–32.

Wild, Christopher P. (2005). 'Complementing the genome with an "exposome": the outstanding challenge of environmental exposure measurement in molecular epidemiology'. *Cancer Epidemiology, Biomarkers & Prevention*, 14, 1847–50.

Wild, Christopher P. (2009). 'Environmental exposure measurement in cancer epidemiology'. *Mutagenesis*, 24(2), 117–5.

Wild, Christopher P. (2011). 'Future research perspectives on environment and health: the requirement for a more expansive concept of translational cancer research'. *Environmental Health*, 10 (**Suppl**), S1–S15.

Williamson, Jon (2005). *Bayesian Nets and Causality: Philosophical and Computational Foundations*. Oxford University Press, Oxford.

Williamson, Jon (2006). 'Causal pluralism versus epistemic causality'. *Philosophica*, 77(1), 69–96.

Williamson, Jon (2009). 'Probabilistic theories'. In *The Oxford Handbook of Causation* (ed. H. Beebee, C. Hitchcock, and P. Menzies), pp. 185–212. Oxford University Press, Oxford.

Williamson, Jon (2011a). 'Mechanistic theories of causality Part I'. *Philosophy Compass*, 6(6), 421–32.

Williamson, Jon (2011b). 'Mechanistic theories of causality Part II'. *Philosophy Compass*, 6(6), 433–44.

Williamson, Jon (2013). 'How can causal explanations explain?' *Erkenntnis*, 78, 257–75.

Wilson, Fred (2014). 'John Stuart Mill'. In *The Stanford Encyclopedia of Philosophy* (Spring 2014 edn) (ed. E. N. Zalta). Available at: <http://plato.stanford.edu/archives/spr2014/entries/mill/>.

Witschi, Hanspeter (2001). 'A short history of lung cancer'. *Toxicological Sciences*, 64(1), 4–6.

Woodward, James (2003). *Making Things Happen: A Theory of Causal Explanation*. Oxford University Press, Oxford.

Woodward, James (2010). 'Causation in biology: stability, specificity and the choice of levels of explanation'. *Biology and Philosophy*, 25, 287–318.

Woodward, James (2013). 'Causation and manipulability'. In *The Stanford Encyclopedia of Philosophy* (Winter 2013 edn) (ed. E. N. Zalta). Available at: <http://plato.stanford.edu/archives/win2013/entries/causation-mani/>.

Woodward, J. and Hitchcock, C. (2003). 'Explanatory generalizations, part I: a counterfactual account'. *Noûs*, 37(1), 1–24.

Worrall, John (2002). '*What* evidence in evidence-based medicine?' *Philosophy of Science*, **69**, S316–S330.

Wunsch, Guillaume (2007). 'Confounding and control'. *Demographic Research*, 16(4), 97–120.

Wunsch, Guillaume, Mouchart, Michel, and Russo, Federica (2014). 'Functions and mechanisms in structural-modelling explanations'. *Journal for General Philosophy of Science*, 45(1), 187–208.

INDEX

5 philosophical questions 237, 259, 265
  epistemology 237, 238, 241, 245, 259, 265
  metaphysics 237, 238, 241, 245, 259, 266, 269
  method 237, 239, 241, 245, 259, 266, 269
  semantics 237, 239, 241, 245, 259, 266, 269
  use 237, 239, 241, 245, 259, 267
5 scientific problems 4, 262, 267
  causal explanation 5, 31, 32, 50, 63, 93, 96, 100, 102, 109, 112, 117, 120, 121, 126, 131, 132, 150, 154, 155, 189, 192, 228, 229, 237, 240, 250, 258, 262
  causal inference 4, 36, 38, 39, 44, 47, 48, 54–58, 61, 63, 74, 76, 79, 86, 93, 96, 97, 99, 100, 106, 112, 116, 121, 133, 135, 136, 151, 154, 155, 157, 161, 168, 171, 177, 179, 187, 189, 191, 228, 229, 237–240, 250, 258, 262, 266, 269
  causal reasoning 5, 27, 33, 36, 38, 39, 44, 47, 58, 61, 74, 76, 82, 86, 88, 90, 94, 97, 100, 103, 116, 124, 135, 136, 151, 154, 157, 161, 168, 171, 174, 177, 189, 196, 228, 230, 232, 237–240, 250, 258, 263, 270
  control 5, 36, 39, 48, 63, 73, 100, 101, 105, 108, 136, 168, 177, 179, 182, 183, 187, 188, 192, 228–230, 237, 240, 250, 258, 263
  prediction 4, 36, 39, 50, 60, 76, 112, 136, 161, 168, 177, 189, 190, 192, 228–230, 237, 240, 250, 258, 263

A

abduction 20
Abrahamsen, Adele 125, 127, 130
absences 30, 94, 117, 132, 147, 150, 215, 217
action-oriented theory of causality 183
activities 123, 124, 126, 127, 133
activities and entities 123, 124, 133
agency theories 178
agent:
  ideal 191
  rational 191
Agresti, Alan 66
AIT: Algorithmic Information Theory 137
algorithmic search 62
Andersen, Hanne 270
Anjum, Rani Lill 152
Anscombe, Elizabeth 139, 202, 250
anti-realism 194
approach to causation:
  analysis of folk intuitions 203
  analysis of language 202
  causality in the sciences 207
Aristotle 153, 251
Aronson, Jerrold 113
asymmetry 185
automated reasoning 62
average response 96

B

background assumptions 229
Bandyoapdhyay, Prasanta 68
Baumgartner, Michael 164
Bayesian degrees of belief 83
Bayesian interpretation of probabilities 83
Beaver, Donald 270
Bechtel, William 125, 127, 129, 130
Beebee, Helen 163, 190, 252
beliefs:
  background 92
  causal 191
  compatible 92
Bell labs 17
BINP: Bangladesh Integrated Nutrition Policy 58
biomarker 260
Bird, Alexander 152
bit string 137
biting the bullet 205
black box 71, 122
blameworthiness 28
BN: Bayesian Network 68
Boniolo, Giovanni 114
Boudon, Raymond 66
Brenner, Sydney 56
Broadbent, Alex 53

Bukodi, Erzsébet 97
Burns, Patrick 94

C

Caldwell, John 167
Campbell, Donald T. 63
Canberra Plan 204, 241, 244
capacities 151, 153
Cartwright, Nancy 18, 58, 127, 153, 186, 250, 253
case series studies 11
case-control studies 11
Casini, Lorenzo 62
category mistake 237
causal assessment 227
causal beliefs 191
causal explanation 5, 50, 63, 93, 96, 100, 102, 109, 112, 117, 120, 121, 126, 131, 132, 150, 154, 155, 189, 192, 228, 229, 237, 240, 250, 258, 262
causal field 31
causal inference 4, 36, 38, 39, 44, 47, 48, 54–58, 61, 63, 74, 76, 79, 86, 93, 96, 97, 99, 100, 106, 112, 116, 121, 133, 135, 136, 151, 154, 155, 157, 161, 168, 171, 177, 179, 187, 189, 191, 228, 229, 237–240, 250, 258, 262, 266, 269
   fundamental problem of 96
causal interaction 113
causal learning 94
causal lines 112
causal Markov condition 68
causal methods 9
   evidence-gathering 254
   structural equations 172
causal modelling 228
causal mosaic 8, 177, 255, 258, 269
causal pies:
   see pie-charts 32
causal pluralism 249, 256, 270
   concepts of causation 251
   methods for causal inference 253
   sources of evidence 253
   types of causing 250
   types of inferences 252
causal process 112
causal production 189

causal reasoning 5, 19, 27, 33, 36, 38, 39, 44, 47, 58, 61, 74, 76, 82, 86, 88, 90, 94, 97, 100, 103, 116, 124, 135, 136, 151, 154, 157, 161, 168, 171, 174, 177, 189, 196, 228, 230, 232, 237–240, 250, 258, 263, 270
   based on variations 174
   counterfactual process view 94
   mechanistic 54
causal relata 38, 231
   as events 231
   as facts 232
   as objects 231
   as processes 232
   as variables 232
   liberalism about 235
   nature of 232
causal relevance 116, 132
causal scepticism 3
causalist perspective 3
causality:
   agency theory 100, 189
   and (in)determinism 82
   and evidence 228, 241, 253, 268
   and explanation 115
   and inference 252
   and language 202
   and levels 267
   and logical necessity 162
   and pluralism 202, 249
   and time 81
   and truth 228
   and truthmakers 228
   as mind-dependent 163, 189, 194
   by absences 30, 94, 117, 132, 147, 150, 215, 217
   by omission 94, 117
   counterfactual dependence 241
   counterfactuals 86
   CPDs 268
   and context 158
   and explanation 155
   and finking 157
   and holism 159
   and interactions 158
   and masking 157
   and truthmakers 156
   evidence of 153, 155, 158
   locality of 154

   reality of 153
   deterministic 81
   difference-making account 88, 100, 174
   epistemic theory 191, 192
   generalist theory 38
   indeterministic 164
   inferentialism 192, 193, 203
   information 268
   manipulation 12, 241
   multifactorial model 164
   probabilistic 33, 77, 83, 164, 175
   production account 112, 268
   projectivism 190
   rationale of 168
   regularity theory 161
   singularist theory 38
   transference account of 113
causation:
   by absences 94, 117, 132, 147, 150, 215, 217
   by omission 94, 117
   general 39
   quasi 117, 217
   singular 39
cause:
   actual 38
   direct 71
   generic 41, 80
   indirect 71
   individual-level 37
   individual-level vs population-level 37
   intervening 28
   necessary 29, 39
   negative 77
   population-level 37
   positive 77
   potential 38
   prima facie 77
   prima facie negative 77
   probability changing 79
   screen off 78
   single-case 41, 80
   sufficient 29, 32
   token 37
   token vs type 37
   type 37
   vs condition 31
causes:
   negative 164
CDA: Categorical Data Analysis 66

CERN: European organization for nuclear research 12
ceteris paribus 159
Chakravartty, Anjan 151, 152, 159, 232
chance 33
change-relating relation 101
children's reasoning 94
circularity 182, 184
CitS: Causality in the Sciences 202, 205–207, 245, 260, 261
Clarke, Brendan 55
classical propositional logic 88
CMC: Causal Markov Condition 68
Cochrane Collaboration 57
cohort studies 11
Collier, John 139
Collingwood, R. G. 180
common cause 54, 79, 121
common cause principle 16, 117
complex system 125
complexity 137
compressibility 137
conceptual analysis 28, 204, 239, 241
conceptual design 243, 270, 271
conceptual engineering 243, 270, 271
condition
 normal vs abnormal 31
confounding 16, 54, 55, 68
conjunctive fork 16
connecting principle 42, 84
conserved quantities 113
constant conjuction 162
context 129
control 5, 17, 36, 39, 48, 63, 73, 100, 105, 108, 136, 168, 177, 179, 182, 183, 187, 188, 192, 228–230, 237, 240, 250, 258, 263
Cook, Thomas D. 63
correspondent (or correspondence) theory of truth 233
Corry, Richard 182
counterexamples 220, 223
counterfactual dependence 241
counterfactual reasoning 86, 92, 94, 95, 97
 imprecision of 87
 usefulness of 97

counterfactual validation 92
counterfactuals 153, 156
 and laws of nature 89
 and modal realism 90–92
 legal 93
 philosophical 88
 psychological 94
 statistical 95
 truth of 89
Courgeau, Daniel 40, 42, 65, 84
CPD: Capacity-Power-Disposition 152, 153
Craver, Carl 128, 129, 131, 217, 259
Crick, Francis 56
cross-sectional studies 10
Cummins, Robert 127

D

Danks, David 95
Darden, Lindley 217
data generating process 72
data mining 84
Davidson, Donald 231
Dawid, A. Philip 97
de Vreese, Leen 142
decision theory 182
deductive arguments 20
deductive method 21
default world 47
determinism vs predictability 82
disease causation 33
dispositional essentialism 152
dispositions 152
 see causality
 CPDs 151
distributed understanding 270
doctrine of four causes 251
Doll, Richard 51, 56
dormitive virtue 155
Dowe, Phil 113, 131
Driessen, Geert 41
Drouet, Isabelle 69, 71
Durkheim, Emile 170

E

EBA: Evidence Based Approaches 13, 56, 57
EBM: Evidence Based Medicine 56, 57

Eells, Ellery 44, 80
EMA: European Medicines Agency 47
entities 123, 124, 133
entities and activities 123, 124, 133
environmental factors 260
EPA: Environmental Protection Agency 38
EPIC: European Prospective Investigation into Cancer and Nutrition 261
epidemiology 29
epistemic theory 191, 192
epistemology 237, 238, 241, 245, 259, 265
EPR: Einstein-Podolsky-Rosen correlations 79, 117, 219
evidence 209
 difference-making 53, 55, 268
 gathering methods 56
 hierarchies 57
 integration 58
 mechanism 53, 54
 observational 108
 production 53
example (science):
 e. coli 150, 152, 154, 156, 235
 helicobacter pylori and gastric ulcer 54
 vibrio cholerae 53
 Hodgkin and Huxley 132
 AIDS 27
 Alzheimer disease 37
 aphids and *buchnera* 129
 apoptosis 60, 62
 atherosclerosis 121, 122
 Bangladesh Integrated Nutrition Policy (BINP) 58
 brain cancer 46
 cerebral hypoxia 147
 cholera in London 136
 clash of civilization thesis 176
 combatting AIDS 188
 cosmic background microwave radiation 17
 CT scans 47
 disease transmission 135
 divorce and migration 79
 DNA photographs 56

INDEX | 305

example (science): (*continued*)
DNA repair mechanisms 158
economic crisis 108
effects of alcohol consumption 48
effects of sunlight 49
Einstein-Podolsky-Rosen correlations 82
Eternit trial 81, 86
exposomics 260
gender discrimination in university admission 68
gender medicine 49
gene knock-out experiments 99, 106, 158, 174
gravitational waves 128
HIV 107, 188, 191, 192, 195
honeybee colony 118
human papillomavirus 76
ideal gas law 102, 151
illiteracy and immigration 41
kinetic theory of gases 133
lac operon 150, 152, 154–156, 158
memory 133
MEND programme: Mind, Exercise, Nutrition, Do It 131
mesothelioma 81, 86
mesothelioma and asbestos 227
migration behaviour 36, 40, 161
mobile phone radiation 46, 48, 53
mother's education and child survival 167
MRI scans 47
myocyte or muscle cell 158
natural and social regularities 161
Newton's laws 179, 183
North Karelia project 106
obesity 131
overexpression experiments 99
personalized diagnosis 50
personalized medicine 35
policy in social science 106
protein synthesis 218
quantum mechanics 82

radiation 46
radio wave transmission 135
regulation of gene expression 138
residual-current circuit breaker (RCCB) 157
resistor 114
rotating laser 111
school composition and educational achievement 41
self-rated health 60, 62, 70, 72, 123, 124, 128, 129
smoking and cancer 75
smoking and heart disease 47, 49, 120, 122
smoking and lung cancer 42, 152, 174, 175
spin-echo experiments 104
supernovae 118, 154, 156
symmetry 179
tuberculosis 29
wealth and influence in the Tobriand islands 124
X-ray crystallography 56
year of the Fire Horse 41
example (toy):
heart 127
soothing a newborn baby 127
accelerating a car 179, 182, 183
airplane shadows crossing 111
aspirin and headache 96, 159, 230
bag of books 133
Big Bang 184
billiard balls colliding 111, 116, 162
Billy & Suzy 194, 206, 216
Billy & Suzy, the next episode 211
capacities of coke 157
car and its shadow 138
coffee drinking, smoking, heart attack 54
contraceptive pills, pregnancy, and thrombosis 69, 84
describing a car with information theory 137
electrical short-circuit 27

exercise and weight loss 55
gender and salary discrimination 103
lack of oxygen causing death 118, 132
light switch and bulb 140
lung cancer 16
melting ice 181
missing the bus 47, 87, 117, 147
not watering the plant 215
plugging in your bluray player 135
removing the sun 87
Schaffer's gun 217
smoke and fire 162
smoking 16
smoking and lung cancer 164
smoking and yellow fingers 78
squirrel and golf ball 37, 80
squirrel's kick 36, 80
storks and birth rates 73, 101
supernova 184
the first philosophy conference 212
yellow fingers and lung cancer 73
example:
science vs toy 211–214
exogeneity 72, 173
experimental methods 12
experimental philosophy 201
expertise 209
explanation 133
and pragmatics 109
D-N model 39, 126
epistemic 115
mechanistic 39, 156
ontic 115
exportable results 18
exposome 261, 267
exposomics 145, 260
meeting-in-the-middle methodology 261, 266
exposure 261
extensive quantities 114
external validity 18, 50
extrapolation 18
see external validity 49

F

Fagan, Melinda 270
Fair, David 113
faithfulness condition 69
fallacies, atomistic and ecological 41
falsificationism 22
Faraldo, Rossella 114
FDA: Food and Drug Administration 47
feedback 138
fictionalism about possible worlds 91
finking 157
Fisher, Ronald A. 13, 61
Floridi, Luciano 142, 271
fragments of causal theory 237, 259, 265
    integration of 243–246
Franklin, Rosalind 56
frequentist interpretation of probabilities 83
Frigg, Roman 63
Frisch, Mathias 126, 185
Frosch, Caren 94
function 125, 127
    contextual description 128
    isolated description 128, 129
    role-function 127
    selected-effects 127
functional individuation 124

G

Gasking, Douglas 181
gene knock-out experiment 99, 106
generalization:
    causal 38, 101
    change-relating 109, 125
    empirical 101
    invariant 102, 125
generalize 18
Gerring, John 52, 78, 172
Gerstenberg, Tobias 95
Giere, Ronald 175
Gillies, Donald 54, 55, 69, 84, 183
Glennan, Stuart 125–127
Glymour, Clark 64, 95
Godfrey-Smith, Peter 252
Goldstein, Harvey 40, 65
Goldthorpe, John H. 97, 254

Good, I.J. 77
Gopnik, Alison 95
Greenland, Sander 65
Griffiths, Thomas L. 95
Guala, Francesco 18
GWAS: Genome Wide Association Studies 37

H

H-D method: Hypothetico-Deductive method 22, 192
Haavelmo, Trygve 65
Hall, Ned 37, 206, 216, 252
Hart, H.L.A. 28–30, 203, 251
Hartmann, Stephan 63
Hausman, Daniel 100, 186
Heckman, James 65
Hellevik, Ottar 66
Hempel, Carl G. 93
Hill's guidelines 51, 52, 57
Hill, Bradford 51, 52
Hitchcock, Christopher 38
HIV: human immunodeficiency virus 107
Holland, Paul 96
Honoré, Tony 28–30, 203, 251
Hood, William C. 65
Howick, Jeremy 54, 56
Hume, David 30, 162, 186, 231
Humphreys' paradox 71
Humphreys, Paul 83
hypothetical populations 175

I

IARC: International Agency for Research on Cancer 38, 47, 75
IBE: Inference to the Best Explanation 22
Illari, Phyllis 53, 55, 115, 122, 125, 130, 132, 143, 146
individual response 75
inductive arguments 20
inductive method 21
inferentialism 192, 193, 203, 252
information 268
information channel 137
information loss 137
information theory 137
information transmission 137

interaction 32, 125
internal validity 18
intervening cause 28
intervention 73, 100, 101
    constraints on 103
    ideal, real, practical, feasible, hypothetical 104
    structure altering 103
    type of 104
INUS: Insufficient, but Non-redundant part of an Unecessary but Sufficient condition 28, 29, 39
invariance 73, 175
    across changes 108
    under intervention 101, 102
it-will-work-for-us 18
it-works-somewhere 18

J

Jackson, Frank 204
Jimenez-Buedo, Maria 63
Joffe, Michael 52

K

Kelly, Michael 56
Killoran, Amanda 56
Kim, Jaegwon 231
Kincaid, Harold 144
Kittel, Bernhard 40
Kleinberg, Samantha 42, 44, 84
Knobe, Joshua 38
Kolmogorov, Andrei Nikolaevich 137
Koopmans, Tjalling Charles 65
Kushnir, Tamar 95

L

La Caze, Adam 56
laboratory experiments 12
Ladyman, James 141
LaFollette, Hugh 49
Lagnado, David 95
Laudisa, Federico 163
laws 150
laws of nature 89, 156
legal causation 28, 93, 117, 251
legal responsibility 28
Leonelli, Sabina 265, 270
Leuridan, Bert 129
levels of causation 30, 42, 50, 80, 84, 165, 267

Lewis, David 88, 163, 241
LHC: Large Hadron Collider 12
liability
  see responsibility 38
linguistic analysis 28
linking 53
Little, Daniel 44, 127
Livengood, Jonathan 95
LoA: Levels of Abstraction 22
locality 154
Longino, Helen 209
Longworth, Francis 252
Lucas' critique 105

M

Mach, Ernst 3
Machamer, Peter 217
machine learning 62
Mackie, J.L. 28, 29, 44, 163, 202
Mandel, David 95
manipulation 12, 96, 100, 174, 181, 186, 241
manipulationism 73, 100
  and explanation 109
  conceptual 107
  methodological 107
mark method 112
Markov condition 68
masking 55, 157
matching 97
MatEH: Mechanisms and the Evidence Hierarchy 55
material implication 88
Mayntz, Renate 127
McCormack, Teresa 94
MDC: Machamer, Darden, and Craver 124, 217
mechanisms 120
  and activities 126
  and causality 126, 131
  and explanation 126, 127, 132
  and recursive decomposition 71
  complex system 125
  contextual description 128
  discovery of 122
  entities and activities 122–124, 133
  function 125, 127
  functional individuation 124
  identification 124
  in biology 122
  in demography 122
  in economics 122
  in neuroscience 122
  in psychology 122
  in social science 126
  in sociology 122
  isolated description 128, 129
  mixed 131
  multifield 131
  mutual manipulability 129
  organization 122–125, 130, 133
mechanistic reasoning 54
MELF: Meaning, Existence, Lawfulness, Fact 92
Mellor, David 232
MEND programme: Mind, Exercise, Nutrition, Do It 131
Menzies, Peter 47, 95, 182
Merton, Robert 127
meso-level relations 44
metaphysics 237, 238, 241, 245, 259, 266, 269
method 237, 239, 241, 245, 259, 266, 269
methods:
  Bayesian networks 64, 68
  case series studies 11
  case-control studies 11
  categorical data analysis 66
  cohort studies 11
  contingency tables 32, 66
  cross-sectional studies 10
  data-gathering 9
  experimental 10, 12, 61, 100, 254
  lab experiments 12
  multilevel 39–41, 65
  natural experiments 14
  observational 10, 254
  potential outcome models 33, 96
  probabilistic temporal logic 84
  qualitative analysis 14
  quantitative 40
  quasi-experimental 96
  quasi-experiments 14
  randomized controlled trials (RCTs) 13
  recursive Bayesian networks (RBNs) 62
  recursive decomposition 71
  regression models 65
  simulation 15
  statistical analyses 40
  structural equations 65
  structural models 65, 97
Mill's methods 169
Mill, John Stuart 30, 169, 170
Miller, Luis M. 63
Millikan, Ruth 127
miracle:
  convergence 90
  divergence 90
Mitchell, Sandra 118
mixed mechanisms 131
modal realism 89–92
model relativity 229
model validation 229
modelling 23
models:
  as fictions 229
  as isolations 229
  as maps 229
  as mediators 229
  as representations 63
  associational 66
  causal 66
  validity of 63, 108
modularity 73, 105
Moneta, Alessio 67
Moore, Michael S. 28, 38, 250
mosaic of brains:
  see distributed understanding 270
Mouchart, Michel 65, 97
multifactorial causes 29, 32, 164
multifield mechanisms 131
Mumford, Stephen 152
mutual manipulability 129

N

natural experiments 14
Neapolitan, Richard 64
necessary cause 29
negative capacities 156
negative causes 164
negative feedback loop 138
negligence 117
NICE: National Institute for Health and Care Excellence 47
noise 137
Norton, John 126

O

objective interpretation of
  probabilities 83
observational methods 10
omission 94, 117
organization 123, 125, 130, 133
Osimani, Barbara 57
overexpression
  experiments 99

P

paradox of material
  implication 88
Pearl, Judea 64, 65, 186, 232
Pearson, Karl 3
peer review 219
Peto, Richard 56
phenomena 122, 132
philosophical problems, origin
  of 208
philosophy, role of 201
PI: philosophy of
  information 209
pie-charts 28, 32
pluralism:
  evidential 57
  in causality 202
Popper, Karl 22, 208
possible world 88, 91
  similarity of 89
possible-world semantics 88
post-genomic era 99
powers 152
  see causality
  CPDs 151
prediction 4, 36, 39, 50, 60, 76,
  112, 136, 161, 168, 177, 189,
  190, 192, 228–230, 237,
  240, 250, 258, 263
Price, Huw 182, 204
principle of causality 162
probabilistic causes 33
probabilistic dependencies 69
probabilistic
  independencies 69
probability raising 80
probability theory 77
probability:
  interpretation of 83
  marginal and conditional 71
problem of confounding 16
process 120
  causal 112, 113, 218

pseudo 112, 113, 218
process tracing 112, 115, 120
  in biology 115
  in social science 115
productive causality 132
projectivism 190
propensity interpretation of
  probabilities 83
propositional logic 88
Psillos, Stathis 164
PSP: philosophy of science in
  practice 209

Q

QM: Quantum Mechanics 82
qualitative analysis 14
quasi-causation 117
quasi-experiments 14

R

randomization 13
randomized controlled
  trials 13
ranking functions 193
Rappaport, Stephen 261
RBN: recursive Bayesian
  network
  see methods 62
RCCB: Residual-Current
  Circuit Breaker 157
RCTs: Randomized Controlled
  Trials 13, 57
recursive Bayesian networks
  (RBN) 62
recursive decomposition 71, 73
reduction 182
reference class 50, 55
regular succession 162
regularism 164
regularity 30, 179
  as frequency of
    instantiation 165
regularity theory 161
Reichenbach, Hans 16, 78, 112,
  138
Reiss, Julian 127, 192, 193, 253
Rescher, Nicholas 92
responsibility 38
  legal 38
  moral 38, 93, 95
Robinson, W. S. 41
Roessler, Johannes 95
Rose, David 95

Rosen, Deborah 36
Ross, Don 141, 142
Rothman, Kenneth J. 28, 29, 32
Rubin, Donald 96
Russell, Bertrand 3, 102, 112
Russo, Federica 42, 44, 55, 65,
  67, 97, 131, 146, 171, 241,
  259
RWT: Russo-Williamson
  Thesis 52, 53, 241, 253

S

Saggion, Antonio 114
Salmon, Wesley 82, 112, 113,
  115, 131, 138, 218
Samaniego, Fernanda 104
same cause, same effect 164
San Pedro, Iñaki 69, 79
Schaffer, Jonathan 217
Schanks, Niall 49
Scheines, Richard 64
Schulz, Laura E. 95
screening off 16, 78
semantics 237, 239, 241, 245,
  259, 266, 269
Shannon information 137
Shannon, Claude 137
Simpson's paradox 41, 54, 68
simulation 15
Sloman, Steven 95
Smith, Martyn 261
Snow, John 136
SNP: Single Nucleotide
  Polymorphisms 37
Sobel, David M. 95
Sober, Elliott 37, 42, 84
social science:
  individualism vs holism 39
  microfoundations 44
sociology of science 209
Socratic gadfly 255
Solomon, Miriam 57
Spirtes, Peter 64
Spohn, Wolfgang 193, 252
Spurrett, David 142
Stanley, Julian C. 63
statistical individual 42
Steel, Daniel 49, 55, 106
Stegenga, Jacob 57
Stella, Federico 38
strength of an effect 32
Strevens, Michael 107

structural equation:
    counterfactual reading 172
    manipulationist reading 172
    variational reading 172
structure-altering
    interventions:
    in biochemistry 106
    in social science 106
structure:
    economic 65
Suárez, Mauricio 69, 79, 82
sub-population 50
Sudbury, Aidan 69
sufficient cause 29
sufficient component
    model 28
Suppes, Patrick 36, 77
SUTVA: Stable Unit Treatment
    Assumption 96
symmetry 185
systems biology 107
Sytsma, Justin 95

T

technology 209
temporal information 18
temporal logic 84
tendencies 153
Tenenbaum, Joshua B. 95
Thagard, Paul 54
theory of reflexivity 105
theory of truth 233, 234
third variable problem 16
time 18
timely philosophy 210
truthmakers 228
    as capacities 230
    as regularities 231

U

use 237, 239, 241, 245, 259, 267

V

validity 171, 229
Vandenbroucke, Jan 51
variable:
    aggregate 65
    aggregate level 40
    individual 65
    individual level 40
    proxy 108
variation 108
    and counterfactuals 174
    and invariance 175
    and manipulation 174
    and probabilities 174
    and regularity 174
verificationism 202
Vineis, Paolo 260
von Wright, Georg Henrik 180

W

Wagenknecht, Susann 270
Waters, Kenneth 37
Weber, Erik 142, 252
Weslake, Brad 182
WHO: World Health
    Organisation 47
Wild, Chris 261
Williamson, Jon 55, 64, 122,
    125, 130, 132, 191, 192, 241,
    252, 253
WITHBD questions:
    'what-if-things-
    had-been-different'
    questions 102
Woodward, James 100, 186,
    232, 241
Worrall, John 56
Wunsch, Guillaume 65, 97